Global Issues in Water Policy

Volume 12

More information about this series at http://www.springer.com/series/8877

John F. Raffensperger • Mark W. Milke

Smart Markets for Water Resources

A Manual for Implementation

Springer

John F. Raffensperger
RAND Corporation
Santa Monica, CA, USA

Mark W. Milke
Department of Civil and Natural Resources Engineering
University of Canterbury
Christchurch, New Zealand

ISSN 2211-0631 ISSN 2211-0658 (electronic)
Global Issues in Water Policy
ISBN 978-3-319-55007-7 ISBN 978-3-319-55008-4 (eBook)
DOI 10.1007/978-3-319-55008-4

Library of Congress Control Number: 2017933677

Printed on acid-free paper

This Springer imprint is published by Springer Nature
The registered company is Springer International Publishing AG
The registered company address is: Gewerbestrasse 11, 6330 Cham, Switzerland

For my grandchildren—
John F. Raffensperger,
Santa Monica.

A mi querida, luz de mi vida—Mark W. Milke,
Christchurch.

Foreword

There are many theoretical and experimental articles in the journals, and some books, on water markets for the allocation of water among competing users. That literature is often technical and difficult, written for academics by academics. This book is unique in offering a manual for staff practitioners charged with dealing with the important nitty gritty details of designing a "smart market" for water users. "Smart" means that the task of users is much simplified by applying computational algorithms to user bids to acquire or to supply water from their location in the supply–demand–transportation network.

Such markets are intended to make decision-making easy for the user, who is assured by the algorithms that he or she will get the best deal possible given his or her bid, the location in the network, the hydrology of the drainage–storage system, and the bids of all others. Bidders need to focus primarily on carefully estimating and calculating the value to themselves of additional water bought from, or sold to, others in the market.

Similarly, this manual is intended to make the task of the staff supporter easier and to help them work with others to assure that any particular application covers all the relevant bases. Toward this end, there is a glossary, along with over 40 figures and lots of examples and illustrations of the key elements that go into the design of an effective and flexible water allocation and transfer market. Every case is different, different in the sense of varying with local technology, water sources, and the preferences of those who are to participate in the market. I believe the authors have done a very credible job in their writing style and in the plan of the book to achieve these objectives.

Markets are often seen as contrary to the consumer-user's interest. Others are often thought to profit at the expense of the consumer user. This book emphasizes throughout that smart water markets are for users, by users, and that users are the direct beneficiaries, the ones that "profit" and are made better off, by their participation in the market.

It is now 35 years since Stephen Rassenti, in his PhD dissertation (1981), pioneered the concept of the smart computer-assisted market and 34 years since we published the first article reporting the results of laboratory test-bedding of a proposed combinatorial auction for the allocation of runway (landing–takeoff) slots for an airline industry facing complex coordination issues after the Carter Administration had moved to deregulate the airlines in the 1970s. From that beginning, we went on to electric power, which became a successful application of the smart market concept to a very complex problem. We also explored water, but found it very difficult to effect meaningful change. Interest was high with drought, but quickly faded when the rains came. These contributions, and far more by the many contributors to the literature, including the authors, are referenced in this book for those interested in the history, the background, and the many sources the authors have drawn upon over the last 35 years. The authors and many others have moved the development of water markets much further than our efforts were able to move the needle of progress.

This book is a testimony to this success and is effective, I believe, in making the essence of water market design easy for those who must be the instruments of new applications.

All this is personally very satisfying to me, and I welcome the reader to the pages that follow.

Chapman University
2002 Nobel Laureate in Economics
Orange, California
October, 2016

Vernon L. Smith

Preface

The world desperately needs new approaches to managing fresh water. At best, today's water allocation mechanisms work just okay. At worst, the water allocation mechanisms provoke perverse behaviors. Most of the time, society's water allocation mechanisms work badly—within the same catchment at the same time, some people waste water while others are desperate for it.

The water resource literature urges that water should be allocated with markets. But the literature is thin on exactly how those markets should be designed and operated, leaving open a long list of questions. How can users trade water in a complex catchment, such as when water flows between a river and the adjoining **groundwater** beneath the earth's surface? How can two users trade water on their own when their trade will affect many other users and the environment too? How can government avoid an expensive check on every water trade? How can a water regulator match water rights, as printed on paper, to the highly variable real thing? How can a "free market" operate while satisfying agreed environmental limits? Is the state of our ability to model the hydrology good enough to create *any* useful market?

In this book, we will show how to answer these questions with a **smart market**. A smart market is a trade-coordinating mechanism in which a market manager uses optimization to allocate products or services. Critically, the optimization makes the allocation decisions, not the market manager. We are not referring to eBay or Amazon, which are simple shops that use the computer mainly for communication. Smart markets enable much more complicated trading. The word "smart" here refers to the use of mathematical optimization to allow the market manager to make trades quickly while managing complexity.

A smart market is not a "free" market, but rather a market that is carefully planned and carefully managed. Researchers and practitioners have been developing this mechanism for over 60 years. The new system depends on a combination of recent technologies: hydrological simulation, computer power, and the Internet.

One of the most important smart markets is the modern electricity market, which we shall examine later. Power generators and users exchange electricity across complex networks of wires and switches. Ensuring that generation matches demand, in near real time, is a complicated problem. Yet active markets are working worldwide, thanks to the ability of mathematical optimization to match generation to demand in a way that respects the physics and the line capacities.

Despite important differences, the analogy of electricity markets to water markets works surprisingly well. Among other things, electricity markets produce spot prices for electricity. Our proposed smart market produces spot prices for water. An active market with spot pricing enables a range of other types of markets based on the underlying spot prices, such as trade in capacity rights for **infrastructure**, and various types of futures and options that can help users lower their risk. The price information will help users and government better plan for infrastructure.

This new approach can reduce the **transaction cost** of water trading while improving the environmental outcomes. The smart market could apply to any kind of water resource, whether groundwater, surface water, or the combination of the two, with or without reservoirs, for consumptive use and non-consumptive use—all of these can work with the same smart market method. The implementers could include, as components in the market, characteristics of water quality (e.g., salinity and nitrate). A single smart market can accommodate many users in a complex catchment.

Setting up a smart market for water will require:

- Specifications of water rights based on actual water availability;
- Maintenance of modern records for those rights;
- Water metering;
- Specification and measurement of the requirements for water flows to ensure environmental protection, including aquifer levels, with enforcement to maintain those flows;
- A hydrological model that describes the water flows with reasonable accuracy;
- The legal ability to trade.

Existing water markets work badly where any of these prerequisites is missing, and few existing water markets have all of them in place. Good water managers should put these prerequisites in place anyway, even if government does not intend to implement a smart market.

The purpose of this book is to explain the mechanism of a smart market for wholesale water, with a focus on large water users such as farmers, industrial firms, and water agencies. While we have not had the opportunity to implement such a market, and we are not aware of one working anywhere as we imagine it, we intend for communities and policymakers to use the book as a manual for implementing such a market. We have written this book for policymakers, water regulators, business developers, large water users, and environmental organizations. Economists, operations researchers, and hydrologists with an interest in water allocation may also find the book of use. We hope to inspire a wide audience to consider a better way to allocate water.

Some sections of later chapters require knowledge of linear programming, and nonspecialists can skip these. To broaden our audience, some sections in early chapters cover the basics of hydrology and of economics—specialists may find that they can skip those parts.

The style is intentionally nonacademic, but the material is technical in many places. We give only a partial survey of water allocation mechanisms in theory and practice; other recent books give excellent surveys and we see no need to duplicate that information. Our examples are pitched at the nonspecialist and involve simplified cases, but the techniques scale to situations with thousands of water users.

We hope to demonstrate that the technology and knowledge exist for society to implement economically efficient and environmentally protective systems for allocating water. Communities that choose to do so will accrue the benefits to users and to their local environments.

Acknowledgments

This work was developed over many years with help from many people.

Thanks most especially to our colleague, E. Grant Read, who served as our intellectual guide for many years, patiently tutoring us in markets, and whose many suggestions we have incorporated into this book. He read drafts of early chapters and made numerous suggestions. We credit many of the ideas to him.

A number of others have made great contributions that we bring into this effort. Ranga Prabodanie studied markets for nitrate runoff. Antonio Pinto studied markets for impervious cover. Rebecca Teasley studied groundwater as a postdoctoral fellow at Canterbury. Thanks to Tom Cochrane, from the Department of Civil and Natural Resources Engineering, University of Canterbury, who worked with us on impervious cover.

Many people gave us comments on this manuscript. Thanks to David Ahlfeld, Eric Garner, Steve Gorelick, Richard Howitt, Xindi Hu, Darren Lumbroso, Michelle Miro, James Murphy, Chris Perry, Ranga Prabodanie, Manuel Pulido-Velázquez, Linus Schrage, Vernon Smith, and David Zetland.

Thanks to Florence Kerns for drawing the cartoons.

Thanks to Janet Bray for her patient, careful, and often brilliant copyediting.

Any errors are our own.

Contents

List of Aside

List of Figures

List of Tables

Glossary

Abstraction The taking of water, whether from above ground (spring, river, reservoir, lake) or below ground (well).

Allocation The amount that each water user is assigned by a water regulator for a given time period. This amount is specified initially by a license and can be changed frequently through regulatory actions or user actions in the market. See also "License."

Aquifer A body of subsurface material that contains and transmits a useful amount of groundwater.

Aquitard A body of subsurface material that retards flow of groundwater through it.

Assessment point A location in a catchment where the regulator has specified with an environmental flow requirement, e.g., where a river's flow must be at least some amount or where an aquifer must have at least a specified head. Usually, the regulator would monitor an assessment point in some way. The smart market model has a constraint for each assessment point and each relevant time period in the auction schedule.

Auction schedule The schedule for operating the smart market, including how often the auction runs, the number of tradable periods between auctions, the number of tradable future periods, and the timing associated with environmental flow constraints. Example: the auction runs every 6 weeks, with users trading water for each week within the 6 weeks, and for 48 weeks into the future, with environmental flow constraints modeled by day.

Bid block A quantity and an associated price provided by a bidder, interpreted as a step in a bidder's demand function.

Catchment A region that is relatively isolated hydrologically from other regions.

Cone of depression A conical drawdown of the aquifer head around a well when the well is taking water.

Confined aquifer An aquifer between aquitards.

Consumer surplus The sum of the differences, over all consumers, between each user's value and the price the users pay. Example: a user is willing to pay $100 for a megaliter of water, but the price is only \$60, so that consumer's surplus is \$100−\$60 = \$40.

Demand function A mathematical formula that describes how a person's value for something, such as water, varies with quantity. A user will pay more for the first unit of water than for the second. With increasing quantities, a user is willing to pay less for each additional unit. A user might even pay to have some taken away, as in the case of a flood.

Drawdown A reduction in head of an aquifer due to abstraction.

Dual price A value associated with a linear programming constraint; the value is the improvement in the objective function if the constraint constant increases by 1. In the smart market optimization model, the objective maximizes the total value of water, so the dual price is the change in total value of water if the constraint were tightened or relaxed by 1 unit. The dual price is the marginal value of the resource associated with the constraint. Thus, the dual price gives the price per unit associated with the constraint at the optimum.

Environmental flow requirements The water that is set aside to protect the natural and cultural environments.

Externality The impact of a transaction on a third party not directly involved in the transaction.

Gross pool market A market design in which users provide bids for the total quantity of water they wish to have, regardless of any initial allocation. Contrast with net pool market.

Groundwater Water beneath the earth's surface.

Groundwater discharge The flow of groundwater out of an aquifer.

Groundwater recharge The flow of water from above ground into the groundwater.

Head The height above sea level to which water, at a specific three-dimensional location, would naturally rise if a well were open in that location. Head can be thought of as the energy density of the water at a particular point.

Hydraulic conductivity A property of material, the ease with which water can flow through pores or fractures.

Hydrological year A yearlong period aligned with the annual precipitation cycle rather than the calendar. Often, the hydrological year starts at the end of the dry season.

Infrastructure Equipment used to store and move water, such as pumps, reservoirs, pipes, and aqueducts.

License Government permission to users to conduct activities related to taking water. This permission includes access rights such as digging wells in a particular place, specification of the allowed uses for the water to be extracted, and the identity of the users. The license also specifies a term (e.g., 5 years, 10 years, or perpetuity) and a nominal quantity that the user can take per period (week, month, season, or year). The nominal quantity is subject to regulatory change, especially in dry periods. See also "Allocation."

Linear program A set of linear equations with a linear objective function. The variables in the equations correspond to decisions, such as the amount of water allocated to each user, and the equations correspond to physical or logical constraints on those decisions, such as allowed limits on aquifer drawdown. The objective function specifies a decision-maker's value for each possible decision, whether to maximize profit or utility or to minimize cost or disutility. Solving the linear program finds the optimum values of the decision variables.

Marginal value The change in total value for a unit change in some variable or amount of resource. If two users of water have different marginal values for water, the users should be able to benefit from trade, where the user with lower marginal value gives up a unit of water to the other user in exchange for money. In a linear program, the dual price of a constraint is the same as the marginal value of the resource associated with the constraint.

Market clearing The process of settling open offers to buy and sell. In the smart market, the market manager chooses allocations and prices so that the selected sellers are willing to sell at the given price and the unselected sellers are not willing to sell at the given price and so that the selected buyers are willing to buy at the given price and the unselected buyers are not willing to buy at the given price. Usually, the quantity sold by willing sellers matches the quantity bought by willing buyers, but the quantities may not match in a complex catchment, because of losses, such as from evaporation.

Market manager A person who has the responsibility of operating the smart market. He or she is responsible for ensuring an orderly market-clearing process, for choosing which buy and sell bids to accept, for setting prices, and for making and receiving the associated payments. The market manager does not make the allocation decisions; rather, the optimization does. See also "Regulator."

Market power The ability of a firm or group to manipulate prices over an otherwise competitive market.

Megaliter A measure of volume, a million liters.

Net pool market A market design in which users provide bids only for changes from their existing initial allocations. Contrast with gross pool market.

Optimization Choosing the values of variables in a math model, such as a linear program, that maximize or minimize a given mathematical objective, subject to a set of constraints.

Over-allocation In a given period, if the regulator finds that an environmental requirement cannot be satisfied when all users take their entire allocation during the period, then the catchment is "over-allocated" for that period. If all environmental requirements are satisfied and at least one requirement is more than satisfied, the catchment is "under-allocated."

Pigouvian tax A tax imposed by government to penalize activity that causes a negative externality. The policy goal is to include the marginal damage in the consumption decision, in an attempt to reach an efficient outcome.

Regulator The government agent responsible for enforcing users' limits, for setting and managing the environmental flow requirements, and for supervising some responsibilities of the market manager. See also "Market manager."

Reservoir A natural or artificial lake or pond from which water is abstracted.

Response coefficient The reduction in head at an assessment point at a given time current or future, for one unit of withdrawal at a given well in the current period. Hydrologists use simulations of water flow to estimate response coefficients.

Smart market An auction where buyers and sellers submit bids and a market manager uses mathematical optimization to match the buying and selling of bids and sets prices. Smart markets are most useful when trades have complex interactions.

Spot market A market with goods for immediate or near-term delivery. A spot market for water allows quick changes in water allocations and contrasts with other markets that, for example, trade the long-term license for water abstraction.

Thin trading Market behavior in which trades happen relatively rarely compared with trades by the same people for other commodities or a market behavior in which the market has a small number of buyers and sellers.

Transaction cost The cost to arrange a trade, including the cost to find a trading partner, to negotiate the price and the conditions, to obtain regulatory approval, and to enforce the agreement.

Unconfined aquifer An aquifer without an aquitard above it, so its pressure is constrained by the atmosphere. The aquifer nearest the ground surface is unconfined, as it has an aquitard or impermeable strata below.

Unsaturated zone The layer of earth between the ground surface and the top of the topmost aquifer.

User In this book, anyone who takes water. This book uses the term to mean a person or firm that abstracts water for commercial use, as in agriculture and industry.

Water intake A place where a user abstracts water from surface water or groundwater. This book considers each combination of user and abstraction point as a different water intake. The total number of water intakes is the total number of user–abstraction point combinations.

Water table The surface of an unconfined aquifer where the head equals atmospheric pressure.

Well A place where a user abstracts groundwater.

Chapter 1
Water Allocation: The Joint Problem of Interaction and Transaction Cost

Abstract This chapter reveals the problem of water allocation as a joint problem of hydrological interaction and externalities. The chapter describes the reason to use markets to allocate water. The chapter explains why most current methods of water allocation have a high transaction cost, and summarizes the problems that need to be solved to improve water allocation.

Keywords Water allocation • Hydrology • Commons • Externality • Transaction cost

1.1 The Complexity of Shared Water

The world is facing a water crisis as demand grows, while in many places supply is falling as a consequence of climate change and environmental damage. Many people want a different or additional allocation for themselves or for another party or natural feature. Each of us holds a range of views about fairness, equity, justice, and sustainability, which would affect our actions if water users' licenses were changed. The collection of voices includes farmers, horticulturalists, hydroelectricity generators, environmentalists, recreation-seekers, local residents, indigenous peoples, and scientists, with many fitting in multiple categories.

When people allocate other goods and services, the discordant voices often find accord via markets. People do not argue much about allocation of dental floss, concrete, or haircuts, because allocation of these commodities is fundamentally simpler. Allocation of commodities from the mall requires only well-known and fairly simple mechanisms of trade, credit, transportation, law, and enforcement. What is the difference between allocating items bought at the mall and allocating our fresh water?

Water is different to other goods and services because water is shared. All users in a watershed—commercial, recreational, or environmental—want water from the same resource. It is all connected. This shared nature is what makes it so complicated, and contentious, and beautiful. Strong feelings rise quickly when people perceive damage to our shared resource.

J.F. Raffensperger, M.W. Milke, *Smart Markets for Water Resources*, Global Issues in Water Policy 12, DOI 10.1007/978-3-319-55008-4_1

The scale and extent of the use make water allocation unique. Because each person's use of water affects many other people, and because so much water is consumed, allocation of fresh water can be nothing like allocation of dental floss.

When anyone takes public water, his or her use modifies the behavior and quality of the water. The extraction will first affect the user's neighbor, and then it will affect people and the environment further away. After many days, the extraction will affect many people and many environmental features, to a varying extent, depending on the hydrology, the user's timing and quantity of take, and how the water is used. The use may be "consumptive" or "non-consumptive." It may be of "surface water" or of "groundwater." It may have some other useful label. Whatever the water is called, wherever it is obtained, however it is used, each human interaction with the environment affects many other people and the environment, in varying degrees, and over different time horizons.

The effects are felt across the catchment, though with decreasing and delayed intensity further away. These effects include lower aquifer levels at neighboring wells and reduced flow in streams. Our uses of most commodities affect mainly ourselves, the seller, and the seller's supply chain, but have little impact on our neighbor, much less a wetland 50 km away. These water uses may help some people, but usually they lead to hydrological changes that hurt other people and the environment, especially people and environmental features nearby. The hydrological changes are **externalities** of the user's operations. An externality is an impact of a transaction on a third party not directly involved in the transaction.

Managing these externalities due to the connected nature of water drives the need for societal vetting of every transaction. If a person requests water, society must ask, "What effect will this take have upon others?" If the effect is wholly bad, society should decline the request. If the effect is not too bad, society would like to approve the request so society can prosper. The externalities represent the trade's social impacts that a simple market ignores.

However, the interactions and effects are complicated, so evaluating these requests is also complicated, and subject to controversy. Most governments therefore have a water regulator. The water regulator's available tools are (1) scientific tools such as hydrological modeling, (2) the political process for allocating water for non-commercial users and the environment, and (3) some kind of license process for commercial users.

In decades past when water was generally abundant, water regulators had an easy time. Water use during a drought might lead to short-term localized depletion, but recharge of the water resource the next year usually brought the resource back to normal. Everyone took much less than the total, so the effects had a modest and short-lived impact on everyone else and on the environment. The vetting process for water use applications was easy.

As demand grew, the water regulator's role became more difficult. Now, government has fully allocated water almost everywhere and contention is common. Users' increasing demand for water puts them in conflict with each other, with the water management authorities, and with people who care about the environment.

Contention is greatest in catchments with **over-allocation**, where government has licensed more rights than there is water.

Making matters worse, full allocation and over-allocation are fuzzy concepts because of rainfall's natural variability. Fixed allocations with variable rainfall results in under-allocated water in wet years, but over-allocated water in dry years. Over-allocation pressures the environment. Commercial users feel they have to fight for every drop, while recreation seekers and environmentalists feel that nature is getting shorted. Government is caught in the middle. Resolution lies with the courts, which is expensive and mean.

To find accord among the many voices with regard to water, societies must find mechanisms that can mediate between users while managing the externalities of water use. Society cannot easily increase the available pool of water, so our best tool to get better value from the water is to incentivize some people to give up water and allow others to take more. Commercial users say they are operating a business, and have to make decisions on that basis. They could be motivated to give up water voluntarily if they were compensated for it. Society is already accustomed to this kind of exchange.

That a market is the most economically efficient way to allocate resources is settled in the scientific literature. Markets usually allocate resources better than quotas, government directives, or legal decisions. Without individual incentives to improve water management, users will waste water and damage the environment. By "waste," we mean something quite specific. For example, take Bob, who is using water and gaining a little from it, as in Fig. 1.1, while Ann could use that same water and gain more from it than Bob could. If water goes for a low-value use when a high-value use needs it, the water is being wasted. Societies need to find a way both (1) to allocate the available water so the commercial for-profit water goes to its highest value use, and (2) to set aside sufficient water for the non-profit and environmental uses. Ann should be able to pay Bob for water, so that both are better off. We need water markets for the same reasons we need markets for dental floss.

The so-called orthodox economic view is that government should get out of the way and let the market decide. Markets maximize benefits to society, if the market is properly implemented and regulated—and that is a big *if*.

Ann	Bob
If I had one more megaliter[a] of water, my dairy yield would improve by $150. If I had one less megaliter, I would lose yield worth $150.	If I had one more megaliter of water, my crop yield would improve by $70. If I had one less megaliter, I would lose yield worth $70.
Ann could buy one megaliter from Bob for $110. Both would gain $40.	

[a]A megaliter is a million liters

Fig. 1.1 The gains from trade

A good water market would account for the externalities, internalizing the impacts as an inherent part of market choices. Many of the externalities damage the environment. To manage these externalities, society needs to consider the **environmental flow requirements**, or the amount of water that should be maintained to protect both the natural and cultural environments. If the cost of water does not reflect the negative effects, then the cost is too low, users consume too much water, and the environment suffers. But once the trading system accounts for the externalities, the public should have no complaint with a transaction for water between commercial users, just as third parties have no complaint with a transaction for tractors between commercial users. As long as business follows the law, and the law protects against the externalities, the trade should be nobody else's concern. Water users are responsible only for adhering to the conditions of their licenses. Water users are not responsible for protecting the environment; the regulator is.

Users may suggest that they themselves can allocate water through trading, implying that they can account for the externalities without government. However, the water regulator, not the user, is responsible for ensuring that water allocation does not adversely affect the environment. Many people are all too tempted to violate constraints. The drive for profits tends to push against every constraint, whether that constraint is an absolute limit such as water in a reservoir, or a legal constraint such as a limit on groundwater rights. In short, government must regulate a water market through the rule of law. We will assume the rule of law, and we will assume that government has the means and will to enforce the law. Yet even with the rule of law, and with conscientious and honest users and good regulation, the water markets that have been tried have not worked well.

Societies want solutions that are simple and predictable, like the solution for allocating dental floss. People have made a lot of progress in improving water efficiency through irrigation technology, industry cooperation, and science, but they have made poor progress with improving systems for the allocation of water. If a simple and predictable solution were available, our clever parents and grandparents probably would have thought of it.

Many places today have water markets, but water is thinly traded—trades happen only rarely, especially in comparison with other commodities of similar status in the market, such as energy and labor, which are widely traded through sophisticated markets of all kinds. Farmers, for example, frequently trade crops, equipment, and land, but they trade water rarely. Some people might think that a lack of information on water availability and reliability causes water trading to be thin. Numerous markets operate fine with little information for buyers and sellers—garage sales for example. Water trading must be thin for some other reason.

1.2 The Transaction Cost of Seeking Water

If two users want to reallocate water by sharing it, presumably one user will take more water and the other will take less. This can be easy if the two users are taking water from a pipe that runs from a shared reservoir, but that convenient situation is rare. More commonly, the users have to find each other, and this searching is difficult and annoying. Then, when willing partners find each other, they cannot account for the effects on other users and the environment. So, the local water regulator reviews the change in detail. The reason for thin water trading is a high transaction cost. Let's see an example.

Ann is a successful dairy farmer. To develop her business, she needs more water for her growing herd, as in Fig. 1.2.

To get more water, she goes to the government water regulator and asks the regulator to increase her water license. She has to submit a formal application listing detailed information about her property and how she plans to use the water, as in Fig. 1.3. Ann's time and expense for this process are what the economists call a "transaction cost." The transaction cost is the cost of trying to make the deal happen.

Because the regulator has already fully allocated the water in Ann's catchment, the water regulator tells her "No."

So Ann decides to try to buy water from someone who already has an approved allocation (Fig. 1.4). She reasons that the offset should cancel out: the seller takes less water from the catchment and she takes an equal increase. The regulator should

Fig. 1.2 Ann needs more water

Fig. 1.3 Seeking approval for more water takes time, and is unlikely to be successful

Fig. 1.4 Negotiating is part of the transaction cost

therefore approve the transaction. To buy this water, Ann must search for someone who is willing to sell. Ann spams all her friends and family with email, places an ad in the newspaper, and even puts up posters down at the local tavern.

Eventually, she finds Bob, her neighbor. Ann has to negotiate with Bob as to the price, quantity, and timing. They have to write a contract, and must work out what to do if Bob doesn't do what he says he will do. Ann and Bob probably have to hire lawyers to protect themselves and to make sure everything is legal.

Weeks after Ann started this process, Ann and Bob go together to the water regulator to get approval for the trade. Ann's transaction cost is getting bigger.

The vetting process is also a large cost to government. Government must decide whether to approve the deal, and will ask for evidence to support the request. The regulator will want to recoup the costs of vetting the transaction, so it will charge an appropriate fee. If the regulator rejects the request, then Ann and Bob find that their invested time and money were fruitless.

Evaluating the trade is more complicated than evaluating a new abstraction. To make this decision, the regulator should get a hydrologist to consider the effects of the proposed change on other users and on the environment, and to consider how these effects change over time, as in Fig. 1.5. Although the deal between Ann and Bob would mean no change in the total amount taken from their two wells, the

Fig. 1.5 Getting government approval to a trade is part of the transaction cost

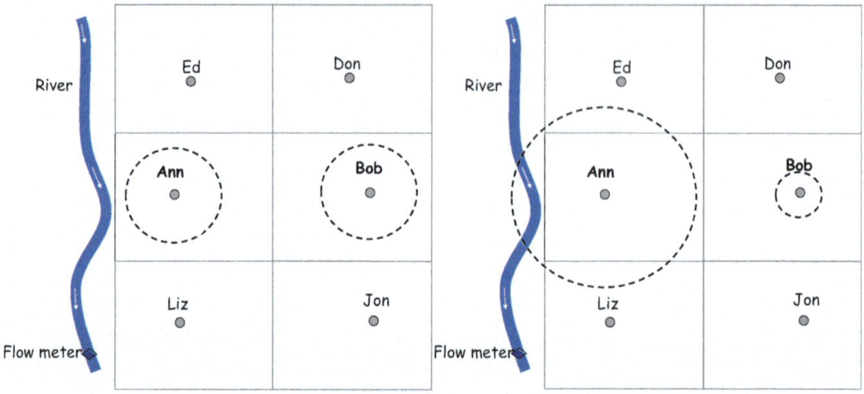

Fig. 1.6 The hydrological effects on the minimum flow in the river are different if Ann and Bob do not trade water (*left*) versus when they do trade (*right*)

water would be extracted at a different point and applied on different land, so the effects of the take would change. Suppose the hydrologist's concern is the minimum flow in the local river, as measured by a flow meter.

As diagrammed in Fig. 1.6, Ann's property is close to the river and Bob's property is far away from the river. Taking water from Bob's well further from the river will result in small delayed effects on the river. Taking water from Ann's well will result in immediate large effects on the river. The trade cannot be one-to-one, because Ann's impact on the river is bigger and sooner than Bob's impact, as measured by the government's river flow meter at the assessment point. Because users such as Ann and Bob are in different locations relative to the river, the effects of their use of water are not the same, so the exchange between them cannot be one-to-one. This contrasts with dental floss—we pay our $1.95 and we get our packet of floss without anyone else caring.

Therefore, government says "No," as Fig. 1.7 shows. However, the government hydrologist observes that if *everyone* in the catchment reduced their allocated take of water, in a way that was carefully timed to take into account how far each user was from the flow meter at the assessment point, the cumulative reduction could balance the effects on the river flow from Ann's additional take. The calculation would require a computer model to work out all the simultaneous effects on the river of all the changes in the various users' takes.

At this point, Ann will have had enough of the whole thing, and maybe she will take the government to court. The transaction cost is now enormous. Moreover, we haven't even said how much water is involved. In some regions, this transaction cost is even worse than our cartoon example. Lest the reader think that our cartoon story is unrealistic, consider Fig. 1.8, which shows that a water transfer can take 2 years in the new "expedited process" for California.

Fig. 1.7 Government cannot easily approve one-to-one trades, because the users' wells have different hydrological effects

The transaction cost comprises up to 70% of the total costs of water in water markets, according to empirical studies by Garrick and Aylward (2012) and Garrick et al. (2013). These figures are for the trades that happened, not the transactions that failed.

A high transaction cost on water trading results in fewer trades, and the few trades are large. Commercial users routinely factor the cost and chance of success of a license application into their business plans. Lund (1993) points out that users' aversion to this transaction risk (the chance that the trade could fail) makes trading even less attractive than it would be if users assessed the transaction purely on its expected value. If the transaction cost is larger than the gain from the trade, they back away. With or without trades, applications for larger water takes are likely to affect more people and more of the environment, so they result in longer vetting processes, with a concomitant larger transaction cost. Faced with such large costs, users make fewer and larger applications, more carefully assess the risk of rejection, and more willingly appeal unfavorable decisions where the potential gain is great.

Imagine paying $5,000 just to get to the grocery store—you would stock up for a whole year at least. Due to this high transaction cost, allocations and trades tend to be for large quantities of water and are fixed for years, decades, or in perpetuity.

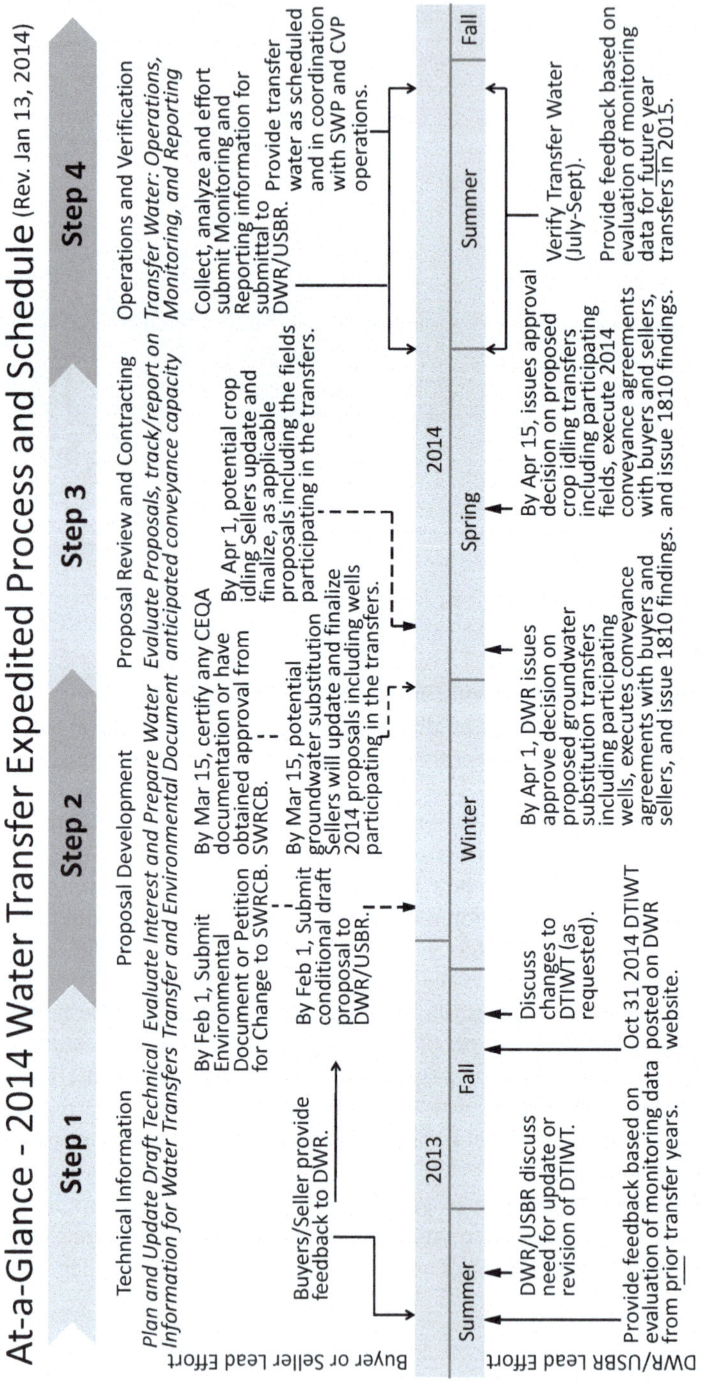

Fig. 1.8 Process for transferring water in California (Modified from CA Water Commission 2016)

1.3 Can We Solve Water Allocation Problems by Working Better with Communities to Have Them Agree on Allocations?

Local communities can often allocate scarce resources with little government regulation. People in a local community often share a limited resource in a way that sustainably maximizes its value. For example, fishermen in a small bay can work together to make sure each family gets enough, while ensuring that the local fish population maintains a healthy level. This "non-market" approach is desirable because the members of the community feel they are looking out for each other and know best how to allocate their local resources. So why not encourage this approach for water?

The Nobel Prize winner Elinor Ostrom described in detail the situations under which community-based allocation works, and the situations where it does not (Ostrom 2009). Briefly, if the resource is relatively simple and predictable and has clear boundaries, and if the community is geographically small, where most people know each other and can communicate easily with each other, then the community can develop processes to manage the resource.

Ostrom pointed out that the community approach will not work when the resource is complicated and unpredictable, or when the geographical area is large and the population using the resource is spread out. Seemingly simple factors can make a critical difference, such as whether a fish's eggs are external and visible, or internal and invisible. Water resources, generally, are complicated and unpredictable, literally flowing over thousands of miles, including underground.

The community, presumably through a committee of some kind, must match people who wish to have more water with those who are willing to give it up. The committee must check each transfer to protect from injury third parties and the environment. The committee must wrestle with the uncomfortable fact that, for many users, water is an input to a commercial enterprise, and is worth real money, so users who give up water will want compensation. This decision problem becomes particularly difficult when the community faces an application for a new use. Entrenched local decision-makers can easily overlook the issue of whether the new use is better for the community as a whole.

Community processes for water allocation can work for small well-defined areas, especially when the water is controlled in a reservoir, but do not work in general. As much as many people wish to allocate water through informal community processes, arranging the allocation is too difficult—the transaction cost is too high—just as it is with almost every water allocation mechanism currently used.

1.4 Can We Solve Water Allocation Problems with More Regulation?

Perhaps society could avoid a market approach by improving the allocation oversight process? Some people are not using water as productively as it *could* be used. The regulator could simply take water from the wasters and give it to those who would use it more productively, while ensuring water for the environment. Alternatively, if water use damaged the environment, the regulator could take water from the damaging use and give it to the environment. Maybe society could spend a lot of money on research and identify which crops *should* be grown on each plot of land, and then allocate water only in the quantity appropriate for that water use. These approaches are likely to end in confrontation.

Neither the current market nor government has an easy way to get one user to voluntarily give up valuable water, even if another user could use that water more productively, and even if one user were willing to compensate the other for giving up water. Nor can government increase water to the environment without taking it from someone. Any successful allocation system must allow for new high-value uses of water to replace existing low-value uses, while compensating the existing users. Current allocation systems have no convenient mechanism for compensating people who give up water, and any proposal that addresses only part of the problem, such as just the farmers, or just the environment, will certainly fail. If the regulator did have such an ability to change allocations quickly, the regulator has no way to know the users' values for water. If the regulator simply asked the users outside of a market, the users would have enormous incentive to exaggerate.

Furthermore, a single user is rarely the only cause of an environmental problem. Rather, the impacts on the environment at a particular place and time are usually caused by many users, spread over a wide area, using water over a long span of time. The regulator has no objective way to choose which user should cut back.

These command approaches to resource use have their own high transaction costs. They lock in specific land and water uses, and they decrease innovation. An underlying benefit of a water market is that, because so many individuals in the market have incentives to use water efficiently, water users will be continuously looking for ways to obtain more value from their water. Any system that mandates how water should be used is pointing down the road to inefficiency.

1.5 Is Infrastructure the Solution to the Problem of Water Allocation?

Water users want reservoirs, pipes, and canals, because this infrastructure gives them control over the water. Storage allows access to water in the dry season, and reduces the risk of unexpected drought. In addition, infrastructure makes both regulation and water trading easier, because the water is more visible and controlled.

Infrastructure lowers the transaction cost for water trading. So should society promote water infrastructure to enable trading? Probably not.

First, infrastructure reduces the transaction cost per trade only somewhat. Users still trade only rarely, even when the water is controlled; sometimes users trade once a season, but more likely only once per year or decade. Other industries, such as petroleum products and electricity, have high infrastructure costs yet an otherwise low transaction cost per trade, with more-active trading than for water even when water is controlled. So something else is going on. Part of the problem is that infrastructure enables control of only some of the water, surface water especially. Infrastructure does not allow easy trading of groundwater, nor trade between surface water and groundwater. Infrastructure, when it is available, does not reduce the transaction cost far enough.

Second, users and government should make infrastructure decisions based on the **marginal value** of water. That means infrastructure decisions should be made based on the price for an additional small amount of water taken at a specific place and time. Society could find the marginal value of water by first paying for infrastructure and then trading water. Unfortunately, the marginal value may be too low to justify the initial investment. Users and government have little information about the value of water, so building infrastructure just to reduce transaction cost is ill advised. Reducing the transaction cost by adding infrastructure can easily lead to waste.

As Australia improved its water trading, billions of dollars' worth of infrastructure became idle. People found that they no longer needed some infrastructure. Trading allowed the water to move to where users most needed it, and the new location was sometimes different to where the infrastructure designers had forecasted.

Because the transaction cost for trading is so high, infrastructure that appears economical would not look so economical if the decision-makers knew the true marginal value of water. Policymakers should instead find ways to reduce the transaction cost per trade sufficiently, and design a market intentionally so that trading is active. Then users will be able to see the correct marginal value of water. After that, users can work out the correct value of new infrastructure. Society wants improved allocation from infrastructure, but society can accomplish this better, and at lower cost, by lowering the transaction cost of water trading.

1.6 Here Are the Problems with Market Approaches that Need Solving

All of us need to see the true price of water, so we all know its value. Some of us complain when our government wants us to switch to low-flow showerheads and low-flow toilets. Nevertheless, we can't have water that is both in ample supply and priceless. We know that if it's free we'll use more than we really need—why fix the leaky tap if the water's free? But if we see a price on the water, if we lose money from the leaky tap, we'll fix it. And we'll put in the low-flow showerhead of our own accord. Most critically, we need the big users—business, especially agriculture—to see this price signal and to use water more efficiently.

Society also needs to price water to protect the commons, because society shares water as it does the atmosphere, a fishery, or the medieval village grazing ground. The commons can be abused as individuals use it too freely. If business people take as much as they wish, without a cap, they will damage the public resource. Who will buy on behalf of the fish and birds? Even if someone wanted to buy water for the environment, as many people want to do, society has no way to enable such a transaction The political process is too inflexible to properly manage either the commercial use of water or the dynamic environmental flows. By imposing a cap, business is constrained—and no one likes living under a constraint. But the economy needs constraints. Business is damaging the commons, and is injuring non-commercial users and the environment. Society has to cap water use to protect the commons.

Users can appeal rejected applications by citing scientific uncertainty, especially when hydrology models do not have the authority of law. "You don't know for sure that the environment will be impacted by my application for extra water because the rainfall is uncertain," they will say, "but the negative economic impacts of denying me water are measurable and certain." At multiple points in the process of evaluating a trade, the principle of protecting the environment can crumble in the face of certain economic benefit and uncertain environmental effect. Many times, the government calculates correctly that the environment will be injured only "rarely" or "to a small degree". But the level of acceptable environmental damage can vary among regulators, and even over time with the same regulator, and the flow of water remaining for the environment tends to get cut in every transaction. The accumulation of these multiple cuts can leave the environment with low flows.

A business's damage to the commons also hurts other businesses. When one farmer draws down the aquifer, his or her neighbor has to spend more to draw water from a deeper well. When one farmer wastes water, his or her neighbor loses the chance to use it productively. Responsible water users benefit from markets that include enforcement and a low transaction cost. Markets provide an incentive for civil behavior by water users and those trying to protect the environment, and deter those who try to benefit from uncivil behavior (Hardin 1972). Users and the environment will benefit from a well-designed water market. Society needs water markets to (1) protect the commons from over-use, and (2) enjoy the gains from trade.

The problem to date has been that no one has worked out *how* to implement a market for water in the general case. While economists stress that a market is the best way to allocate resources, they too rarely describe in detail how to set up the market. A very few researchers and organizations have developed trading systems for water, but those have shortcomings, a high transaction cost in particular. Consequently, too many water markets around the world operate badly, having few trades while damaging the environment. As far as we can tell, no one has a market for wholesale fresh water in complicated hydrology with protection for the environment and third parties, and with correct price signals to users.

It should be clearer now why water management is so difficult. One person's use of water directly affects many other people and the environment, which results in contention. The connectedness of water creates the need for government supervision of water use. To try to resolve the contention, society has a vetting process, and the need for vetting is understandable and appropriate. This government supervision is currently society's only way to manage the many-to-many effects of water use.

The vetting process needs improvement. No matter how good the decision was about the license, the resource will approach full allocation over time. With full allocation comes more contentious decision-making, and so the vetting process becomes more complicated and expensive. In addition, with thin trading, water users have less incentive to introduce more efficient uses, and little economic penalty for poor use of their water. Over time, the allocations will become economically inefficient. To solve the problem of water allocation, policymakers must find a way to lower the transaction cost for trades and facilitate entry of new water users. Setting up such a market will take time and care, and its prerequisites may be difficult to implement.

The goals of water allocation are, first, to steward the environment for future generations, and, second, to maximize the value of the resource to society. Looking at water use around the world, it is easy to get discouraged about the potential for sensible water markets. The many parties involved—farmers, large industries, municipalities, governmental water regulators, environmental groups, and the public—engage in gamesmanship, grandstanding, and uncivil behavior. The tension between players and the long history of contention and lawsuits are not conducive to the collaboration and compromise needed to develop a good water market. Any market needs to have earned trust between the various parties. Let's reflect on what policymakers must do to build trust so water markets can strengthen. The policymaker wants to:

- protect the environment from damage due to commercial use of water. Taxpayers don't want to pay commercial users to take less water simply to protect the environment. Society would like a method to allocate more water to the environment without public demonstrations and court actions.
- have more productive and competitive agriculture and industry through more-flexible water allocation, with users sharing water in the most productive way. This requires lowering the cost to obtain or change a water allocation.
- compensate users who give up water.
- match water allocation to its actual availability on a more real-time basis. Over-allocation should never happen.
- settle these arguments in this generation, and to leave a better society and a better environment for our children.

People have traded for millennia. Yet, with all the knowledge available to researchers and government, people have still not worked out a good way to buy and sell water. It seems that society needs an entirely new idea. Whatever the new idea, it needs to have a low transaction cost. It must coordinate each user's taking of water with that of other users, over time and space. It must move water to its highest value, because anything less is waste. And the new allocation system must protect the environment.

1.7 Overview of the Smart Market

Suppose a hydrologist built a detailed hydrological optimization model of a catchment, which included all the relevant users, and many constraints to ensure that environmental flow requirements were satisfied. The model would balance the flows of water at each point, while accounting for expected inflows, reservoir capacities, etc. The optimization would find the optimal amount of water to give to each user, taking into account each user's demand for water, including the user's different values for water in each time period and the decreasing marginal value for more water. The hydrologist could then click a button and solve the optimization model to maximize the value of available water, while taking into account the quantity of available water at every location and the environmental flow requirements.

This optimization model would depend on the hydrologist knowing the value of water to each user. But these values are private. The hydrologist could interview each the user, asking about the user's value for water, but the users would not have the incentive to tell the truth and probably couldn't be bothered to think about the question anyway.

Instead of guessing users' water values, the hydrologist could ask the users, via a web page, to bid for the water directly. By having their money at stake, users would have an incentive to tell the truth. The optimization model would determine the allocations and prices, while ensuring that the environmental requirements were satisfied. That is the smart market in a nutshell.

The smart market would be operated by a central market manager. The market manager would run the hydrology model, accept bids from users, and publish the final allocations and prices. These prices would vary with space and time, with a price for each user's location, for each environmental assessment point, and for each time period. To users and policymakers, the prices would be a rich source of information about the value of water.

Under the smart market, users trade time-based leases for water. Their ongoing rights are unchanged. They merely rent out their water allocations for specific time-limited periods, e.g., a week or a month—January water is a different commodity to July water. The water user bids only for water at his or her location. They do not need to worry about who supplies the water or who is selling. The trading partner is always the market manager. Nor does the user have to worry about environmental flows. As long as the user follows the rules, the environmental requirements will be satisfied, because constraints in the optimization require those flows to be met.

The smart market design is not a laissez-faire free market, but a managed regulatory framework that enables voluntary market operations. Instead of a one-to-one market, it is a many-to-many **spot market**. In this spot market for water, users can frequently trade short-term leases for water, while not changing the long-term or permanent rights. The smart market enables users to adjust quickly to changes in the economy and the environment that affect their use of water, because the market allows users to quickly adjust their allocations. Users who do not wish to trade would still be subject to regulation, including adjustment of their water allocations to match actual water availability, and the requirement for metering. Users who do wish to trade would have the benefit of a simplified mechanism for selling or buying water allocations; one that considers interactions of uses and has a low transaction cost.

References

CA Water Commission. https://cwc.ca.gov/Documents/2014/02_February/February2014_Agenda_Item_10_Attach_1_TransferProcess.pdf. Accessed 13 Mar 2016.

Garrick, D., & Aylward, B. (2012). Transaction costs and institutional performance in market-based environmental water allocation. *Land Economics, 88*(3), 535–560.

Garrick, D., Whitten, S. M., & Coggan, A. (2013). Understanding the evolution and performance of water markets and allocation policy: A transaction costs analysis framework. *Ecological Economics, 88*, 195–205.

Hardin, G. (1972). *Exploring new ethics for survival*. New York: Viking Press.

Lund, J. R. (1993). Transaction risk versus transaction costs in water transfers. *Water Resources Research, 29*(9), 3103–3107.

Ostrom, E. (2009). Social-ecological systems: A general framework for analyzing sustainability of social-ecological systems. *Science, 325*, 41–49.

Chapter 2
Water Markets: Theory and Practice

Abstract The purpose of this chapter is to review some economic fundamentals to market approaches, to point to the extensive literature on markets by water policy analysts, and to examine applications of water markets worldwide.

This chapter first reviews the problem of hydrological externalities, the Coase Theorem, and why the Coase Theorem does not usually apply to water resources. The next section observes that many researchers have shown the improvements that society should expect from water markets, noting especially the high transaction cost associated with almost all water markets. The final sections examine actual experiences in various countries that have attempted to encourage water trading.

Keywords Water markets • Externality • Transaction cost

2.1 The Problem of Hydrological Externalities

2.1.1 What Is an Externality?

We have defined an "externality" as the impact of a transaction on a third party not directly involved in the transaction. Since this chapter is about the theory and practice in general, the reader should interpret each component of this definition in a general way—the impact, the transaction, and the third party.

By "impact," we mean any significant effect on someone else or on the environment. A farmer taking water changes the hydrological flows. These changes could be both a lower river flow and a higher run-off from land. The same take of water may raise or lower the aquifer, add nitrate to the aquifer and to local streams, and decrease or increase risk of flood. A wastewater treatment plant discharges materials into a river, increasing the phosphorus in it and thereby causing hypoxia in local and distant waters. The increase in phosphorus is an impact, as is the hypoxia. A hydroelectricity generator taking water from a reservoir changes the hydrological flows, too, though in different ways than the farmer. With hydroelectricity generation, the upstream reservoir loses water, while the downstream river gains water, possibly diluting the farmer's nitrate run-off. Some of the impacts are positive, but most are negative.

J.F. Raffensperger, M.W. Milke, *Smart Markets for Water Resources*, Global Issues in Water Policy 12, DOI 10.1007/978-3-319-55008-4_2

By "third party," we mean any person who has no direct voice with the person making the impact. Such people could be downstream farmers, people who need clean drinking water in downstream towns, fishermen, hikers, boaters, etc. The hydrological impacts result in loss to these third parties. The losses could be financial, as when a lowered aquifer prevents a farmer from irrigating, or emotional, as when the contamination in a river inhibits recreation. We will also treat the environment as a third party, as a proxy for the many people who will be directly and indirectly affected by environmental changes. Determining how environmental impacts affect people is complicated and contentious. So governments usually treat the environment as a third party, such as by specifying quantitative standards for minimum river flows and maximum daily loads for nitrate. The quantitative environmental standards serve as proxies for people's needs and wishes.

By "transaction," we mean any human behavior that has significant hydrological impact, including drawing down an aquifer or reservoir, or changing the quality of water through a change in temperature or alteration of substances in the water. We intentionally exclude from this discussion actions that have no significant impact. The actual transaction was the direct behavior—taking water to irrigate crops, processing wastewater, or releasing water from a reservoir to generate power. The agent doing the behavior could be an individual land user, a non-profit organization, a corporation, or a government agency. The person doing the transaction usually has no specific desire to impact the third parties. We could assume that people may be vindictive or mischievous, but that assumption is not necessary here.

Thus, a hydrological externality is the hydrological impact of a water-related transaction on another person or the environment. This helps to define the limits of the system that we aim to address. Outside of this system is the market into which the farmers sell their crops, the houses that flush into the local sewers, and the power network that receives the generated electricity. Those behaviors can be coordinated by market processes, for crops, for sewage treatment, for electricity, outside of the water allocation process. They affect water allocation, in that high crop prices increase demand for irrigation, higher urban water consumption tends to increase sewage flows, and high prices for electricity increase demand for hydroelectricity. But the actors involved with those markets come to the water market seeking water resources specifically. Our interest in externalities is about water, not about crop prices per se, or toilets, or power bills.

These externalities represent social impacts: tangible and intangible costs to people and the environment from users taking water resources. The water users gain value, often direct and financial, from the water resources, but they do not pay for the externalities. That is why the impact is called an externality—someone has a profitable transaction with the system of hydrological resources, others have costs as a result, and the cost remains external to the person with the profitable transaction: they do not compensate those who lost.

A "good" market will account for these externalities, where appropriate, and factor in the externalities as a part of the wider market choices. Those who are impacted by the externalities are thus compensated in some way, at which point the externality becomes "internalized" to the market, and the transaction still has impact but the externality is greatly reduced.

2.1.2 The Coase Theorem

Societies, to varying extents, are accustomed to tolerating and managing externalities. A farmer's production, the wastewater plant's operation, and electricity generation contribute good things to society, so governments allow some impacts for these goods. A person usually has the right to use his or her own property even when that use causes reasonable irritation to others.

When some people find the irritation to be unreasonable, they may go to their policymaker or court to seek relief. What should the policymaker do? Let us assume that the policymaker is benevolent and seeks a policy that maximizes total welfare.

To assess whether the water user's impact is reasonable, the policymaker may compare the advantage to the user with the cost to the third party. If the policymaker finds that the user, say, Ann, gains $100 by irrigation, and a third party downstream, say, Bob, loses $80 from additional saturation, then the policymaker may allow Ann her action, as society benefits by $20.

This situation looks unfair to Bob. If Bob had some legal ability to prevent Ann's behavior, either by stopping the outflow of a pipe, or through an enforceable right not to be flooded, then Ann should be willing to compensate Bob $80 for Bob's loss. Ann could carry on, though with only $20 profit, and Bob should be indifferent. If Bob had no such right, then he must bear the loss (Fig. 2.1).

To the Nobel Prize winner Ronald Coase (1960), this problem is reciprocal. That is, if Bob stopped up Ann's outflow pipe, Ann would be injured by $100 (ignoring for simplicity Ann's ability to substitute). Coase observed that preventing Ann's action would impose harm on Ann, just as allowing Ann's action would impose harm on Bob.

Coase also noted that as long as the right (e.g., either to flood or not be flooded) were initially assigned either to Ann or to Bob, then Ann and Bob could bargain to find the social optimum. If, in fact, Ann's gain were only $75, and Bob's loss were still $80, then one expects Ann to restrain her action, whether because Bob paid her $5 for the restraint, or because Ann did not want to pay $80 to Bob. The policymaker could favor either Ann or Bob, awarding the right based on careful analysis, established precedent, or caprice. According to Coase, however the initial right is allocated, the market will provide the socially optimal solution. Thus, the critical step for the market to maximize social welfare is that the rights are in fact allocated, not that they are allocated in any particular way. One way to interpret Coase is that "the market will provide," laissez-faire.

When the Coase Theorem assumptions are true, it works well. In an ideal case, government need not intervene except to help clarify and enforce the rights. The avoided interventions are government-imposed taxes, user charges, and expensive regulations. This ideal case is the one espoused by libertarians, free marketers, and

Fig. 2.1 Bob and Ann could bargain more easily if the initial right were assigned

conservative politicians. Wearing the hat of the benevolent policymaker, we wish for this ideal case.

2.1.3 The Failure of the Coase Theorem for Water Resources

When the assumptions for the Coase Theorem are false, the theorem fails. The theorem fails not because Coase was wrong, but because the conditions under which he said it would be true do not hold. He pointed out such conditions in his article, but still the statement of the Coase Theorem drives many wrong-headed political speeches and ill-formed policies. Governments seized on Coase's argument, as it relieved them of the work of governance. Businesses also approved, as it gave them the justification to spew first and see if anyone complained later. In particular, the Coase Theorem does not hold in general for water resources. Instead, the commons suffers a tragedy.

For design of a water market, the most critical assumption in the Coase Theorem is that the transaction cost is zero. That means that Ann and Bob can communicate easily, perhaps because they are neighbors. They can negotiate and agree on the price conveniently, perhaps because Ann can see the effect of excess water on Bob's property, and Bob knows the value of Ann's crop. They each can enforce the contract on the other, whether through the courts or physical control of the water. In general, for water resources, this assumption of a zero transaction cost is abjectly false. The transaction cost associated with a significant use of water resources is almost always high.

With the case of Ann and Bob, the parable assumes that bargaining takes place in a broader market that provides some price information; but in fact it is only a bilateral negotiation at best. Ann and Bob are unlikely to be able to observe the cost to the other. Both have incentive to misrepresent their costs, insisting that they are more injured than the other is. Ann and Bob will not have equal bargaining power; Bob could represent a multinational corporation, while Ann operates her own dairy farm. Bob may not even know that Ann is to blame for his inundation. The transaction cost is big—far above the zero cost required for the market to find its frictionless optimum.

Here is an example more typical of water resources: Consider a major farming community in a large catchment. The farmers take water from a river and from local groundwater to irrigate grass, which the farmers also fertilize; they then feed it to cattle. Much water is lost to evaporation. Some water runs back to the stream, now laden with nitrate and phosphorus from the fertilizer. Downstream users, both farmers and non-farmers, receive much less water, and what they can get is contaminated. The downstream farmers must switch to some activity that does not require irrigation, towns must buy expensive water treatment facilities, and fishermen lose their livelihoods.

One user's water abstraction today has a great many externalities, with impacts on many neighbors near and distant, and on the environment, both immediately and (with increasing uncertainty) in the future. Individuals downstream have great difficulty in identifying those upstream actors who are contributing to an impact—because the actions causing the impact can be occurring on private land out of sight of neighbors and regulators—and, when they do, they may find that the contribution of each actor to the impact is only small. The many farmers each impose a small fraction of the externality, and if any single farmer were to stop farming, the resulting improvement to the third parties would be small. Each user would have to bargain with virtually everyone else in the catchment (ideally using good hydrological data), which would raise an even larger transaction cost, problems of free riding, budget constraints, and negotiating with precious little information on the costs. The problem is a huge many-to-many transaction, completely unlike the tidy scenario with Ann and Bob.

Consider another example of water pollution. When a factory discharges into a catchment, they have incentive to hide the fact of their discharge and to play down its effect; the neighbors may experience the effect (say, an illness) without knowing the cause, much less the culprit. Thus, the neighbors face a large transaction cost in finding the cause. Eventually, they may realize their illness is due to polluted water. After finding the cause, the neighbors would have difficulty in proving their case. Thus, the neighbors face a large transaction cost in asserting their rights. Having found the culprit, ordinary citizens rarely have the budget to sue a large corporation. Thus, because of a budget constraint, they cannot act. Eventually, someone might obtain sufficient resources to file a lawsuit, but citizens with small budgets would be unlikely to outlive a long battle with a deep-pocketed firm. This is a probabilistic budget constraint, where a poor player with an apparent high chance of winning will almost certainly lose to the wealthier player. Thus, in this case, the transaction cost is astronomical and budgets are paralyzing.

Further, despite existing water license systems, rights tend to be poorly specified. In many places, users can take all the groundwater they wish, with no restraint. Even when the license is limited somehow, water managers assume users will not use their full water license, leaving users to think they do have those rights; but regulators retain authority to shut down wells on short notice in drought. This is not from anyone's lack of attention or diligence, but simply an attribute of the license system. Licenses are poorly specified because adjusting the allocations is a long and expensive process, and this large transaction cost means that neither users nor government want to do it very often.

Perfect information of the catchment—perfect hydrological models—will not solve the problem of poorly specified licenses. Good hydrological models are easy to make; given the expected inflows, they can accurately estimate the effects of users' abstractions. But in many places users' abstractions are not metered. From the regulator's point of view, much of the uncertainty about the catchment is due to this lack of measurement and lack of information about users' plans, not primarily the uncertainty of the weather (though that is important), and even less so inaccuracies in hydrological modeling.

Thus, the Coase Theorem fails for water resources under current systems of water allocation. The unregulated market will *not* provide. Laissez-faire fails to allocate water resources efficiently. Scientists saying this include McAfee (1997) and Baliga and Maskin (2003), Maskin being a 2007 Nobel Prize winner.

Consider this example paraphrased from Baliga and Maskin: in winter, my neighborhood has too much smoke from wood burners, and the smoke is making people sick. If enough of us bought heat pumps, the smoke would decline to a level such that no one got sick. If I buy a $3000 heat pump, everyone in the neighborhood benefits without having to pay anything; they get a free ride. I decide to wait for someone else to buy the heat pump. And therefore no one will buy a heat pump. When this situation happens, society does not reach the optimum where the cost of heat pumps needed to give reductions in emissions equals the health benefits from lowering emissions. Ideally, everyone would pay for the illness caused by their smoke to others, which would correctly incentivize people to buy heat pumps. Following this logic, Baliga and Maskin concluded that government is necessary to help achieve the least-cost solution.

Business people are not responsible for the externalities per se; they are responsible only for adhering to the conditions of their own permits. Government is responsible for rationing access to the commons, so a market in hydrological resources in any form, whether one-to-one trades, or the proposed smart market, probably should not be operated purely privately (such as by a users' cooperative).

A market without regulation doesn't work. Further, the market needs more than regulation; it needs help.

If the local water regulator went to any commercial user and showed the user a price for water, say $2.25 per megaliter, for which the user could at that price have a choice—just for this week—of selling a quantity or buying a quantity, the user will almost certainly react either to buy or sell. If the water were worth more than $2.25 to the user, the user would buy; if it were worth less the user would sell. But there's no use waving a just-this-week gain in front of the user if they have to pay thousands of dollars and wait months or years for approval from the water regulator. The catch is that the transaction cost must be low enough to make the transaction worthwhile.

2.2 Water Allocation by Pricing Mechanisms

In his article "The Tragedy of the Commons," Hardin (1968) observed that users tend to despoil public resources if users are not charged for their damage to those resources. The price actually paid by water users almost everywhere in the world is less than the true marginal cost to society of additional water use, and hence is too low. Price signals can make environmental protection more effective. Current allocation systems are prone to criticism from both those who want more water for commercial use and those who think that more water should be allocated to the environment. Many commercial users might believe that environmental protection is too tight, and that government underestimates the negative economic impact of increased environmental protection. Many other people might believe that government allocates water to commercial users without full study of the environmental impact. The introduction of prices would improve the discussion of benefits and costs of increased or decreased water use within a catchment.

Of course, researchers have exhaustively studied this charging problem, often with the phrase "price mechanism." A long line of water policy analysts have argued persuasively that some pricing mechanism is desperately needed (Dinar and Subramanian 1997).

The phrase "price mechanism" is a catchall unspecific term. The implication is that the policymaker wants some kind of **Pigouvian tax** in which users pay for the externalities they impose on others. For example, a regulator could charge all water users a fee per unit of water used. But because of the transaction cost and the complexities of hydrology, determining a good Pigouvian tax is difficult. Since the hydrology varies and users' impacts vary, a given tax may overcharge some users and undercharge others. A better approach would take into account each user's marginal value for the water, but the regulator has no way to discover each user's marginal value for water.

True marginal cost prices maximize **consumer surplus**, which is the sum of the differences, over all consumers, between each user's value and the price the users pay. Marginal cost prices would signal users where and when to use water, where to install augmentation (e.g., dams), and how much to pay for the water. Users can best discover the marginal cost prices through a market.

The economics literature about water specifies conditions for a water market. Strong markets are characterized by a low transaction cost, frequent small trades, government actions and regulations that promote markets rather than hinder them, enforcement of those regulations, public information, clear property rights, protection of those property rights against externalities (such as unintended effects of others' trades), and prices that reflect the marginal cost to society.

We shall be brief in our discussion about existing water markets: other authors have written recent and extensive volumes on these (e.g., Easter and Huang 2014; Garrick 2015). The short story is that past markets have not worked well, because of the high transaction cost. Water markets work especially badly where rights are poorly specified, where the rules have perverse incentives, such as "use it or lose it", or where water is badly controlled.

Hirshleifer et al. (1960) wrote a comprehensive book, *Water Supply*, that seems dated, but over 50 years of additional literature seems to have added surprisingly little. They begin with a strong endorsement of allocating water with market forces, then they describe the hydrological cycle, the economics of allocation, investment in new supplies, and water law in the U.S. They include case studies of New York and southern California, describing expensive public policy mistakes resulting in over-investment. They give several recommendations towards setting up a market. In their words:

1. Establish a system of law that permits a clear definition of the extent of property rights in water, particularly adapted to eliminate commonality problems which may pertain with particular force in the case of water.
2. Provide certainty of tenure for these rights and establish a clear basis for their transfer between individuals and private or public agencies under voluntary contractual agreements of purchase and sale.
3. Provide a procedure whereby rights to yet-unowned water may be secured by individuals, private or public agencies, or political entities.
4. Establish rules and procedures for the protection of outside parties against the spillover effects caused by the actions of water-right owners. These rules would be similar to zoning in the case of land use and would protect against harmful effects such as pollution, flooding, and the creation of drainage problems, by water-right owners. Alternatively, procedures could be set up whereby injured parties can obtain compensation for damages.
5. Develop and provide information regarding the extent and quality of water resources and the technical and economic factors relating to the various possible sources of supply.

Hirshleifer, DeHaven, and Milliman point out some of the technical difficulties with establishing a market for water. "Hydrologists have argued that the distinction between underground waters in definite channels and diffused or percolating waters is highly tenuous because nearly all ground water is percolating, that is, moving by laminar flow as contrasted with the turbulent flow of most surface waters." *Water Supply* seems to have only modest concern for the environment. Perhaps this would have been different if the book had been written more recently than 1960.

Smith (1982) proposed a two-part deed mechanism for groundwater rights. The first deed conveys the right to draw a fraction of the natural inflow, and the second deed conveys the right to draw recoverable water from the aquifer. The rights would depend on actual water available, and would be tradable like real property. Wells would have to be metered. Smith then described the positive outcomes associated with the tradability.

Despite the promising title, "Formal Water Markets: Why, When, and How to Introduce Tradable Water Rights," Thobani (1997) reviewed the problems caused by centralized water planning, such as inefficient use and overcapitalization in dams and hydraulic infrastructure, and that government has not always provided enough water to poor people. He points out some of the practical problems of implementing water pricing, including difficulties in measuring water flows (including return flows) and determining opportunity costs.

Dinar (2000) gives an excellent discussion of the issues from a political economy perspective, describing water markets around the world, and clearly analyzing the strengths and shortcomings of those markets. The book includes discussions about the economics of groundwater.

Bate (2002) examines water availability globally, and points out that irrigation projects are routinely beset by corruption and mismanagement. He observed that, without the ability to calculate prices, "quota systems have proven insufficient to constrain demand." The difficulty of creating a water market lies in the transaction cost, including the costs of creating, monitoring, and enforcing property rights. He lists the success of water markets in Pakistan, India, Chile, and Spain, while observing how the inadequately developed markets in Mexico and the U.S. have caused drainage of aquifers and inefficient water use. However, he observes that, "on balance, water trading countries in arid areas have a better than average environmental performance." He concludes by pointing to international tensions due to water scarcity, but adds that "water markets can stop water wars."

Similarly, Howitt (2000) wrote that "... efficient transfers rely on the detailed local knowledge of water use and value, known only to the ultimate user of the water ... The price offered for water condenses a wide range of values and preferences into a single signal that is readily understood and usually persuasive ..." It is understandability and persuasiveness that water markets need.

The papers in Easter and Huang (2014) describe the benefits of water and markets generally, and the environmental and economic damage from mispriced water. They discuss climate change and war as related to water. They distinguish formal from informal markets. In informal water markets, neighbors trade following informal community rules, often for small quantities. Formal water markets trade over larger regions, usually for larger quantities. Water markets reduce the ability of government employees to extract rents, because they have less control over allocations. The authors recognize and describe the fundamental problem of third party effects.

Partially based on Coase, the best science for water allocation is to use tradable rights, as suggested by Tietenberg (2006). In Tietenberg's words, "Tradable permits address the commons problem by rationing access to the resource and privatizing the resulting access rights." Government rations the total quantity, but, within that limit, users can trade. Tietenberg cites carbon emissions trading and New Zealand's fishery quota system as examples. Smart markets are a step beyond simple markets in that they can help manage complex markets involving common resources, thereby improving economic efficiency within sustainability constraints.

A properly functioning market has three requirements. First, users should not have much **market power** (an ability to control the price). If they do have market power, the market will not be efficient. Empirically, market power quickly falls as the number of users increases (Montero 1998; Deshel 2005); as few as ten can be enough to severely limit market power. Water users in most catchments would number in the hundreds; hence, market power is unlikely.

Second, the market must have sufficient monitoring and enforcement, as any market requires. For the commons, this monitoring and enforcement role falls to government. In general, management of the commons requires government intervention, some kind of market supervision. As Gordon (1954) put it:

> The blade of grass that the manorial cowherd leaves behind is valueless to him, for tomorrow it may be eaten by another's animal; the oil left under the earth is valueless to the driller, for another may legally take it; the fish in the sea are valueless to the fisherman, because there is no assurance that they will be there for him tomorrow if they are left behind today. A factor of production that is valued at nothing in the business calculations of its users will yield nothing in income. Common-property natural resources are free goods for the individual and scarce goods for society. Under unregulated private exploitation, they can yield no rent; that can be accomplished only by methods which make them private property or public (government) property, in either case subject to a unified directing power.

Third, trades in the market must have a low transaction cost. Users need to be able, with relative ease, to find each other, negotiate a price and a contract, close the deal, and enforce the contract. "Market failure" occurs when these conditions are not met, which explains why water markets worldwide trade so little.

2.3 Experiences with Water Markets

2.3.1 Chile

People advocating for water markets often cite Chile's experience, but researchers have also mentioned Chile to bolster arguments against them. Augusto Pinochet's military government in Chile aggressively pursued market-oriented policies. In 1981, the government allowed water resources to be privatized and it eliminated legal restraints to water markets. The goal was to allow users to own water like any other property, and to encourage private investment in water infrastructure. This was an extreme "free market" approach. Perhaps surprisingly, little trading takes place.

Chile has abundant water resources, but they are spread unevenly. The country has dry places, and irrigation plays an important role in agriculture. Some trading occurs in areas that have controlled water in reservoirs and canals.

Thobani (1997) writes positively of the Chilean experience, indicating that water markets there have improved economic growth and protected the environment. He lists common arguments against water markets, including objections that "life-sustaining water should not be bought or sold," that the poor will be excluded, that water transfers will damage the environment, and that those who stand to lose will obstruct change. His final reasons seem most compelling: that setting up new institutions is expensive and complicated, implementing initial water allocations is complicated, and protecting against externalities is challenging. Thobani believes that the best way to assign initial water rights is based on historic use, admitting that this would provide a windfall to some farmers. He goes on to recommend an information campaign and auctioning of new rights.

We cannot do justice to the full range of water trading around the world, so we focus on a few places where governments have encouraged water trading.

Good reports on water resources around the world are widely available. See, for example, Garrick (2015), and Salman (1999), which includes an article by Jacque Emel with a proposal for groundwater rights, and reviews of groundwater management in Bangladesh, Jordan, Mexico, Nepal, and Yemen.

Along with his supportive comments on the Chilean experience, Thobani mentions some problems: (1) Private hydroelectricity companies are in conflict with farmers where the companies were given water rights with conditions about releasing water to farmers at prescribed times. "This experience underscores the need to specify weekly releases of non-consumption rights [such as the right to release water for hydroelectricity generation]—whether the rights are tradable or not is irrelevant." (2) Chile does not enforce its environmental law. (3) To prevent competition, a single power company has managed to obtain rights to most rivers.

Bauer (1998) describes in detail the Chilean experience with water markets from 1976 to 1995. In contrast to Thobani, Bauer notes that users do not often trade water, and when they did, they usually traded water rights with land.

> ... [D]espite the illusion of laissez-faire, markets are not simple, automatic, or self-maintaining mechanisms: how they operate depends on wider legal and institutional frameworks, political and economic conditions, and geographic context. In the case of

water these are often unfavorable or at least constraining. A balanced view of water markets would aim to take advantage of their benefits without exaggerating them or trying to wish away their flaws.

The flaws in the Chilean system are instructive. The system still relies on one-to-one trades, with each individual trader having specific property rights. The prevention of third-party effects still requires regulatory oversight that inhibits completely free trade.

Another flaw is that Chilean law does not always give certainty of title. In theory, all water rights have had to be registered with the title offices, but the law also recognizes unregistered rights under earlier legislation, and users cannot trade unregistered rights. As Bauer noted, "Thus in many areas there are an unknown number of legally valid rights which in theory could be asserted at any time."

Record keeping of water property rights is uncoordinated and incomplete. No agency reviews all trades, so complaints between traders must be resolved through cumbersome and slow court processes. Bauer observed that "... price signals remain uncertain, ambiguous, or contradictory in much of the country, indicating that water is simultaneously cheap and valuable. ... [T]here is no guarantee of being able to buy rights at an affordable price in the future." In particular, potential buyers have not offered prices high enough to compensate owners for the risk of drought.

According to Bauer, groundwater is "abundant and relatively untapped in central and southern Chile." As a result, users see no benefit to trading in many parts of the country, and the lack of trades between users of groundwater and surface water limits the ability to find mutually beneficial trades in other parts. The military government interfered to prevent auction of water rights in several cases, particularly to protect hydroelectricity projects, and this history has further inhibited trust in water trading.

Another difficulty that Bauer found was that the "mere fact of an abrupt legal change—whether market-oriented or not—caused problems for people who were slow to be informed or marginalized from the legal and governmental system."

Chile has continued to try to improve its markets. In 2012, a promising Internet-based system for spot transactions was begun for the Limarí basin in an attempt to reduce the transaction cost (Easter and Huang 2014, 117).

The experience in Chile shows that a "free market" solution is not sufficient to produce a more efficient system of water allocation, notwithstanding the tidy supply–demand graphs in the economics textbooks. Simply legally allowing a free market for water does not make a market happen, unless the water is controlled. Society may as well remove the legal restraints to brain transplants. They may be legal, but a host of practical issues must still be resolved.

2.3.2 The U.S.

The U.S. has 50 states, and water laws can vary between states, within each state, and sometimes by county. The legal frameworks in the various states hinder development of a common pool market, and result instead in attempts at water markets between a few similar users.

Most U.S. states base water allocations at least partly on appropriative doctrine. This means that users acquire rights to water by use, and the earliest user has senior rights to later users. In dry years, senior users get water first. Many states also have riparian doctrine for water rights. Appropriative doctrine contrasts with riparian doctrine, in that, in the latter, ownership of land confers the right to use water on or adjacent to the land, while all upstream and downstream users can have equal rights. In both of these doctrines, users are required to use water "reasonably," and users argue about what "reasonably" means, especially in dry years, with resolution through the courts.

Texas has at least four different water doctrines, from highly controlled water with well-specified rights in the Rio Grande basin, to completely uncontrolled rights in groundwater. Texans can take as much groundwater as they wish, which naturally affects surface water. In spite of this, surface water users have no recourse in the courts if excessive groundwater use denies them water.

Another difficulty with the growth of water markets in the U.S. has been the lack of water metering and regulators' lack of visibility of total water used, and sometimes even of where water is used. For example, California does not require farmers to meter their water, despite the current severe drought and depleting aquifers.

Water markets in the U.S. have helped meet some water allocation goals. Researchers have analyzed water markets in the U.S. since at least 1995 when Lund and Israel published their work on the topic (Lund and Israel 1995). One type of trading method that seems to work in a limited way in the U.S. is the "bulletin board" system where traders find partners via electronic means. Water authorities developed these with participation of water users in the Rio Grande basin and the northern Colorado River basin. This method can lower the need for vetting when users are taking water near each other from a surface source (lake or river), so they have a similar impact on the environment, making these simpler one-for-one trades acceptable.

California has used a "water bank," in which the state served as the main broker for water trades (Israel and Lund 1995), originally as a response to the 1977 drought. The Department of Water Resources (DWR) offered users a price of $125 per acre-foot, regardless of source. Sellers could fallow farmland, use groundwater instead of surface water, or transfer water from local reservoirs. DWR bought 820,665 acre-feet. DWR then offered the water for sale at $175 delivered at the entrance to the California Aqueduct. DWR sold only 389,970 acre-feet. Crudely speaking, the market did not clear because the prices were not market-clearing prices. California also operated water banks in 1992, 1994 and 1995. Other researchers have written in more detail about the California water bank mechanism (Howitt et al. 1992; Howitt 1997).

Many American policymakers and users want more market-based water allocation, and researchers have spent considerable effort studying the potential for it. These researchers, in varying degrees, look at ways to modify water laws or policy to allow more market-focused water allocation. Harou et al. (2009) gives a good review of hydroeconomic modeling. However, 20 years of advocacy and research have produced surprisingly limited results in trading behavior.

So the U.S. water policy varies regionally. Every region is different, hydrologically, legally, and culturally. Some areas will be more amenable to the smart market approach than others, but only people steeped in the local knowledge will know. Many good reviews are available. A recent thorough review of U.S. water law and markets is Lachman et al. (2016).

2.3.3 *Australia*

The Australian government supports trading and has helped enabled trading by eliminating restrictions. Drought has motivated the Australians to make water markets work, especially in the Murray–Darling Basin, which has the most active water markets in the world. As in Chile, most Australian trading is in reservoir and other surface waters, though, increasingly, there is also trading in groundwater basins.

To allow markets to function effectively, the Australian government made institutional changes. They found that they had to improve record-keeping of water rights. They had drawn down their aquifers, so they started licensing groundwater as well as surface water, and they have implemented rules requiring the two to be managed as one resource. The Australian states are relatively independent, so Australia worked to allow trade between states. In 2014, the federal government introduced new rules intended to harmonize conflicting rules between states.

The improvements in water trading in Australia have benefited the country considerably (National Water Commission 2010). The country's water markets are active because the government has aggressively eliminated restrictions on trade. For example, a user can usually trade a water access right without any restriction, based on the water use, historical or anticipated level of use (thus avoiding the use-it-or-lose-it dilemma), even when the water might be transported outside the basin, or if the person has some other water right. A trade can occur within a given area even if the water is over-allocated; the authority's logic is that a trade will not cause further over-allocation (Murray–Darling Basin Authority n.d. 1). Traders must disclose traded prices. Regulators may restrict trades because of physical constraints, the need to address hydrological connections, and the need to protect the environment. Private brokers arrange the trades, generally through the Internet. Figure 2.2 shows the steps in the trading process. Users can trade groundwater within a basin, treating the aquifer as a single reservoir. The Australian

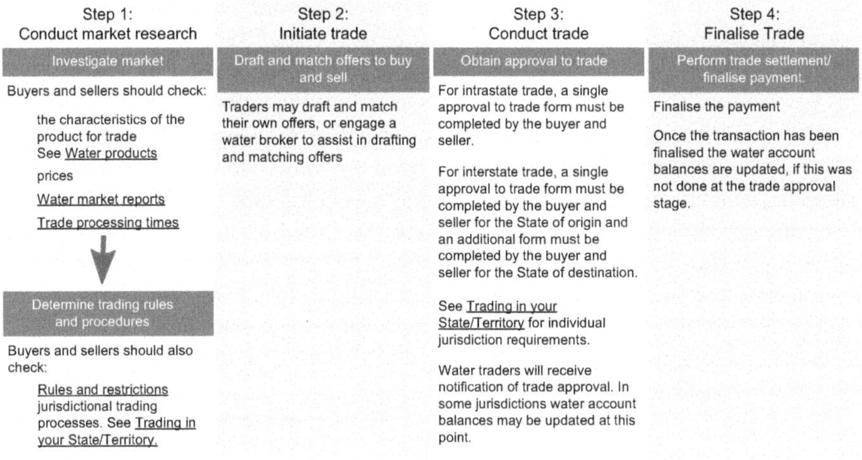

Fig. 2.2 Steps for an Australian trade of water allocation (Reproduced from National Watermarket 2015, © Commonwealth of Australia (Bureau of Meteorology) 2015)

government publishes considerable information about water trading on its many websites.

Australian water markets have been using sealed bid auctions and bulletin board systems for water markets for years (Bjornlund 2003). These systems eliminate the need to search for trading partners, reduce the costs of bargaining and contracts, and presume enforcement. These methods can work in situations where users have roughly the same environmental impact and take from a relatively uniform water body at a similar time (a river, canal, or lake). The auction systems show the potential for trades and they develop trust in trades, which together have led to markets that are more active.

Water trading up until now has been somewhat successful where the water is physically controlled, as in the Maipo River Basin in Chile (Rosegrant et al. 2000), the Texas Rio Grande basin, and parts of the Australian Murray–Darling basin. For the most part, though, water markets throughout the world suffer from thin trading, where trades are rare and big. Simply allowing trading is insufficient to create a successful market, because of the many-to-many effects.

An especially good case of Australia's water trading is the privately operated Waterfind platform, at www.waterfind.com.au. Waterfind's trading mechanism could be viewed as a type of smart market, in that trades are computer-checked against a large complex set of rules. The operators of the Waterfind system have worked to ensure that the list of rules agrees with the local authorities' rules. The computer system automatically submits the trades to the local authorities for checking, but, by that time, the approval is straightforward, because Waterfind has already checked it. Figure 2.2 shows the steps to trade. Those steps are mostly automated. As of Aug 2015, Waterfind hopes to start operations in California (ABC 2015).

When active water trading began in Australia, many people immediately sold unused rights, which gave them cash and "freed up" water on paper. But the unused water had benefited the environment, and now was being taken for irrigation, which then damaged the environment. In some cases, the "freed-up" water rights exceeded the actual water available in a typical year. The result was that the Australian government had to buy back rights to protect the environment. Whatever system society uses to allocate water, it should have convenient knobs and dials, where regulators can adjust quantities upwards and downwards to manage effects on individuals and the environment.

Hanak et al. (2011) examined Australia's water markets. They pointed out that California law requires flows for the environment, and requires that the market can trade only real water. Australia water rights holders can trade "paper water," rights without actual water associated; they can also sell the gross water, rather than only the net water, from irrigation improvements. These regulations can result in greater injury to the environment. From their page 341:

> The clarity of property rights and lack of environmental limitations on water sales have allowed the Australian water market to evolve quickly and with lower transaction costs than in California. However, this market efficiency comes at a high environmental cost, which became increasingly apparent during the past record drought. To address this problem, the national government plans to spend over $3 billion in the next decade to purchase back environmental water and to invest over $5.5 billion in water savings whose yield will be shared between irrigators and the environment.

They then mention researchers who have called for change so that Australian water allocations reflect the value of environmental water.

References

ABC. *California's millennium drought turns attention to South Australia for expertise and experience*. http://www.abc.net.au/news/2015-05-03/california-millennium-drought-turns-attention-to-sa-expertise/6438004, 3 May 2015.

Bate, R. (2002). Water—Can property rights and markets replace conflict? In J. Morris (Ed.), *Sustainable development: Promoting progress or perpetuating poverty*? (Chap. 15). London: Profile Books.

Baliga, S., & Maskin, E. (2003). Mechanism design for the environment. In K. G. Mäler & J. R. Vincent (Eds.), *Handbook of environmental economics* (1st ed., pp. 305–324). Amsterdam/London: Elsevier Publishing.

Bauer, C. J. (1998). *Against the current: Privatization, water markets, and the state in Chile*. New York: Spring Science+Business Media.

Bjornlund, H. (2003). Efficient water market mechanisms to cope with water scarcity. *Water Resources Development, 19*, 553–567.

Deshel, M. D. E. (2005). Optimal allocation of tradable pollution rights and market structures. *Journal of Regulatory Economics, 28*(2), 205–223.

Coase, R. H. (1960). The problem of social cost. *Journal of Law & Economics, 3*(1), 1–44.

Dinar, A. (Ed.). (2000). *The political economy of water pricing reforms*. New York: A World Bank Publication/Oxford University Press.

Dinar, A., & Subramanian, A. (1997). *Water pricing experiences: An international perspective*. World Bank technical paper, no. WTP 386. Washington, D.C.: The World Bank. http://documents.worldbank.org/curated/en/580151468768597034/Water-pricing-experiences-an-international-perspective

Easter, K. W., & Huang, Q. (Eds.). (2014). *Water markets for the 21st century: What have we learned?* (Global issues in Water Policy 11). Dordrecht: Springer Science+Business Media.

Garrick, D. E. (2015, July 31). *Water allocation in rivers under pressure: Water trading, transaction costs and transboundary governance in the Western US and Australia*. Cheltenham: Edward Elgar Publishing, Cheltenham, UK.

Gordon, H. S. (1954). The economic theory of a common-property resource: The fishery. *Journal of Political Economy, 62*(2), 124–142.

Hanak, E., Lund, J., Dinar, A., Gray, B., Howitt, R., Mount, J., Moyle, P., & Thompson, B. (2011). *Managing California's water: From conflict to reconciliation*. San Francisco: Public Policy Institute of California.

Hardin, G. (1968). The tragedy of the commons. *Science, 162*, 1243.

Harou, J. J., Pulido-Velázquez, M., Rosenberg, D. E., Medellín-Azuara, J., Lund, J. R., & Howitt, R. E. (2009). Hydro-economic models: Concepts, design, applications, and future prospects. *Journal of Hydrology, 375*(3–4), 627–643.

Hirshleifer, J., De Haven, J. C., & Milliman, J. W. (1960). *Water supply: Economics, technology, and policy*. Chicago: University of Chicago Press.

Howitt, R. E. (1997, September 28–30). *Water market based conflict resolution*. Rosenberg International Forum on Water Policy, San Francisco. http://ciwr.ucanr.edu/files/186749.pdf

Howitt, R. E. (2000). Is California's future hydraulically sustainable? *California Agriculture, 54* (2), 10–15. doi:10.3733/ca.v054n02p10

Howitt, R., Moore, N., & Smith, R. T. (1992, March). *A retrospective on California's 1991 emergency drought water bank*. California Department of Water Resources. Sacramento.

Israel, M., & Lund, J. R. (1995). Recent California water transfers: Implications for water management. *Natural Resources Journal, 35*, 1–32.

Lachman, B., Resetar, S., Kolra, N., Schaefer, A., & Curtright, A. (2016). *Water management, partnerships, rights, and market trends*. Santa Monica: RAND RR933, RAND Corp.

Lund, J. R., & Israel, M. (1995). Water transfers in water resource systems. *Journal of Water Resources Planning and Management, ASCE, 121*(2), 193–205.

McAfee, R. P. (2010). Four issues in auctions and market design. *Revista de Análisis Económico – Economic Analysis Review, 13*(1), 7–24. http://www.raeear.org/index.php/rae/article/view/117/226. Accessed 18 Mar 2017.

Montero, J. P. (1998). A simple auction mechanism for the optimal allocation of the commons. *The American Economic Review, 98*, 496.

Murray–Darling Basin Authority. (n.d.1). *Guidelines for water trading rules, introduction*. http://www.mdba.gov.au/sites/default/files/pubs/01_WTG-REFERENCE_final.pdf. Accessed 30 Apr 2016.

National Water Commission. (2010). *The impacts of water trading in the southern Murray–Darling Basin*. National Water Commission, Commonwealth of Australia. http://www.nwc.gov.au/__data/assets/pdf_file/0019/10783/681-NWC_ImpactsofTrade_web.pdf. Accessed 30 Apr 2016.

National Watermarket. (2015). http://www.nationalwatermarket.gov.au/about-trade-delete/index.html. Accessed 30 Apr 2016.

Rosegrant, M. W., Ringler, C., McKinney, D. C., Cai, X., Keller, A., & Donoso, G. (2000). Integrated economic–hydrologic water modeling at the basin scale: The Maipo river basin. *Agricultural Economics, 24*(1), 33–46.

Salman, S. M. A. (ed.). (1999). *Groundwater: Legal and policy perspectives*. Proceedings of a World Bank seminar. World Bank technical paper, no. 456. World Bank, Washington, DC.

Smith, V. L. (1982). On divestiture and the creation of property rights in public lands. *The Cato Journal* (Winter), *2*, 563–629.

Thobani, M. (1997). Formal water markets: Why, when and how to introduce tradable water rights. *The World Bank Research Observer, 12*, 161–179.

Tietenberg, T. H. (2006). *The evolution of emissions trading: Theoretical foundations and design considerations* (Vol. 230). Washington, DC: RFF Press.

Chapter 3
How Water Flows, How It Is Modeled, and How We Can Optimize It

Abstract The purpose of this chapter is to give the reader an understanding of water, water simulation models, and optimization. We briefly discuss different types of hydrological optimization, which will lead the reader to the smart market.

After that, the chapter introduces the operations research method of linear programming. The point of this section is that linear programming can find the best solution to a particular type of problem and that it can calculate prices. We describe the use of hydrological optimization in the existing research literature, and then we show how to combine the hydrology model and linear programming to create the smart market for water.

Keywords Groundwater • Hydrology • Optimization

Aside: Groundwater Terminology

Terminology varies by country and discipline, so we shall define terms. Generally, we shall follow convention from hydrological science.

A "cell" in a hydrological model such as MODFLOW is a three-dimensional block, within a three-dimensional set of cells representing the aquifer. Thus, a stack of cells will be under any point on a surface map.

A "cell location" is a point on the ground surface above the center of a cell.

Head is defined as the height above sea level to which water, at a specific three-dimensional location, would naturally rise if a well were open in that cell. Thus, for wet cells, head measures the level of the water in the cell, plus the water pressure at that level. If the stack of cells contains no physical barrier to vertical water flow, head will equalize within the stack. The head need not be equal at all cells at one cell location: head in lower cells can be either higher or lower than in those above. Head can be thought of as the energy density of the water. The use of head in water discussions is similar to its use in the phrase "head of steam."

Drawdown is a reduction in head due to abstractions determined by the market.

(continued)

J.F. Raffensperger, M.W. Milke, *Smart Markets for Water Resources*, Global Issues in Water Policy 12, DOI 10.1007/978-3-319-55008-4_3

Abstraction is the act of taking water from any source.

A **user** is someone who abstracts water.

A **well** is a place where a user abstracts groundwater. The top of the well is at ground level. MODFLOW assumes that each well extends from the top of the top wet cell (where water is abstracted) to the bottom of the same or a lower cell. Thus, a well has a location on a two-dimensional surface map, but can join a series of cells vertically. We assume that each well does not draw water from below the well bottom, and that the user can always abstract at the same rate, irrespective of head.

An **assessment point** in terms of groundwater is a cell at which head must be measured, and possibly constrained. The regulator may define some assessment points in response to environmental requirements, such as the need to ensure that flow in a stream is maintained, or to prevent saltwater intrusion from a coast. The regulator may specify other assessment points to prevent wells running dry. Assessment points may be viewed as boundary conditions on the market.

The wells and assessment points are within one **catchment**, which is a region of land that is relatively isolated hydrologically from other regions.

3.1 Groundwater Flow

Smart markets for groundwater will be more challenging to develop and explain than smart markets for surface water, because of the complexity of groundwater flow. Many readers may find it useful to learn more about the complexity of groundwater flow. Figure 3.1 shows the basics of the water cycle, giving names to the various water flow processes.

If the groundwater concepts get baffling later in the book, refer back to this section. The take-away point from the hydrology section is the concept of the response coefficient—the amount that the groundwater level falls at a particular place and time when someone takes water out of a well.

The term water cycle, while reflecting the circular nature of water processes, glosses over how differently water flows through surface water (lakes, rivers) and groundwater. Table 3.1 summarizes the world's water resources and the time water molecules spend in various parts of the water cycle. Even though evaporation and precipitation are common processes, very little water exists in the air; a water molecule typically spends only a few days in that part of the system. Water at the surface of the planet, in rivers, lakes, reservoirs, and wetlands, is also a relatively small proportion of the total fresh water: a water molecule that falls as rain over land, and does not enter groundwater, might spend only weeks or years at the

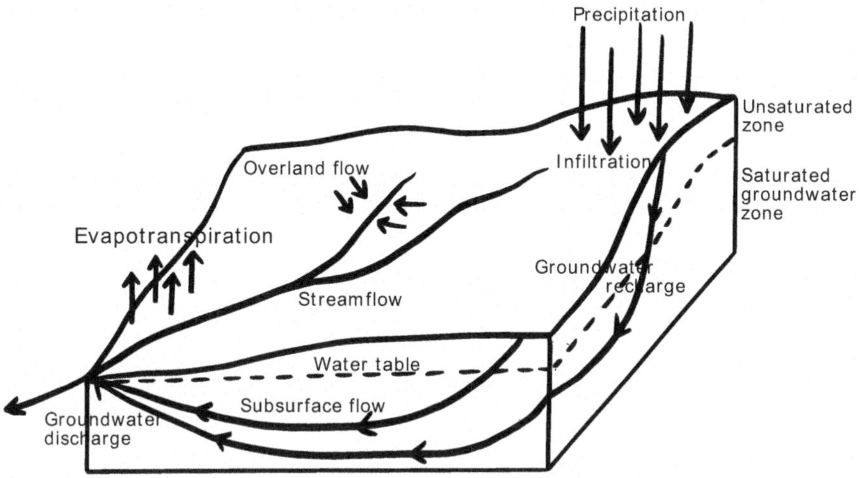

Fig. 3.1 Schematic representation of the hydrological cycle (Modified from Freeze and Cherry 1979)

Table 3.1 Estimate of the world's water balance

Reservoir	Volume ($km^3 * 10^6$)	Volume (%)	Average residence time
Oceans	1370	97.25	4000 years
Ice caps and glaciers	29	2.05	50 years
Groundwater	9.5	0.68	100 (shallow) to 10,000 (deep) years
Lakes	0.125	0.01	70 years
Soil moisture	0.065	<0.01	1 month
Atmosphere	0.013	<0.01	10 days
Streams and rivers	0.0017	<0.01	1 month

Sources: Pidwirny (2006), Freeze and Cherry (1979)

surface before returning to the atmosphere or becoming seawater. Most non-frozen fresh water on our planet is groundwater. This water moves slowly; a water molecule that becomes groundwater might spend centuries or millennia in an aquifer.

From using bathtubs and garden hoses, everyone develops some intuition about the flow of water. It flows downhill because of gravity and because water molecules slide smoothly over each other, unlike molecules of a rock. Water flows more quickly in a channel when the sides and bottom of the channel are smoother and straighter, because less energy is lost overcoming friction. And deeper water is under more pressure, endangering the submariner. A patch might not leak when a bathtub is nearly empty, even though the patch is below water level, but the water gushes out when the bathtub is full. The area of the bathtub does not matter; the depth of the water it contains does. The depth of water is proportional to the pressure, and that pressure represents an energy potential that can break open the patch.

The key to describing water flow, then, is the water's level of energy. Energy is the potential to do work, to move a mass over a distance. Water has both potential energy and kinetic energy; groundwater has little kinetic energy because it is moving so slowly. Water has potential energy for two reasons: either because it is at an elevation (and so could fall and spin a wheel), or because it is under pressure (and so could burst a patch in a bathtub or a tank).

At the bottom of the bathtub, the energy of the water at the outlet is all pressure energy and no elevation energy: the water is at the same elevation as the outlet, and so the water in the tub has no elevation energy. The water's pressure energy (above atmospheric pressure) is proportional to its depth above the outlet. At the top of the bathtub, the energy of the water is all elevation energy. It has no pressure except for the atmosphere above it. The water's elevation energy is proportional to the distance to the outlet, which is the depth of the water above the outlet.

The water's level of energy at a specific point is always proportional to the depth that a column of water would rise to if taken from that point. The water levels indicate water energy, and because water energy helps us figure out whether water will flow, the levels give us a simple way of evaluating the ability of water to flow. Consider a water tank with the top of its water level at 100 m above sea level, and another water tank 1 km away with water at a height of 90 m above sea level, with the two connected by a pipe. One can estimate the flow between them and the friction that the water would have to overcome.

What happens when water flows? Let's use your bathtub (not ours) for a little experiment. At the bottom end of the tub, put a plug in that can be opened and closed from below, and put enough water in it at the top end of the tub to cover the base. The water level is the same throughout the length of the bathtub, and the water does not flow. If you put more water in at the top end of the bathtub, the level of water at that end increases: that increases the head. The difference in head between the top end and the plug end of the bathtub drives water to flow, and the length of your bathtub is important in figuring how quickly it will flow. If you stop putting water in at the top, water flow will slow down as the head difference decreases until both ends of the bathtub have the same water level (or head) and water flow stops.

If you could put a fabric over the surface of the water during this adjustment of water levels in the bathtub, you would see that the fabric starts out with a steeper slope and then the inclination decreases with time until the water stops flowing and the fabric is flat. This flat nature of the water levels with time has led to the use of the term **water table** to describe the level of a water column over an area.

Groundwater is not that mysterious. Groundwater and groundwater flow follow the same physical laws that govern surface water and its flow as in the bathtub. Only the physical setting differs. If you take your partly filled bathtub of water, put a good screen over the outlet, and start shoveling soil into the bathtub (this is why we are using your tub and not ours!), the soil will fall to the bottom and you will have a little lake. If you keep putting in soil, you will eventually not see any water, but the water is still there. If you scraped off the top layer of soil where you poured the water, the soil would be moist. It would have water attached to the soil grains, partly filling the spaces between soil grains – the soil pores. If you dug further down, you would find a soupy mess. That is where the soil pores are completely filled with water. The water that completely fills the pores is called groundwater. Although the water that partly fills pores is below ground, it is difficult to extract and so is not a practical water resource and is not considered groundwater. This zone where the soil pores are partly full is called the **unsaturated zone**.

To measure the head at various points in your bathtub, you cannot simply measure the depth of water, because the water is below the soil. You could put a straw into the groundwater at the top end of the tub and another at the plug end. Water will rise in the straw to a level that reflects where the saturated pores begin. If you wanted to, you could get a measured string, dip it into the straw and figure the depth to the water level, thereby finding the depth that the water level is above sea level. You could then congratulate yourself on having just measured the hydraulic head! The head in the straws will be the same at both ends of the tub. If you pour water into the top end of the bathtub, it will go into the soil and seemingly disappear. It will become groundwater (except for a bit that stays in the unsaturated zone), and you will see the level of water in the straw at the top end of the bathtub go up. Over time it will go down, while the level in the straw at the plug end of the bathtub will go up—until both levels are equal. Water will have flowed, underground, from the top end to the plug end of the bathtub. If you open up the plug of the bathtub for a while and then close it, the plug end will lose water, and water will again flow down to that lower level.

Outside of your bathroom, similar processes go on. Groundwater flows from high head to low head. Water going into groundwater is called **groundwater recharge**, while water leaving groundwater is called **groundwater discharge**. Groundwater recharge can happen from rainfall, but also from a river that leaks water downwards, and from excess irrigation. Groundwater discharge can go into a spring, a lake, a river, or the ocean. The level of groundwater at any time is the result of two processes—groundwater recharge and groundwater discharge. Groundwater is extracted at wells. The water levels in the bathtub straws define the surface of the water table, just as the fabric did with your unsoiled bathtub. A pumped groundwater well is like a straw in your soil-filled bathtub that you suck on (not recommended at home!).

The rate of groundwater flow, like that of the water flow between tanks, is proportional to the difference in hydraulic head: if you tip up the top end of your bathtub, water will flow more quickly from top end to plug end. In contrast with your clear bathtub, water flowing through your dirty bathtub instead has to overcome a lot more friction as it moves around the solids. The smaller the solids are, the more friction there is. Because the water loses more energy flowing past smaller particles, it has less hydraulic head, and the flow is slower. Similarly, the longer your bathtub is, the more particles the water has to flow past, and the slower the flow.

Groundwater flows through pores that vary in size over orders of magnitude and that are proportional to the particle sizes. Clay particles are of the order of 0.001 mm in size, silt particles 0.004 mm, sand particles 0.4 mm, gravel particles 4 mm. The flow rate in groundwater is still proportional to the difference in hydraulic head, but is greatly affected by the size of the pores. For the same difference in head, groundwater could travel 100 m/day in gravel, 1 m/day in sand, 1 m/year in silt, and 1 m/100 years in clay. Materials that transmit water more readily have a higher value of **hydraulic conductivity**.

Pumping groundwater from the subsurface creates a **cone of depression** around the well. This is like the depression in the water level that you observe around the plughole in your clean bathtub after you pull the plug. The cone of depression reflects the decrease in head around the well because of groundwater pumping. Figure 3.2 shows the problems that can result. For the initial case in Fig. 3.2a, a home use creates a small cone of depression, but, because this is far from a nearby stream and the amount of water extracted is small, the effects are minor. In Fig. 3.2b, a new groundwater user is pumping a greater amount of water and is closer to the stream. This cone of depression is deeper, and leads to water moving out of the stream into the well and a lowering of the water table. This could lead to significant environmental impacts. In Fig. 3.2c, the use is even greater, creating a very large cone of depression. The effect is so great that the stream is drained dry, and the groundwater level has gone to the bottom of the well of the house on the left. This means that the house on the left must drill a deeper well to continue to tap the groundwater.

This example shows how relatively small changes in the groundwater level (whether due to pumping, greater groundwater discharge, or reduced groundwater recharge) can lead to large impacts on the environment or other water users. Looking back at this example helps explain why Ann's groundwater use in Chap. 1 causes greater environmental damage than Bob's. Ann is closer to the stream, and pumping there will lead to a greater decrease in head at the stream, and

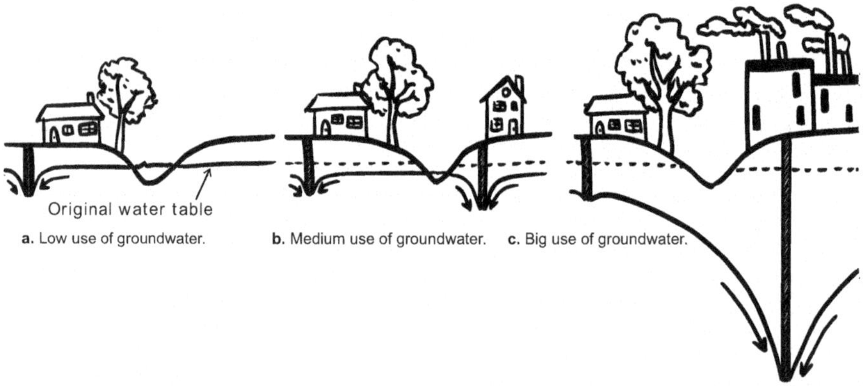

Fig. 3.2 Change in water table from increasing demand for groundwater (Modified from Hughes 2016)

so more loss of stream flow into groundwater. Because Bob's well is further away, pumping the same amount of groundwater has less effect on the stream.

The impact also varies with time. In a clean bathtub, the cone of depression forms quickly when the flow out of the plughole changes. For the dirty bathtub, the friction of the soil particles slows down this response. Looking at the decrease in water levels (head) near a well after it begins pumping, the heads decrease most near the well and soon after pumping increases, with less strong and later effects further away, and further into the future. A brief period of pumping will have an effect that becomes insignificant at some point in time limiting the need to consider water use effects indefinitely. The head at any point will recover with time as more recharge occurs elsewhere in the system.

Because of this time effect, a model used to evaluate the impact of groundwater use needs to consider how much water someone wants, where they plan to take it from, and the future demand. This makes managing groundwater resources significantly more challenging than managing water use from a tank, a lake, or a river. To help manage the problem of timing of water use, the manager needs to consider the decrease in head at other locations in future periods for a given amount of groundwater abstraction.

Figure 3.3 shows these effects for a hypothetical well and three assessment points, A, B, and C. In this example, pumping 1000 m^3 of water from Well 1 for 1 day will lead to a decrease of 0.66 m in the head at assessment point A 18 days later (denoted as $R_{1,A,18} = -0.66$), and a decrease of 0.09 m after 120 days. The greatest impact at the nearby assessment point A occurs in 18 days. The greatest

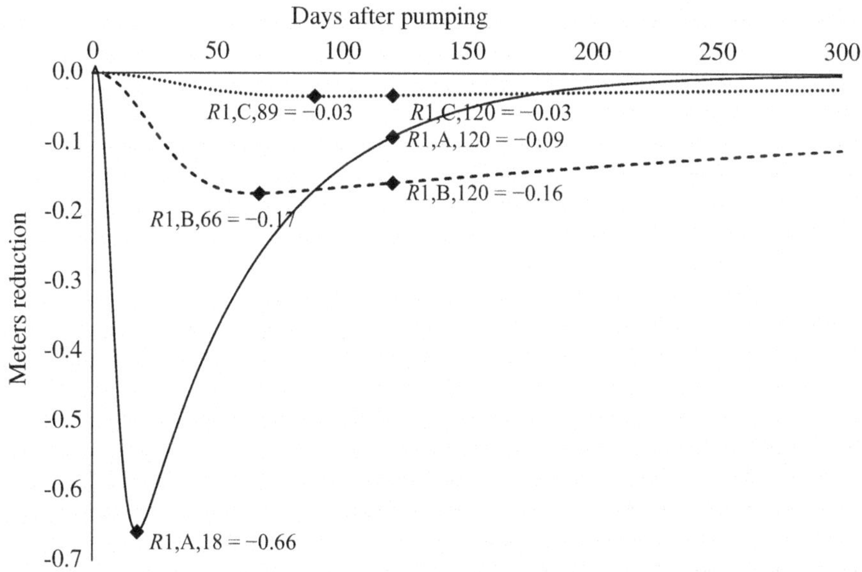

Fig. 3.3 Decrease in aquifer head at assessment points *A*, *B*, and *C*, as a function of time and distance from a given well for each unit of water pumped. For example, if Well 1 takes a unit of water at time 0, $R_{1,B,120} = -0.16$ gives the effect 120 days later on assessment point *B*

impact at assessment point B further away occurs about 66 days after the pumping event.

These R values are termed **response coefficients**. They are key to the working of a smart market for groundwater use.

Figure 3.3 assumes that the terrain has only one layer of groundwater. The ground beneath our feet is usually a series of layers of porous materials above a rock of either sedimentary or igneous origin. Groundwater can flow through rocks, though the flow rate can be slow if the rock's pores or fractures are small. Commonly, geological processes have created layers beneath us, with more-permeable layers (e.g., sand) alternating with less permeable layers (e.g., clay). Because water flows more readily in sand than in clay, these less permeable layers form semi-barriers to flow. When groundwater recharges in a sand layer and then flows down between two clay layers, the water in the sand layer will build up pressure (or head) because more water is coming in than can be pushed out.

There is no need to go into detail about the implications of groundwater flowing in different layers under different pressures, but, to appreciate later discussions, we introduce a few terms. First, the layers where water flows relatively readily are called **aquifers**, and the layers where water does not flow relatively readily are called **aquitards**. What makes one layer an aquifer and another an aquitard is not so much the ability of water to flow through it per se, as the water's ability to flow through that layer compared with the adjacent layers. A silt layer in between sand layers will be an aquitard, while a silt layer between clay layers will be an aquifer.

The top aquifer will behave like the groundwater in the bathtub because the air in contact with the top of the groundwater, having no water above it, is at atmospheric pressure. This limits the potential for pressure to build up in the aquifer. This topmost aquifer is called an **unconfined aquifer**. The lower aquifers, where pressure can build up, are called **confined aquifers**. Figure 3.4 shows the mixture of unconfined and confined aquifers in the Canterbury Plains of New Zealand.

In an unconfined aquifer, the water level is the same as the head. This is not the situation in a confined aquifer, where the water could be under a variable amount of pressure. Because of this, we will use the term head rather than water level when discussing the effects of pumping on groundwater. The groundwater examples used in this book focus on unconfined aquifers because they are easy to explain. The same concepts apply to confined aquifers.

In summary, groundwater, like surface water, flows in response to differences in water energy. These differences in energy are commonly called hydraulic head. In contrast to surface water flow, groundwater flow is slow, and is complicated by the layers of materials that it flows through. Describing the flow of groundwater is more difficult than describing surface water flow, because the ability of ground materials to transmit water cannot be readily observed or measured. The response coefficients of groundwater abstraction quantitatively capture the many-to-many effects of water use. Ideally, a market engine would account for all these different effects, and encapsulate the effects into something that users could manage via a trading system. But how can users trade when every transaction has hundreds or thousands

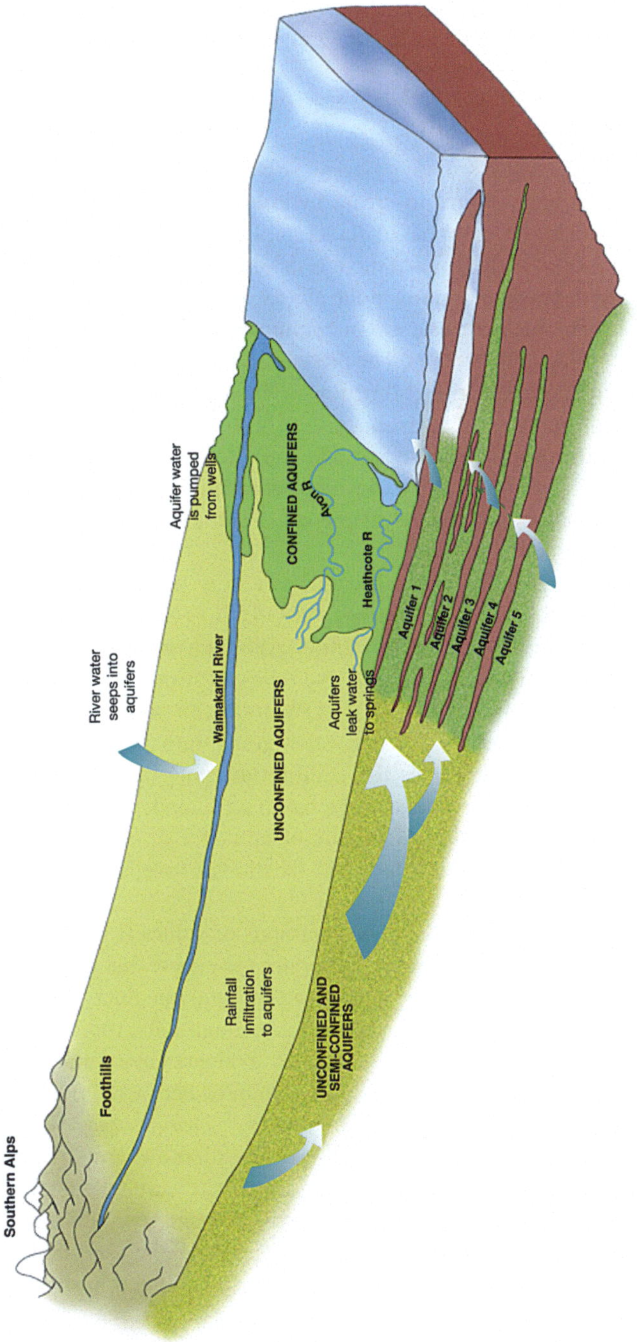

Fig. 3.4 Cross section through part of the New Zealand Canterbury Plains groundwater system (Reproduced from Weeber 2008)

of different environmental requirements? To answer this question, we first analyze the tools of simulation and optimization.

3.2 Hydrological Modeling

Hydrologists have worked hard to create hydrological models to estimate water flow. The models vary from simple to elaborate. For the past 40 years, researchers have developed increasingly detailed models to take advantage of faster computers, while other researchers have improved the ability to estimate the model input data. Sophisticated hydrological models are in common use today in many countries as part of the water allocation process. The models allow regulators to assess required flows for ecological or recreational functions (Kirk 1991), to set maximum allocations on water resources, and to evaluate the short-term or long-term impacts of multiple water extractions. When water use and allocation become more contentious, the regulators are motivated to develop robust models. Research into hydrological processes has combined with improved computational capacity to give regulators strong modeling capacity.

The industry standard model for groundwater flow modeling is MODFLOW. Developed by the U.S. Geological Survey (USGS), it is open-source and has been widely used and modified (Harbaugh 2005). Although it does not use the most elegant approach to modeling, it is robust and relies on strong computational power to make the modeling effective. MODFLOW treats an aquifer as thousands of variously-sized cubes. It evaluates the water going into and out of each cube simultaneously, and then describes the equilibrium state of water in each cube. The hydrologist provides data for each cube based on an understanding of the actual catchment, such as type of land and amount of rain (if a surficial cube), hydraulic conductivity (if a below-ground cube), and locations of wells and their pumping rates.

MODFLOW can require hundreds or thousands of values from the modelers as inputs, and a complex process of model calibration to ensure that predictions match what is measured. Today's advanced computers can evaluate finely detailed models much more quickly than in the past and, given enough data, models can be both accurate and quick. Regulators can now sensibly evaluate environmental effects of proposals for water use and water trades using groundwater models, where 30 years ago the evaluations would have been much cruder.

Surface water modeling of hydrological processes has also greatly advanced in recent years, with improvements in computational hardware, computational software, our understanding of water flow processes, and in geographic information systems that allow for better visualization of results. Better and less expensive sensors and monitoring methods improve hydrologists' ability to put the best inputs into models.

Today's hydrologists can determine the effects of water use, as in our story of the government hydrologist with Ann and Bob's application. But this technical sophistication is not enough to determine the economically efficient allocation of water. Given a specific decision, such as to release a certain amount of water from a reservoir or to take a certain amount of water from a well, the model answers the question, "How will that action affect the environment and the ability of other people to take water?" This "what if" capability is useful in decision-making. Researchers, government water agencies, farms, and hydroelectricity companies regularly use such simulations. Still, the simulation gives little guidance as to what the regulator *should* do, because the simulation cannot optimize for environmental requirements and users' values.

3.3 Optimization with Linear Programming

People often want to know the best decision to make. Given a large set of possible decisions, which one maximizes welfare? Simulation can help with this, but scientists have a more powerful tool with **optimization**. We will focus on an especially useful type of optimization called linear programming. A **linear program** is a specific kind of mathematical problem. "Program" in this context means a plan.

In 1938, Leonid Kantorovich was given the problem of helping the Soviet Union plan plywood production. As a result, he developed the basics of linear programming techniques for optimal allocation of resources. He proved that an economy requires prices for proper allocation. In 1947, George Dantzig, when faced with the problem of planning U.S. military logistics, reinvented some of Kantorovich's ideas with the help of John von Neumann and other brilliant people, and Dantzig took those ideas further.

The linear program has an objective, which is the goal or aim. For example, a group of buyers and sellers will try to maximize their profit. The objective was one of Dantzig's insights. This objective is the scorecard, e.g., "Maximize total economic good."

The linear program has a set of decision variables, which represent the answer. For the buyers and sellers in our example, the decision variables represent the quantities bought and sold. Each variable corresponds to a decision, such as, "How much water should be given to each user in each week of the year?"

The linear program has a set of constraints. These constraints represent logical relationships and limits. For example, a seller cannot sell more than he or she has available, and material cannot be created out of thin air. The constraints prevent the solution from breaking known rules. Each constraint corresponds to a requirement, such as, "The flow in the river must be maintained above X liters/second."

If someone proposes a solution, a complete plan, the constraints ensure the plan is technologically feasible, so that, for example, water doesn't magically appear or disappear and the plan doesn't exceed pump capacities. The objective can calculate how good the solution is. If someone else proposes an alternate plan, and it is feasible, the plan with the higher total economic good can be easily determined.

So, a linear program is ideal for finding a solution to a complex problem in which the component decisions interact and the decision maker must choose them simultaneously.

One of Dantzig's main contributions was to develop an algorithm to solve a linear program quickly. One feature of Dantzig's algorithm is that "price information"—the marginal value of another resource—pops out of the algorithm as part of finding the optimal solution. This is called the **dual price**. In the linear program, each constraint results in a dual price. For example, when a constraint ensures that material bought equals the material sold, the price on the constraint can be interpreted as the market price at which the material is traded. In fact, the optimization model gives marginal prices for water in a surprisingly rich way—by location and by time period, if the model has constraints by location and time period.

Samuelson et al. (1958) showed that a linear program could model trade. Their insight was that an optimization model behaves similarly to a market, in that the optimization finds the best allocation, simulating Adam Smith's "invisible hand" of the market finding the best allocation. Given a group of buyers and sellers, the analyst wants to know what trades will occur and at what price. The solution to the linear program specifies the trades that people would likely want to make, because the linear programming model has the objective of maximizing the gains from trade, which is the same as that of the people in the market. This assumes that the model's objective function contains people's true values for the trades.

Linear programming, and its difficult cousin integer programming (in which objects are indivisible), are now used widely throughout the world for optimizing resource allocation problems. These techniques are used by power companies for reservoir management, by airlines, and by many other organizations.

Linear programming is also used in smart markets for electricity, to match generation to demand. This last application is quite different to those used by the firms mentioned in the previous paragraph. What is special about this application is that the linear program is serving as a real world market engine. Section 4.2 looks more at the background of this smart market.

Aside: A Two-User Optimization of Water Use with Linear Programming

Let us examine a numerical version of the story of Ann and Bob.

Ann and Bob each have a water quota of 3 units. Both Ann and Bob would buy an additional unit for $1, and would sell their unit for $2.

The local regulator requires that a nearby stream maintain a certain flow. The total decrease in flow allowed by a combination of Ann and Bob's water use must be less than 12 *L/sec*.

Every unit that Ann uses decreases the flow by 3 *L/sec*.

Every unit that Bob uses decreases the flow by 1 *L/sec*.

(continued)

Without any trade, the existing water use is at the environmental limit. Because each uses 3 units, the combined effect from the two water users is 12 *L/sec*. Without a trade, neither of them can use more water and neither can be compensated for using less water. We will define this allocation of water as having no consumer surplus—users are not receiving money to compensate for a reduction in water use and are not receiving extra economic value for receiving extra water.

We found the optimum through linear programming. If a trade is allowed, the water distribution between Ann and Bob that optimizes economic surplus is:

Ann sells 1/3 unit for $0.67 or at $2 per unit. Ann's final allocation is 8/3 units.

Bob buys 1 unit for $0.67 or at $0.67 per unit. Bob's final allocation is 4 units.

After the trade, the stream is no worse. Ann's water use has an impact of 8 *L/sec* (8/3 units * 3 *L/sec*) and Bob's water use has an impact of 4 *L/sec* (4 units * 1 *L/sec*).

Let's see what the trade implies. Suppose Bob, the buyer, went to the water regulator to increase his water quota from 3 to 4 units. The water regulator would evaluate the request and conclude that this change would decrease water flow in the stream by 1 *L/sec*. To make up for this, the water regulator would have to take away 1/3 of a unit from Ann to guarantee the stream flow. The water regulator would have to pay Ann $2 * (1/3 unit) = $0.67. So the marginal cost to the economy of Bob getting 1 more unit is $0.67. Bob is happy with this, since he would pay up to $1 for the unit but pays only $0.67. A trade where Bob gets more water would work.

Suppose Ann went to the water regulator to get a water quota for another unit of water. Would this trade work? Ann's extra water would hurt the stream by 3 *L/sec*. To make up for this, the water regulator would have to take away 3 units from Bob to guarantee the stream flow. Ann is willing to pay a maximum of $1 for 1 unit of extra water. This means that the most she can pay Bob is $1 for 3 units or $0.33 per unit. This is far below the minimum price that Bob wants of $3 per unit. A trade where Ann gets more water would not work.

Let's look at the economic surplus. It is zero without trading. At the optimum, Bob pays $0.67 for 1 unit when he was willing to pay $1 per unit, so this generates consumer surplus of $0.33. Ann receives $0.67 for 1/3 m^3 at $2 per unit, but she would sell only if the price were as high as $2 per unit, and so no economic surplus is generated from Ann, giving a total economic surplus of $0.33. Trading has led to a better economic outcome for society than no trading.

(continued)

With no trading, each uses 3 units, and the total water consumed is 6 units. After trading, Ann's use decreases to 8/3, while Bob's increases to 4, for a total consumption of 6 and 2/3 units.

This example shows how a physics-based allocation mechanism can maximize the value of a resource without damaging the environment. After the trade, the stream is just as healthy as before.

For the case with no trade, Ann's marginal value of water is $2 per unit, while for Bob the marginal value is $0.67 per unit. This leads to a solution where Ann sells water to Bob, and the trade generates consumer surplus. The implication is that society needs a way for the price to be higher when Ann takes an extra unit of water than when Bob does, and to allow trades where Ann sells water to Bob. Because Ann's water use has three times the impact on the stream flow, Ann's extra water needs to be three times more expensive than Bob's to give the optimum social allocation of water.

These effects are the crux of the problem in trading water. Everyone is affecting everyone else and the environment somewhat differently. An efficient system needs to have prices that vary between users. Even if users have the same use for water and the same value for water, the environmental/social constraints that society needs to put on water use are not constant. To buy or to sell, Ann's water needs to be more expensive than Bob's.

3.4 Hydrological Optimization

3.4.1 Previous Applications

To put the new smart market approach into context we provide here an overview of research in water use optimization. See Haimes (1977) for a still relevant overview of how optimization applies to water resources.

Researchers can combine hydrological simulation models with optimization to try to optimize allocation of water. In hydrological optimization, the manager seeks to answer a question about how to manage the water. Governments and companies use hydrological optimization for a range of water resources. Here, we discuss its use in reservoir management, contaminant remediation, and groundwater abstraction.

Modeling methods to optimize reservoirs have a long history (Forsund 2007; Read 1989). The approaches sometimes consider the value of different crops, and a lot of literature describes optimization applied to irrigation. The hydroelectricity firms have invested in reservoir management tools, which use operations research methods. The firms want to know how to schedule generation to maximize profit, given their physical system with many constraints and requirements. Specifically, when should the reservoir operator release water from the reservoir? The release could produce electricity while supplying irrigation to farms and water for the environment, or it could flood local properties. The reservoir operator wishes to manage the release of water now with the trade-off of saving water in case the weather turns dry in the future. Although these methods use costs, each model's user manages only his or her own operations, so the models are not being used to allocate between users.

Consultants use hydrological optimization to help clean up groundwater contamination. Imagine someone spilled a toxic mess on farmland above an aquifer. Left alone, the toxin would spread down through the earth and then into the aquifer, where it would contaminate surrounding water supplies. To remediate the contamination, engineers could dig wells surrounding the spill, and pump water out of the ground so the contaminated water could be treated. If the engineers put enough pumps in the right places, and pump enough water, then they could contain the chemical. To contain the spill at minimum cost, where should the engineers put the pumps, and how fast should they pump? This remediation problem has been examined by Greenwald (1998), Ahlfeld and Mulligan (2000), Ahlfeld et al. (2005), and others. Greenwald developed a software package called MODMAN. More recently, Ahlfeld developed GWM. These programs work with MODFLOW to solve optimization problems involving groundwater flows, providing a way to optimize pumping from many wells, while constraining the combined effects at various places in the groundwater system. For contaminant spills, hydrologists use the models to choose where to place the wells and how fast to pump to minimize the cost of containing the spilled chemical.

In a catchment with many users taking groundwater, as in our example with Ann and Bob, usually each user has a different effect on the aquifer and relevant environmental flow requirements, such as in a nearby river or wetland. How much water can the farmers sustainably take from the aquifer? With a hydrological simulation like MODFLOW, hydrologists can evaluate the impact that pumping from any well will have at any other place in the groundwater system. They can calculate the impact of additional pumping from each well. And then, using MODMAN or GWM, the hydrologists can optimize the pumping from each well, so that the optimized flows satisfy the various physical and environmental constraints.

Hydrological optimization would be a powerful tool for water regulators. They could use it to ask questions such as, "What is the most water that users can take from these wells, while maintaining required environmental flows?", and, "Given that all these wells are taking water, what is the most we could give to this new license applicant while maintaining required environmental flows?" Fortunately, researchers and regulators are increasingly using optimization models of combined groundwater and surface water, though not necessarily for the economic optimization of water allocation. Use of multi-criteria optimization for water resource decisions, for example, is increasingly common.

Thus, scientists can develop physically correct methods for incorporating the economic value of pumping from each well, to optimize the total value from pumping. "How much water should each user take to maximize total economic value, while maintaining required environmental flows?" This critical insight helps make the leap to a smart market for groundwater.

Many researchers (some referenced by Schoengold et al. 2006) have tried to calculate the gains from water trading, and have tried to offer advice to government about how much water users should take for particular activities. We will call the researchers in this category "economic optimizers." Most of the economic optimizers depend on mathematical models that combine the hydrology and the economics, to varying levels of sophistication (Becker 1995; Stubbs 2000; Rosegrant et al. 2000; Draper et al. 2003; Mahan et al. 2002; Yu et al. 2003; Mulligan et al. 2014). Typically, the researchers develop a simulation of the hydrology, whether for surface water only, for surface water and groundwater, or for a developed irrigation system with or without hydroelectricity generation. The researchers then develop an optimization model that prescribes how much water each user should take to maximize total economic good, assuming no transaction cost. Unfortunately, the researchers never know users' true water values, and so the researchers must estimate those values. Harou et al. (2009) reviewed hydroeconomic models. Gorelick and Zheng (2015) give a good survey, focusing on the applications related to the worrying challenges of groundwater management; their paper lists others that discuss techniques for coupling simulation with optimization. These techniques will be the core of the smart markets for water.

Researchers have attempted to estimate the most economical use of water for various land users. These researchers have combined the well-known techniques of irrigation optimization with catchment models to optimize the economic use of water in a catchment with many water users. These models require as input the saleable value to each land user of each potential water use (e.g., wheat, corn), as well as enough parameters to estimate the yield of each crop on each unique parcel of land, and a function allocation that describes how yield increases with water use. Because each land user might be a different distance to market or produce in different volumes, one distorts the model when assuming that a constant value for each crop applies to all users.

These models produce as outputs a list of which land use is appropriate for each land user, the economic value that each parcel of land could produce, and the amount of water each user should receive in order to reach this optimum situation. It also produces an estimate of the catchment-wide marginal value of water for the catchment.

Becker (1995) takes a similar approach of using estimates of crop values by type and region and then solving a linear program to simulate trade among competing uses. He found that tradable water rights would make farmers better off, though the model did not allow for bidding. He observed that the dual prices correspond to market prices for water. In a smart water market with bids, these dual prices on the flow constraints serve as the spot market prices.

While these models may be helpful for identifying the potential for trade, they are impractical for water allocation. The best of the economic optimizers use good hydrological models; they just have not used users' actual values. Even with perfect data, hydrologists have no way to say who should get water, nor when. The reason is that they do not have access to users' water bids, which are private. To our knowledge, regulators never use "economic optimizer" models for actual water allocation.

3.4.2 Cost–Benefit Analysis

Government often attempts to manage water based on what appears to be the greatest good to society. If the catchment would be better off with more water allocated to agriculture than to hydroelectricity, the government can try to adjust the allocations. These studies of the value of water to various sectors are amenable to incorporation into a model that uses hydrological optimization to derive water allocations. Of course, if the market functioned properly, government would not need to do these analyses, nor pick which users get the water. Since existing markets work poorly, the government buys studies to learn how to allocate water better.

To conduct a cost–benefit analysis on a project involving water resources, the regulatory analyst needs to know the benefits both of taking water and of leaving the water. Taking more water from a catchment means less water for kayakers, less ability for indigenous people to continue their traditional practices of food gathering, less recreational fishing, less likelihood of protecting endangered species, less aesthetic value, and so on.

The analyst has a challenge placing a value on these uses of water. The analyst can infer the value by looking at how water use shapes people's decisions in aggregate. For example, the analyst can compare land values for homes with a view of a high-quality natural river, with the value for homes next to a canal used for irrigation water. This method has difficulties in controlling for the numerous factors that might vary between locations, but can be helpful for some water uses.

Unfortunately, many people value water for recreational and cultural purposes, to varying degrees, and so costing these values present greater challenges than commercial uses do. One approach is to account for the expenditures people make in order to pursue a recreational activity. The analyst could add total expenditure that recreational fishers expend on rod and tackle, fishing permits, transport costs, etc., and how much this might decrease after a change in the water use because of less fishing could be estimated. This can help, but does not reflect the economic consumer surplus for users; in other words, how much more they would be willing to pay to enjoy the uses they currently enjoy.

Imagine that the water regulator had a good hydrology model of the catchment. The model accurately estimates the quantity of water available at any moment, and it shows the quantity that the regulator should save for environmental and cultural purposes.

With the regulator's good view of water flow, if a commercial applicant requests a new water license, the regulator can make a decision by simulating the new license. If the change leaves enough water for the environmental and cultural uses, then the regulator can license that surplus to the commercial applicant.

However, if two applicants come along with conflicting requests, the regulator has no rational basis to grant a license to one or the other. The regulator must fall back upon rules, e.g., which applicant arrived first, or which had a relevant prior right, or some political process. So the regulator has no means to transfer water between two commercial uses without the users' agreement. Even if the regulator wants to maximize the value of water to society as a whole, while ensuring that the environment is not short-changed, the regulator has no information about different users' economic values for water.

So, at best, the regulator may be able to say "Yes" or "No" to one application, based on whether granting the license would harm the environment or other users too much. But the water regulator has no basis to take a license from one user and give it to another user based on economic value, because the regulator has no knowledge of users' economic values. Those values are private, in the users' heads, and the users are not keen to share that information with government.

A common approach to discovering these values is to survey people and ask them how much they would be willing to pay to retain a water use or how much they would be willing to receive in order to give up a water use they currently enjoy. Asking people to value an environmental feature will generally turn up positive numbers. People want a nice environment. Even the worst polluter, while having a huge value on the right to pollute, would still place a positive value on having a good environment. Unfortunately, people cannot easily buy a good environment, because no mechanism exists for it; the transaction cost is too large. The lack of a market means that people's assessment of the value of non-extractive uses of water is not constrained by the limitations and trade-offs faced when making other economic decisions.

Valuing an environmental feature by asking people their feelings about it is unreasonable for another reason: it simply doesn't give the right value. If the analyst wants to know the price of a loaf of bread, the market price is a good estimate because individual buyers can shop around for the lowest price on the sort of bread they want. The market tends to push the price of bread to its lowest. However, the analyst cannot find society's optimized cost for the environment. The true value on the environment should be the *lowest cost the society needs to pay* to have the environment at a specified state. However, finding that lowest cost would best be done by having some mechanism to conveniently trade rights related to the feature, and in a precise way.

Young and Loomis (2014) give a book-length treatment of the non-market methods to value water generally, not only for environmental water. They mention the use of mathematical programming models in valuation of water. "Analysts sometimes give more attention to the mathematical formulation of the overall allocative and its solution than to the less interesting but equally important work of specifying the correct valuation model and calculating the appropriate empirical parameters." Toward getting appropriate empirical parameters, they cite Howitt (1995), who developed a robust method for estimating farmers' values for resources using optimization. The authors briefly consider the use of market prices, but they note that this price information is scarce.

Methods to estimate the non-use values of water are helpful only because water markets are so poor. With smart markets for water, the weaknesses of these estimation methods should make them rarely worth the cost of the study.

3.4.3 Catchment Management Models

While regulators do not use optimization to reallocate water, researchers use optimization to help evaluate the potential benefits of water re-allocation proposals. One prominent model of this type is the Water ReAllocation Model (WRAM) used in Australia (Yu et al. 2003). Another is the California Value Integrated Network (CALVIN) (Newlin et al. 2002; Draper et al. 2003; Jenkins et al. 2004). Both of these models can indicate the potential value of water reallocation through trading. They can also indicate the value of water where no market exists, as when planning water infrastructure.

Some of these models simplify or exclude groundwater resources, and, more significantly, none includes user bids for water. Yu et al., for example, find a single price for water in a region.

WRAM, CALVIN, and similar efforts cannot substitute for a proper water market, or even work as a framework for constructing such a market. Though the model will say that Land Parcel A should be planted in rice and be allocated x m^3/ha, while Land Parcel B should be planted in soybeans and allocated y m^3/ha, any attempt to mandate the allocation to each land user would run into problems. The model might have poorly represented the ability of the land to yield that crop, how efficiently water could be used, or the value of the crop for that land user. The model might not represent a new hybrid of oats that promises high yields and strong demand; should the user be given the extra water to grow this new hybrid? The real knowledge and values inside users' heads are not in the model. While this is an economic optimization, the regulator should not use it for clearing a market, because users should make the land use decisions independent of the market, with their land use values reflected in bids, not in the model.

Because of their relevance as precedents to the smart market approach, the accompanying boxes describe WRAM and CALVIN in more detail.

Taking this a step further, suppose the regulator could get users' water values (or bids)? The regulator could then use hydrological optimization to maximize total good to society while maintaining all required environmental flows. What would be the significance of this, and how could the regulator actually use such a tool? We will examine this further in Chap. 6 and later chapters.

Aside: The WRAM Program

The purpose of WRAM (Water ReAllocation Model) is to determine optimal water allocation and reallocation in terms of crop planting decisions and irrigation water requirements. As a by-product, WRAM simulates trading of water entitlements between irrigation areas, and generates water accounts for economic impact analysis. In addition, WRAM performs input-output analysis to evaluate the impact of water trading and reallocation on regional economy (Yu et al. 2003).

Yu et al. (2003) described other literature in water trading: "While all the economic models used in these studies require agronomic and hydrologic data for water demand and water availability, none interacts dynamically with hydrologic models." They then gave a preliminary report on their water reallocation model, WRAM, in which they noted that "WRAM is based on an economic optimization algorithm. At its core is a powerful linear programming (LP) solver that maximizes the net benefit subject to a series of land, water, capacity, and crop constraints."

The developers' intention is to extend WRAM to include a water trading model to permit simulation of the economic consequences of reallocating water resources through trading. Currently, they can evaluate the effectiveness of water markets in Australia by comparing the current water use on each land parcel after trades with the WRAM-optimized water use on each land parcel, and estimating the economic value of each scenario.

The WRAM linear program is as follows.

Indices

$i = 1, \ldots, n$, location,
$j = 1, \ldots, m$, crop.

Parameters

$Alloc_i$ = water allocation for location i,
$Cost_{i,j}$ = \$/megaliter water charge at location i,
$IsGrowable_{i,j}$ = 1 if growing crop j is feasible at location i, else 0,
L_i = hectares of irrigable land at location i,
L_j = hectares of crop j in the basin,
λ = % increase in land use for crop j allowed in the basin,
$Price_j$ = \$/ton price of crop j,
$Rotation$ = crop rotation parameter (currently set to 0.6),
$VarCost_{ij}$ = \$/hectare variable cost other than irrigation to grow crop j at location i,
$Water_{ij}$ = water requirement to grow crop j at location i,
$Yield_{ij}$ = tons/hectare yield of crop j at location i.

Decision variables

$hectares_{ij}$ = hectares of crop j to grow at location i.

(continued)

Model

(1) Maximize basin-wide net benefit,

$$\text{Maximize} \sum_{i,j} IsGrowable_i \; hectares_{i,j} \; (Yield_{ij}Price_{i,j} - Cost_{i,j}Water_{i,j} - VarCost_{i,j}).$$

(2) Total land at location i,

$$\sum_{\text{crop } j} IsGrowable_i \; hectares_{i,j} \leq L_i, \; i = 1, \ldots, n.$$

(3) Control the percent increase in land use for crop j,

$$\sum_{\text{crop } j} IsGrowable_i \; hectares_{i,j} \leq (1 + \lambda)L_j, \; j = 1, \ldots, m.$$

(4) Total water available,

$$\sum_{\text{location } i} \sum_{\text{crop } j} IsGrowable_i \, Water_{i,j} \, hectares_{i,j} \leq \sum_{\text{location } i} Alloc_i.$$

(5) Force crop rotation to prevent monoculture,

$$0 \leq hectares_{i,j} \leq Rotation \cdot L_i, \; i = 1, \ldots, n, \; j = 1, \ldots, m.$$

WRAM estimates crop yields and prices, water charges, and variables costs based on available literature. The solution of the model gives the amount of water traded at location i as $TradedWater_i = Alloc_i - \sum_{\text{crop}j} IsGrowable_{i,j} \cdot Water_{i,j} \cdot hectares_{i,j}$. The solution gives the trade value as $TradedWater_i s$, where s is the shadow price of water ($/megaliter). The authors do not specify how they find the shadow price s, but based on standard duality theory, s would be the dual price of constraint 4. Thus, they do not have a price by location, but only an overall price for the whole basin.

Aside: The CALVIN Program

One well-developed water allocation model is CALVIN (Newlin et al. 2002; Draper et al. 2003; Jenkins et al. 2004). In Newlin et al. (2002), the authors define some terms (italics in the original).

> …an *ideal market* in Southern California is defined as a market without transaction costs or risks, where decision makers have perfect knowledge and foresight of hydrologic and economic information and only environmental and physical constraints prevent water delivers. Theoretically, such economically ideal water management can arise from a variety of mechanisms (including various cooperative or centralized forms, or an economically-enlightened despot), but some form of market mechanism seems among the more likely approaches for approximating economically ideal operation and allocation.

They also define "scarcity cost" as the economic loss to a user because water is not abundantly available, and "maximum economic use" as the maximum quantity demanded if price is zero and if water availability is unlimited. They describe CALVIN as follows.

CALVIN combines year 2020 economic values of water use from agricultural and urban demand models (monthly-varying willingness to pay functions for water deliveries) with projected 2020 infrastructure, unit operating costs, and hydrologic information to suggest economically optimal water operations and allocations.

CALVIN maximizes the "economic benefits from operation of a time series of hydrologic inputs through a network given physical infrastructure and other constraint on a monthly time-step …" The benefits are not found from bids from water users, but rather "CALVIN prescribes operations and allocations based on economic value functions derived for urban and agricultural demands."

The Draper et al. (2003) paper describes the calibration process for CALVIN, and gives the following model.

Parameters

A_{ijk} = gain or loss in water flow on arc k from node i to node j, where nodes are confluences, users, and reservoirs.

$Cost_{ijk}$ = cost per unit flow on arc k from node i to node j.

$Inflows_j$ = external inflows to node j.

L_{ijk}, U_{ijk} lower and upper bounds on flow across arc k from node i to node j.

Decision variables

$flow_{ijk}$ = flow on arc k from node i to node j.

Model: minimize $\sum_i \sum_j \sum_k Cost_{ijk}\, flow_{ijk}$,

$$\sum_i \sum_k flow_{jik} - \sum_i \sum_k A_{ijk} flow_{ijk} = Inflows_j \text{ for all nodes } j,$$
$$L_{ijk} \le flow_{ijk} \le U_{ijk}.$$

3.4.4 Key Ideas from Past Work on Hydrological Optimization

The past research on hydrological optimization has a couple provocative precedents to a smart market for water.

First, scientists can combine hydrology simulations with optimization, and optimization is a good stand-in for a market. That is, given users' values for water, an analyst could calculate the best allocation. The economic optimizers still do not have the users' true water values, so they have to estimate those values. Their models only simulate a market, as if the market had a zero transaction cost. But the germ of the idea for the "market engine" is there.

Second, these papers demonstrate that water markets will be good for society, using quantitative experimental data rather than only the sanitized theory of the economics textbooks. The papers do not, however, suggest how those markets should be designed. The papers do not propose ways to eliminate the transaction cost. They show only that if the transaction cost were zero, society would be better off.

So, the regulator needs more than a hydrologically omniscient model. The regulator needs some kind of market engine. This market engine would understand the local hydrology and required environmental constraints, and the engine would accept the users' values for water. Then the engine would operate as a calculator, and allocate the water in the best fashion, while ensuring that environmental and non-commercial needs were satisfied.

This ideal system would help commercial users share water between them. They could exchange water and money, so each is better off. Instead of a slow thinly-traded market, an active market would incentivize many commercial users to join. The commercial users would put their information into the computer, saying how much water they want, and how much they're willing to pay. The computer does the calculations, so the water is shared among everyone, and each user gets the mix of water and money that makes everyone best off, while satisfying the environmental requirements.

The regulator's model then, is a hydrological optimization system. As mentioned above, such a system would be of immense value in vetting license applications, apart from a water market. In an ideal market for water, users would bid for water through a web page, similar to eBay. The regulator feeds the bids into the hydrological optimization, which contains all the relevant knowledge of the local catchment and required environmental flows. The optimization solves, and then the regulator notifies users of the quantities bought and sold, and of prices, all in the space of a few minutes.

References

Ahlfeld, D. P., & Mulligan, A. E. (2000). *Optimal management of flow in groundwater systems.* San Diego: Academic Press.

Ahlfeld, D. P., Barlow, P. M., Mulligan, A. E. (2005). *GWM – A ground-water management process for the U.S. Geological Survey modular ground–water model (MODFLOW–2000).* (U.S. Geological Survey Open–File Report 2005–1072). http://pubs.usgs.gov/of/2005/1072. Accessed 15 July 2006.

Becker, N. (1995). Value of moving from central planning to a market system: Lessons from the Israeli water sector. *Agricultural Economics, 12*, 11–21.

Draper, A. J., Jenkins, M. W., Kirby, K. W., Lund, J. R., & Howitt, R. E. (2003). Economic-engineering optimization for California water management. *Journal of Water Resources Planning and Management, ASCE, 129*(3), 155–164.

Forsund, F. R. (2007). *Hydropower economics.* New York: Springer Science+Business Media.

Freeze, R. A., & Cherry, J. A. (1979). *Groundwater.* Englewood Cliffs: Prentice-Hall.

Gorelick, S. M., & Zheng, C. (2015). Global change and the groundwater management challenge. *Water Resources Research, 51*, 3031–3051. doi:10.1002/2014WR016825.

Greenwald, R. (1998). *MODMAN: An optimization module for MODFLOW Version 4.0.* Freehold: GeoTrans.

Haimes, Y. Y. (1977). *Hierarchical analyses of water resources systems.* New York: McGraw-Hill.

Harbaugh, A. W. (2005). *MODFLOW-2005, the U.S. Geological Survey Modular Ground-Water Model—the Ground-Water Flow Process.* U.S. Geological Survey Techniques and Methods 6-A16. http://pubs.usgs.gov/tm/2005/tm6A16/PDF.htm. Accessed 15 July 2006.

Harou, J. J., Pulido-Velázquez, M., Rosenberg, D. E., Medellín-Azuara, J., Lund, J. R., & Howitt, R. E. (2009). Hydro-economic models: Concepts, design, applications, and future prospects. *Journal of Hydrology, 375*(3–4), 627–643.

Howitt, R. E. (1995). A calibration method for agricultural economic production models. *Journal of Agricultural Economics, 46*(2), 147–159. Wiley Blackwell.

Hughes, S. (2016). http://geology.isu.edu/wapi/EnvGeo/EG7_water/EG_module_7pt2.htm. Accessed 22 Oct 2016.

Jenkins, M. W., Lund, J. R., Howitt, R. E., Draper, A. J., Msangi, S. M., Tanaka, S. K., Ritzema, R. S., & Marques, G. F. (2004). Optimization of California's water supply system: Results and insights. *Journal of Water Resources Planning and Management, 130*(4), 271–280.

Kirk, R. M. (1991). River-beach interaction on mixed sand and gravel coasts: A geomorphic model for water resource planning. *Applied Geography, 11*, 267–287.

Mahan, R. C., Horbulyk, T. M., & Rowse, J. G. (2002). Market mechanisms and the efficient allocation of surface water resources in southern Alberta. *Socio-Economic Planning Sciences, 36*, 25–49.

Mulligan, K., Brown, C., Yang, Y.-C., & Ahlfeld, D. (2014). Assessing groundwater policy with coupled economic-groundwater hydrologic modeling. *Water Resources Research, 50*(3), 2257–2275.

Newlin, B. D., Jenkins, M. W., Lund, J. R., & Howitt, R. E. (2002). Southern California water markets: Potential and limitations. *Journal of Water Resources Planning and Management, 128*(1), 21–32.

Pidwirny, M. (2006). The hydrologic cycle. In *Fundamentals of physical geography* (2nd ed.). http://www.physicalgeography.net/fundamentals/8b.html. Accessed 18 Sept 2016.

Read, E. G. (1989). A dual approach to stochastic dynamic programming for reservoir release scheduling. In A. O. Esogbue (Ed.), *Dynamic programming for optimal water resources system management* (pp. 361–372). New York: Prentice Hall.

Rosegrant, M. W., Ringler, C., McKinney, D. C., Cai, X., Keller, A., & Donoso, G. (2000). Integrated economic–hydrologic water modeling at the basin scale: The Maipo river basin. *Agricultural Economics, 24*(1), 33–46.

Samuelson, P., Dorfman, R., & Solow, R. M. (1958). *Linear programming and economic analysis*. New York: McGraw-Hill.

Schoengold, K., Sunding, D. L., & Moreno, G. (2006). Price elasticity reconsidered: Panel estimation of an agricultural water demand function. *Water Resources Research, 42*(9), W09411.

Stubbs, C. M. (2000, June). *Hydrologic-economic modeling of irrigated agriculture in the Lower Murrumbidgee Catchment: Investigations into sustainability*, PhD thesis, MIT.

Weeber, J. (2008). *Christchurch groundwater protection: A hydrological basis for the zone boundaries* (Technical Report U08/21, 60 p.). Christchurch: Environment Canterbury.

Young, R. A., & Loomis, J. B. (2014). *Determining the economic value of water: Concepts and methods* (2nd ed.). New York: Routledge Publishing.

Yu, B., Tisdell, J., Podger, G., & Salbeb, I. (2003, July 14–17). A hydrologic and economic model for water trading and reallocation using linear programming techniques, MODSIM 2003: International Congress on Modelling and Simulation.

Chapter 4
Introduction to Smart Markets

Abstract The purpose of this chapter is to help the reader make the transition from optimization to the smart market. The chapter reviews some of the history of smart markets, with early examples of and proposals for smart markets. We describe the modern electricity market in some detail. The final two sections describe early research on smart markets for water, and characterize the smart market for water in comparison with other types of markets and auctions.

Keywords Smart market • Precedents • Complex commodities • Electricity • Natural gas

4.1 Analogies of Water Markets to Markets for Other Complex Commodities

Water is not the only complicated commodity. Many commodities that previously suffered thin trading are now traded actively, with billions of dollars in transactions, by means of smart markets. While smart markets differ from the markets most of us see every day, some industries actively use them all the time all over the world.

A smart market is an auction where buyers and sellers submit bids, and a market manager uses mathematical optimization to match the buying and selling of bids and sets prices. Smart markets are most useful when trades have complex interactions. It is a trade-coordinating mechanism. What makes the market "smart" is the application of optimization to manage complexity.

Everyone is familiar with an ordinary auction, often called an English auction. Bidding starts low, and rises as people continue to bid. The hammer drops when bidding closes, and the item goes to the high bidder, whether the item is a masterpiece painting, a beach house, or a season-long right to water in an Australian reservoir. Theory for these simple auctions has been around for many years (Vickrey 1961). Such auctions work well for one-off indivisible items that are exchanged between one buyer and one seller.

Those simple auctions lack any machinery to handle the many-to-many effects of electricity or water. But a linear program *can* handle these many-to-many effects. In the 1980s and 1990s, a few economists realized that this powerful operations

J.F. Raffensperger, M.W. Milke, *Smart Markets for Water Resources*, Global Issues in Water Policy 12, DOI 10.1007/978-3-319-55008-4_4

research tool could be used to run a special kind of auction, in which market complexities were managed within the market-clearing function.

Depending on the initial rights, the markets may be one-way auctions, one-way procurements, or two-way trading; they are "smart" because an optimization is used to choose the allocation under a set of constraints.

McCabe, Rassenti, and Smith (1991) pointed out that a smart market has many advantages over an ordinary pair-wise market. A smart market serves as the place where buyers and sellers can find each other. A smart market is operated by an auctioneer or manager, who has responsibility to ensure good market operations, enforce contracts and guarantee payments, provide anonymity where appropriate, enforce market rules, and provide overall management of the market. An ordinary auction has those advantages, too. But a smart market is cleared by a constrained optimization model; the constraints ensure market limits (e.g., transmission line capacity in the electricity market) and protect against externalities. Thus, a smart market allows trade of highly complex commodities, lowering the transaction cost for trades that would otherwise be difficult.

Smart markets are now in active use all over the world for a wide range of commodities and services. A smart market for water builds on the experiences of these other smart markets. The wide variety of previously "untradeable" goods are now bought and sold through smart markets. Learning more about these other smart markets will help us explore the issues associated with smart markets for water.

As pointed out by Gallien and Wein (2005), Stanley et al. (1954) developed the first smart market proposal to help the U.S. military buy garments. Vendors offered maximum and minimum quantities: "We'll sell you no more than *X* units," or "We'll sell you only batches of at least *Y* units." Vendors offered quantity discounts, and made complicated offers that depended on acceptance of other offers. Some vendors had scheduling problems, such as in "We'll give you a discount if you can wait six weeks." So, the government could not simply take the lowest-priced bid. The lowest-cost solution had to be found with some kind of computer search algorithm. Stanley et al. chose George Dantzig's linear programming, which was perfectly suited to the task. Vendors turned in their bids, and the computer selected the winners.

This example conveys much of the flavor of smart markets, in that a set of market participants bid on price and quantity bundles, and an optimization model selects the trades. The optimization procedure ideally includes all relevant stakeholders and their physical and economic connections within a system. These connections take the form of model constraints.

While much of the theory and machinery for smart markets was available as soon as Dantzig finished his work in the 1950s, the term "smart market" seems to appear first in Rassenti et al. (1982). That paper, developing work started by Grether, Isaac and Plott (1979), showed how to use integer programming to operate an auction for airport landing slots (though the idea appears earlier in Jaikumar 1980). The authors proposed to reduce airport congestion by letting airlines bid for take-off and landing slots. The problem is that take-offs and landings have to be coordinated. It's no good asking a plane in the air how much they're willing to bid to land, especially if they're low on fuel!

Rassenti et al. wrote a math model that found the best set of take-offs and landings, ensuring that every take-off matched a landing, while keeping flights within the capacity of each airport's runways. Svorencík and Maas (2016) quoted Vernon Smith as saying, "The whole idea of combinatorial auctions in particular and smart computer assisted markets in general, came out of Steve's thesis and our joint papers. The first paper was Rassenti, Stephen J.; Vernon L. Smith and R. L. Bulfin." The U.S. government is now attempting to implement such a market for three airports: LaGuardia, John F. Kennedy, and Newark Liberty (FAA 2015).

Allocating landing slots has similar features to allocating water and electricity in the complexity of the interactions, but is harder because landing slots are indivisible. The airport cannot allocate 35% of a landing slot to an Air New Zealand flight and 65% to Singapore Airlines, as jets cannot take up the same space at the same time. Optimization models with integer variables do not naturally provide dual prices, so Rassenti et al. wrote additional models to find approximate prices. By contrast, fractional units of electricity (e.g., 0.6 kWh) and water (like 0.3 ML) *can* be allocated, so the optimization models used for those commodities can automatically give the correct prices.

4.2 Electricity Smart Markets

A close analogy to our market design is the modern electricity market design. An independent system operator (ISO) manages all the flows simultaneously, processing users' bids for the commodity in a periodic auction. The ISO ensures that the physical flows work, that system constraints are satisfied, and that users pay for their purchases and are paid for their sales. The smart market for electricity was proposed at length by Schweppe et al. (1988), in a book called *Spot Pricing of Electricity*, and implemented for the first time in New Zealand in 1995 (Read and Chattopadhyay 1998; Alvey et al. 1998; Hogan, Read, and Bing 1996).

As an example, consider the Midwest ISO in the U.S. (Carlson et al. 2012). Until 2005, the Midwest had 27 different electricity firms. Each firm managed power as a regulated monopoly in its own areas. Each firm managed its own equipment, and planned its generation. Best practice was for each firm to use mathematical optimization in planning its power generation, to make sure that the generation met demand, and that the power transmitted over the high-tension lines stayed within the capacity of those lines. But the American federal government saw the New Zealand success, and required the electricity firms to connect their separate power grids, and to allow competition. The idea was that every generator would compete to sell power at lowest cost, and electricity would go first to the wholesalers who bid highest. The power grid would be shared. Of course, no one could track who sold electrons to whom, so pairing buyers and sellers would be impossible. When a generator in an electricity market put power into the national grid, it has no way to know which user is buying the power. This is one of many ways that the electricity market is like a water market. The generator is simply charging up the whole grid. Many generators together need to be coordinated, to make sure that the supply of power meets demand, and to make sure that the wires don't overload.

The solution was to use the New Zealand smart market design. The modern electricity market works by having an ISO as market manager, who acts as auctioneer, and who accepts bids from generators (sellers) and wholesalers (buyers). The market manager solves a math model that coordinates generation and demand, ensuring that all system constraints (such as line capacities) are within acceptable limits. The market manager reads the best solution from the computer model, and then tells the generators and wholesalers what to do. The participants will accept this solution because the prices produced by the linear program ensure that each generator either sells at its bid or above, or sells nothing, and each wholesaler buys at its bid or less, or buys nothing. Many times, the ISO pays generators more than their bid, and many times the ISO charges buyers less than their bids. Those are gains from trade.

Most of the developed world is now using smart markets for electricity. These systems run as often as every 5 min, with the equivalent of millions of U.S. dollars trading hands in robust and highly reliable systems. For operation of the New Zealand electricity market, a linear program contains a model of New Zealand's high voltage network. The objective function comes from the consumer surplus resulting from users' bids. The decision variables are the amount of power that each generator should produce and the power flows along each wire. The constraints (such as "Don't melt the wires") maintain security, and power flows must balance.

Linear programming duality theory proves that the dual prices of the solution conform to the principles of marginal cost pricing. These marginal costs then inform economic decisions about expanding capacity of the system. The computer finds a solution that minimizes the total cost of producing the power. As with the airport landing problem, a math model does the coordination.

Generators bid to produce power: "If you pay me $1, I'll produce a little power; if you pay me $2, I'll produce a lot of power." Users (wholesale power distributors) bid to buy power: "If you charge me $1, I'll buy a lot of power; if you charge me $2, I'll buy only a little power." Because of line losses, the generator of a megawatt will deliver less than a megawatt elsewhere, and so these bids cannot possibly be matched one-to-one, just as water often cannot be traded one-to-one.

The ISO has an optimization model of the power network. The model matches the supply and demand bids, ensuring that the power flows correctly over all the wires. If one generator makes power, several users must to be ready to take it, and the power flow cannot exceed the capacity of the lines. So, it's not a free market, but, rather, a closely regulated smart market, managed with the help of a computerized math model that the ISO programmed to include the physics of the power network.

Table 4.1 summarizes the similarities and contrasts between electricity markets and water markets, and the following sections discuss them further.

Table 4.1 Comparison of electricity and water markets

Electricity markets	Water markets
Markets cover a large region or a country.	A market would develop by catchment. A market could work within a water district encompassing a sub-catchment, or it could work between water districts across a larger catchment.
Environmental effects of generation (flows through a hydroelectricity facility, fossil fuel emissions) are somewhat less complex (though no less serious) compared with those from water use.	Water use has complex environmental effects, with users' demands affecting other users and the environment in complex ways, immediately and over a long time.
The market manager generally treats hydrological uncertainty as exogenous to the market.	The market manager may need to treat hydrological uncertainty as endogenous to the market model, which has environmental constraints, and water use has inter-temporal effects.
Participants are specialists in relatively large sophisticated organizations.	Participants' sophistication varies from farmers and small industrial firms, to urban managers and regional water planners.
Capital costs for generators and lines are large.	Capital costs for dams and waterways are large.
The regulator limits access to power lines; taking power for free is dangerous. Given access, users get rights to power through purchases in the market.	In many regions, though not all, users obtain access to an aquifer or surface water by ownership of land. Given access, users may have immediate rights, or may need to obtain them some other way.
Users can increase supply by adding generators.	Supply is highly variable, and can be increased only by storage and expensive desalination.
Transmission is instantaneous, and electricity is not easily stored.	Transmission takes hours or days. Some effects are not apparent for weeks or even decades. Water is easily stored in modest quantities.
Hydroelectricity generators face uncertainty of water supply and uncertainty of demand, and demand can vary on a minute-to-minute basis.	The main uncertainty in water use is of supply; storage ensures that short-term variability of demand is less of a problem. Lower supply correlates with increased demand during droughts, as farmers irrigate more.
Electricity has been bought and sold only within the last 130 years or so.	Water has been bought and sold for thousands of years.
Electricity is seen as critical to modern life, but with no spiritual or cultural significance.	Many people regard water as having deep significance to spiritual and cultural traditions.
A generator is generally recognized as deserving payment for producing electricity; in effect, the generator initially owns the produced electricity.	Water is not usually owned; only the administrative permission to use it is owned.

Adapted from Raffensperger, Milke, and Read (2009)

The analogy of water to electricity is not the environmental effects of generation, but rather the complex network effects due to instantaneous flow; demands and generation affect each other in complex physical ways. Water use has complex environmental effects, with users' demands affecting each other and the environment in complex ways, immediately and over a long time.

The electricity spot market works short term, within minutes, because electricity is generally not stored on a mass scale. In fact, while the uncertainty of electricity demand is important to the electricity market, generation can respond sufficiently fast that the optimization models can ignore the uncertainty. Because of the near-real-time nature of electricity operations, the modern electricity market operates with a far higher level of security and reliability than a water market would require. By contrast, water can be stored in reservoirs and aquifers, and this storage should be managed over weeks and months, even years. When a catchment has storage, short-term variability of demand is less of a problem.

The entire power network is heavily instrumented, and electricity users are closely metered. The modern electricity market is operated by specialists in relatively large sophisticated organizations. Traditionally, electricity networks were operated centrally. Now, the smart market for electricity allows management to decompose into buyers and sellers. The system operator still ensures that the flows satisfy network constraints. Thus, managers have excellent ability to manage the electricity system and its market.

Water networks are not well instrumented, and many users are not metered at all. Wholesale water is managed in a highly dispersed—often fiercely independent—way, with farmers, city water managers, commercial firms, and hydroelectricity generators involved. Thus, water managers have much lower ability to manage catchment hydrology than electricity managers do. The smart market would allow the management to remain decomposed into buyers and sellers, but the market manager would have the ability to specify that the flows are physically possible and to satisfy the environmental requirements. The market interface will have to accommodate a variety of users, and those users will have to be able to give evidence about the water they take. Thus, by moving toward a system like the one for management of electricity, managers would have much improved ability to manage hydrological flows, while enabling its market. Overall, the differences mean that politics have played a greater role in access to water than in access to electricity.

Despite their differences, electricity and water have strong similarities relevant for market design. A water market would have separate entities bidding to buy and sell water allocations through a common pool, as with electricity. Just as the electricity market requires constraints to ensure security of supply, a water market would need constraints to prevent overexploitation of the resource.

Markets for electricity and water share the many-to-many effects of supply and demand. These effects hinder pair-wise trading. Electricity generators and water sellers cannot easily determine the buyer. As a result, a water market would operate best as electricity markets do now, as a common-pool. But these markets differ in key ways.

Most electricity markets clear as single periods because electricity has not been easily stored, lacking battery technology. By contrast, storage is central to water management, so a common-pool water market would usually optimize temporal and spatial interactions. Those interactions would be determined by an external simulation that computes the relevant derivatives.

The inter-temporal nature of water complicates modeling. While some electricity markets have inter-temporal optimization of controlled stored water, water is naturally stored and transported underground. As a result, uncertainty of projected conditions impacts current prices, and this uncertainty will concern users. The market mechanism may need to manage this uncertainty explicitly, as Chap. 11 discusses.

4.3 Natural Gas Smart Markets

A gas market is complicated, because buyers and sellers are connected by a complicated pipe network. Buying and selling cannot happen at the same time, because of the compressibility of gas and the time it takes for a pulse of the stuff to move through the pipe. When someone injects gas to the pipe network, the system operator has no way to match that gas to a single buyer, because one seller's gas mixes with gas from other sellers. Similarly, when someone takes gas out, the system operator has no way to identify a seller, because each seller's gas mixes with gas from other sellers. A gas seller would want to sell to the highest bidder, but this is not always possible. The seller might try to sell more gas—or the buyer might try to buy more gas—than the pipe can hold, in which case the pipe could break. The pipe could be over- pressurized in one place, while under-pressurized in another.

To run the gas market of the state of Victoria, Australia, engineers and operations researchers have worked together to build a smart market (McCabe, Rassenti, and Smith 1989, 1990, 1994; EMRG 2002; Vencorp 2007). During the gas auction, a linear program chooses which bids to take (both sell bids and buy offers) in a way that keeps the pressure within appropriate bounds. The linear program takes the best bids, ensuring that injections and withdrawals satisfy pipe capacity constraints. The model manages the timing and movement of gas around the pipes, and makes sure that capacity is available when sellers are ready to inject and that gas is available when buyers have agreed to take it. The gas does not go to the highest bidder per se. Rather, it goes to the highest bidder who can take the right amount of gas at the right time, while ensuring that the network is stable. Finally, the linear program calculates prices at each connection.

A gas smart market operator needs modeling that is in many ways similar to the modeling needed for a water smart market. Gas flows occur in constructed pipelines rather than in rivers with variable dimensions and the porous media of groundwater aquifers, but many of the same issues arise. As for water, gas flows between adjacent network locations both determine and are determined by pressure differences between the adjacent locations at that time. Similarly, "line pack" in the gas pipeline provides a significant storage component, much like water storage, and unlike electricity storage. Experience has shown that management of the storage inherent in a gas pipeline is central to gas smart market operations.

Being compressible, gas behaves somewhat differently to water, but the equations are similar. In the gas smart market, much like the electricity smart market, the market-clearing formulation explicitly incorporates the flow equations and participant offers. The procedures needed to make the gas smart market work could be applied to a water allocation smart market, though a catchment may be more complicated physically than a pipe network.

4.4 Precedents for Smart Markets for Water

To our knowledge, the first suggestion of a smart market for water was by Hoffman et al. (1993). From that work, those authors received a 1995 National Science Foundation grant (Smith, Hoffmann, and Rassenti 1995), resulting in a string of interesting research related to water markets (Dinar et al. 1998; Murphy et al. 2000, 2006, 2009). This team used experimental economics to study how people might behave in a smart market for water, at least in an academic laboratory setting. They established that the people in such a market would likely behave as hoped.

Murphy et al. (2000) developed a smart market concept for surface water, simulating a simplified wholesale spot water market for California. They cleared the simulated market using a sealed-bid uniform price double auction mechanism, meaning the auction has both buyers and sellers, for allocation of water and its transportation through waterways and pipelines. In this type of market, the auctioneer fills all accepted bids at a price equal to the lowest accepted bid price. This price clears the market by balancing supply and demand.

They assumed simple hydrology—deterministic, single-period, corresponding to a stylized version of southern and central California. Murphy et al. (2006) addressed the environment in more detail, envisioning an agent that purchased water on its behalf.

They assumed a simple network structure for the physical flows, allowing the following linear program.

Maximize total surplus,	max $\sum_{\text{arcs}(i,j)} C_{,i,j}\, flow_{i,j}$ subject to
balance of flow,	$\sum_{\text{arcs } i \text{ to } j} f_{i,j} = \sum_{\text{arcs } j \text{ to } k} f_{i,k}$ for all nodes j, and
conveyance capacity,	$MinFlow_{i,j} \leq flow_{i,j} \leq MaxFlow_{i,j}$ for all arcs (i,j).

Murphy et al. ran 13 experiments, with varying rights to transportation capacity, and over high, medium, and low supply periods. Participants were undergraduate students at the University of Arizona. Their experiment included about ten players, with one simulated time period in each game. As with Becker (1995), WRAM, and CALVIN, Murphy et al. obtained prices from the optimization: the dual prices on the balance of flow constraints serve as the spot market prices at each node j.

Despite having only a few players, the market quickly found a competitive equilibrium with efficient water allocations. From their paper:

> Economists have long advocated the use of water markets to help mitigate these disparities between water availability and demand. Easter et al. 1998, as well as countless others, provide evidence indicating that the development of markets can improve water allocation by permitting the compensated, voluntary transfer of water to its highest-valued use. However, the allocation of private property rights can be successful only if there is an efficient mechanism to allocate and trade the rights. Although the advantages of water markets are well-prescribed, economic theory says little about how different market institutions can affect allocations. Experimental research clearly indicates that the rules governing trading play a vital role in determining the market outcomes and the realized gains from trade... For example, it is well known that posted-offer and continuous double auctions display dramatic differences in capacity to track changing market conditions.

Murphy et al. developed a model that simplifies the hydrology, and their game ignored some institutional requirements for a real water market. But the authors focused on users' behavior. A common difficulty mentioned frequently in the economics literature is the lack of knowledge about users' willingness to pay, when no free market is in operation. These authors realized that the simplest way to obtain this information is to ask the users. Their paper is one of the few that have examined any market system for water with explicit bidding. The authors urged the use of experiments in the form of games to observe the effects of different economic policies as a way to test the policies before implementation, and the value of computers as central market-clearing devices.

Murphy et al. (2000) described the desired endpoint of their early research into smart markets for water resources:

> With the advent of high-speed communication networks and low-cost computing resources, the landscape of exchange system architectures has changed dramatically over the past decade. Smart computer-coordinated markets have recently come to the fore. These markets allow decentralized agents, who best know their own circumstances, to submit virtually unlimited messages (bids to buy, offers to sell, logical and budgetary constraints) to a computer dispatch center. The center can then compute prices and allocations by applying a set of rules, often a sophisticated optimization algorithm that maximizes the possible gains from exchange, while observing constraints.

This is exactly the intention of a common-pool smart market for water.

Setting up a smart market is more complicated than simply solving an optimization with the correct hydrology. Setting up a smart market also requires careful development of institutional arrangements—the "rules of the game"—which will satisfy participants' business concerns, ensure a reasonable level of competition, and satisfy regulators. Policymakers need an implementation path. Such institutional arrangements are relatively simple for a government procurement of conventional commodities, but will be more complicated for water systems. Later chapters explore these matters.

This proposal—to clear a water market using optimization, with water values that are not estimated but come directly from the users—was an important conceptual step towards a smart market for water.

Tisdell, Ward, and Capon (2004) in Australia observed that water extraction imposes a common reduction in environmental benefits. Using an experimental water-trading program called Mwater, the authors designed games, played by university students, to explore an environmental levy and community involvement to allocate water in a complex heterogeneous environment. Participants acted as farmers faced with monthly demands for water, uncertain rainfall, possible crop loss, and water trading. The different games allowed users to receive information about the environment with different timing and different methods of communication. The environmental levy served to pay the cost of altering the natural flow to satisfy demand. The game imposed the levy equally on all players. The authors found (1) that trade would raise traders' incomes, and (2) that trade would raise the cost of the environmental levy, implying that trade would result in more damage to the environment. However, they also found (3) that communication among traders resulted in significantly lower cost of the environmental levy, so communication could reduce environmental damage. They concluded that the level of extraction, rather than the market, primarily determined the environmental levy. Thus, Mwater has no explicit constraints to protect the environment, and it overlooks the matter that water abstractions vary in terms of the amount of recharge they contribute.

Holt et al. (2012) conducted economics experiments for a canal, in which water flows in only one direction. They found that the optimal fee, calculated beforehand, would result in the highest efficiency; a uniform price auction would be nearly as efficient, with the advantage of discovering the usage fee via market operations.

The Easter, Rosegrant, and Dinar (1998) book includes an article by Richard Howitt that recommends an independent system operator (ISO) for California's water. (While his article did not reference the Murphy et al. work, he was a co-author on it.) The ISO would serve as the market manager in a smart market. He points out that water markets suffer from "sequential objections," in which a proposed trade is subject to one criticism after another, and the rules for an ISO would have to prevent this.

Hanak et al. (2011) examined the problems and policy options for California water management. A long section titled "Modernizing California's Water Grid" proposed an ISO for California, echoing Howitt's earlier article in Easter et al. This material first describes California's electricity market and its ISO, and then describes analogous options for a limited ISO and a fully integrated ISO for water. From their book, page 340:

> The second, more ambitious, alternative would make the water ISO much more like California's electricity ISO, which finds market clearing prices for all electricity that moves through the grid. Under this model, the ISO would not simply operate the CVP and SWP systems and a voluntary market; it would have authority to change how water is priced and allocated among the contractors, by establishing a market clearing price for the use of conveyance. Each year, the ISO would set the amount of water available for distribution after environmental requirements are satisfied. Rather than assert their water or contract rights to a specific water quantity and price, the participants would bid for delivery of available water (or available conveyance space). The ISO would then allocate water based on the highest to lowest bids over various time horizons.

Hanak et al. do not discuss the operations research methods that would be required or the detailed hydrological issues. Instead, they focus on the institutional and legal changes that would be required for California.

4.5 What Kind of Market Is a Smart Market for Water?

People trade in many different types of markets, with everything from retail groceries to high-speed financial assets. Before describing the design of a common-pool smart market for water, let's consider the type of market it is.

It would be a *common pool market*. In a common pool market, trades do not occur on a one-to-one basis, but instead, to and from the common pool. This auction does not clear or create bilateral transactions between particular participants. Users will not usually know who was on the other side of the trade; 300 people could be buying from 400 other people, and the waters are mixed in between. Users will bid only for a change, either an increase or a decrease, in water allocation at their own location.

Book exchanges work on this principle, but without a financial compensation. Commodity trades work in many ways like common pool markets. A market manager sets certain rules to establish what grade a commodity meets, and sellers who can supply a commodity that meets the grade can sell into the market, while the market manager assures buyers that the commodity that they buy will meet certain quality criteria. People do not necessarily sell wheat to a buyer; they can sell it to a commodity market. The market manager then arranges for buyers, and sets rules needed to make everyone respect the setting of market prices. Despite the obvious common pool nature of water, most water markets operate on a one-to-one basis, not as a common pool market. For common pool markets (as with both electricity and gas), the smart market provides the trading location, so users need not search for each other. More importantly, a linear program automates the vetting process, so the transaction cost falls considerably.

The smart market for water is an *auction*. That is, an auctioneer solicits bids from people, and awards the goods to those who bid highest. It is not a market with fixed prices, like a grocery store is.

The market is a *spot market*. That is, the auctioneer offers goods for immediate or near-term delivery. The smart market for water must also have future allocations traded at the same time.

The market is *periodic*. It is not running all the time, like a grocery store or financial market. Rather, it would run according to a specified **auction schedule**, such as once per week, once per month, or once per season. The best frequency depends on the local conditions and the participants' needs. Whenever the market manager holds an auction, all potential buyers and sellers need to have the opportunity to bid, in order to maximize the value for all bidders.

The market is *double-sided*—it has both buyers and sellers. People often think of an auction as having one seller and many bidders offering to buy. But anyone with initial rights to water could offer to sell. It cannot be an auction where all the buyers buy from the common pool at one auction, and then all the sellers sell to the common pool in a separate auction, because the market manager could not ensure that the water balances.

The market can be multi-round. Bidders have the opportunity to see tentative results, and can change their bids, subject to rules, before the final round. Allowing multiple rounds helps improve the bidding process. Many bidders are unlikely initially to know their true demand function, so they may provide a poor set of bids. By viewing the results of a test auction, they will have the ability to adjust their bids after seeing whether they would be buyers or sellers, and after they evaluate the likely economic impact of that trial bid.

The market can have multipart bids, in which bidders can make multiple offers in each round. For example, a bidder may offer to sell a quantity of water if the price is high, while also offering to buy a quantity of water if the price is low. Each user can offer several blocks in their bid, e.g., "I will sell 2 ML if the price is at least $1000; I will sell only 1 ML if the price is $500–$999.

The market is not an open, or "outcry," market, where everyone hears the bids. Rather the bids are *sealed*, so participants cannot see each other's bids. The final price should be public, and the amount of water that each user takes can be public.

The market manager sets prices for water at marginal cost, not the highest bid. Roughly speaking, this is akin to a second-price auction. The market may be viewed as a uniform price auction (unlike some auctions, where people pay what they bid, or what some other party bids), adjusted for the local hydrology. Two identical users at a given location face a uniform price, even if they have bid differently; these two users face the same price because, by assumption, their water takes are from the same part of the hydrological system and so have the same impact on the environment. They need to receive a consistent message about society's value on that constraint at that time. The auction may be viewed as using marginal cost pricing, as the market manager charges traders at the rate of the dual price found from the linear programming solution.

As described here, it is not a combinatorial auction, because all variables are continuous. (However, some cases can require integer variables, depending on the nature of the trading).

This auction method falls in the class of smart markets: an auction assisted by a computer, with an optimization to solve the allocation problem associated with a set of bids (McCabe, Rassenti and Smith 1991). Linear programming duality provides equilibrium prices under modest assumptions. Prices clear the market as both buyers and sellers are satisfied with their trades, and no better allocation is possible, given the data. Some existing water markets have rules that prevent inter-sector trading, e.g., between agriculture and cities (Griffin and Characklis 2002). Such trade restrictions may be due to politics, or concerns about differing hydrological or environmental impacts. We do not address the former. As for the latter, the common-pool smart market model does not require such restrictions on trade, because the hydrological response matrix can account for the differing hydrological impact of different users.

References

Alvey, T., Goodwin, D., Ma, X., Streiffert, D., & Sun, D. (1998). A security-constrained bid-clearing system for the NZ wholesale electricity market. *IEEE, 13*, 340–346.

Becker, N. (1995). Value of moving from central planning to a market system: Lessons from the Israeli water sector. *Agricultural Economics, 12*, 11–21.

Carlson, B., Chen, Y., Hong, M., Jones, R., Larson, K., Ma, X., Nieuwesteeg, P., Song, H., Sperry, K., Tackett, M., Taylor, D., Wan, J., & Zak, E. (2012). MISO unlocks billions in savings through the application of operations research for energy and ancillary services markets. *Interfaces, 42*(1), 58–73.

Dinar, A., Howitt, R. E., Murphy, J. J., Rassenti, S. J., & Smith, V. L. 1998. *How might future water markets look: The use of experimental economics to design markets for water*. World Congress of Environmental and Resource Economists, Venice.

Easter, K. W., Rosegrant, M. W., & Dinar, A. (Eds.). (1998). *Markets for water: Potential and performance*. Norwell: Kluwer Academic Publishers.

EMRG. (2002). *1998–2002 final activity report*. Energy Modelling Research Group (EMRG). http://www.mang.canterbury.ac.nz/research/emrg/report9802.shtml. Accessed 5 Aug 2007.

FAA. (2015, Jan 6). *Fact sheet – Notice of proposed rulemaking: Slot management and transparency for LaGuardia Airport (LGA), John F. Kennedy International Airport (JFK), and Newark Liberty International Airport (EWR)*. https://www.faa.gov/news/fact_sheets/news_story.cfm?newsId=18054. Accessed 14 May 2016. See also http://www.faa.gov/news/updates/?newsId=81144

Gallien, J., & Wein, L. M. (2005). A smart market for industrial procurement with capacity constraints. *Management Science, 51*(1), 76–91.

Grether, D. M., Mark Isaac, R., & Plott, C. R. (1979)..*Alternative methods of allocating airport slots: Performance and evaluation*. Prepared for the Civil Aeronautics Board, Polinomics Research Laboratories, Inc., Pasadena.

Griffin, R. C., & Characklis, G. W. (2002). Issues and trends in Texas water marketing. *Water Resources Update, 121*, 207–217.

Hanak, E., Lund, J., Dinar, A., Gray, B., Howitt, R., Mount, J., Moyle, P., & Thompson, B. (2011). *Managing California's water: From conflict to reconciliation*. San Francisco: Public Policy Institute of California.

Hoffman, E., Rassenti, S., Smith, V., Dinar, A., & Howitt, R. (1993). *Designing a smart, computer-assisted market for water allocation in Arid Lands*. 63rd annual conference of the Southern Economic Association, New Orleans.

Hogan, W. W., Read, E. G., & Ring, B. J. (1996). Using mathematical programming for electricity spot pricing. *International Transactions in Operations Research, 3*(3–4), 243–253.

Holt, C. A., Johnson, C. A., Mallow, C. A., & Sullivan, S. P. (2012). Water externalities: Tragedy of the common canal. *Southern Economic Journal, 78*(4), 1142–1162.

Jaikumar, R. (1980). Mathematical programming approaches to the Design of Auctions: A study of the auction of landing rights at congested airports. *Economic Forum, 11*, 29–47.

McCabe, K. A., Rassenti, S. J., & Smith, V. L. (1989). Designing 'smart' computer-assisted markets: An experimental auction for gas networks. *European Journal of Political Economy, 5*, 259–283.

McCabe, K., Rassenti, S., & Smith, V. (1990). Auction design for composite goods: The natural gas industry. *Journal of Economic Behavior and Organization, 14*, 127–149.

McCabe, K. A., Rassenti, S. J., & Smith, V. L. (1991). Smart computer-assisted markets. *Science, 254*(5031), 534–538.

McCabe, K., Rassenti, S., & Smith, V. (1994). Designing a real time computer assisted auction for natural gas networks. In W. Cooper & A. Whinston (Eds.), *New directions in computational economics* (pp. 41–54). Holland: Kluwer Academic Publishers.

Murphy, J. J., Dinar, A., Howitt, R., Rassenti, S. J., & Smith, V. L. (2000). The design of 'smart' water market institutions using laboratory experiments. *Environmental and Resource Economics, 17*(4), 375–394.

Murphy, J. J., Dinar, A., Howitt, R. E., Mastrangelo, E., Rassenti, S., & Smith, V. L. (2006). Mechanisms for addressing third-party impacts resulting from voluntary water transfers. In J. A. List (Ed.), *Using experimental methods in environmental and resource economics*. Cheltenham: Edward Elgar Publishing Ltd.

Murphy, J. J., Dinar, A., Howitt, R. E., Rassenti, S. J., Smith, V. L., & Weinberg, M. (2009). The design of water markets when instream flows have value. *Journal of Environmental Management, 90*(2), 1089.

Raffensperger, J. F., Milke, M. W., & Read, E. G. (2009). A deterministic smart market model for ground water. *Operations Research, 57*(6), 1333–1346.

Rassenti, S. J., Smith, V. L., & Bulfin, R. L. (1982). A combinatorial auction mechanism for airport time slot allocation. *Bell Journal of Economics, 13*, 402–417.

Read, E. G, & Chattopadhyay, D. (1999). Electricity market models: Lessons from Australasian experience. In *ORSNZ 99 (Proceedings of the 34th annual conference, Operational Research Society of New Zealand)*. http://orsnz.org.nz/conf34/PDFs/Chatterjee.pdf

Schweppe, F. C., Caramanis, M. C., Tabors, R. D., & Bohn, R. E. (1988). *Spot pricing of electricity*. Boston: The Kluwer International Series in Engineering and Computer Science, Kluwer Academic Publishers.

Smith, V., Hoffman, E., & Rassenti, S. (1995). *Environmental change and adaptive resource markets: Computer-assisted markets for water allocation*. Award Abstract #9513406, start date 1. http://www.nsf.gov/awardsearch/showAward?AWD_ID=9513406. Accessed 14 May 2015.

Stanley, E. D., Honig, D. P., & Gainen, L. (1954). Linear programming in bid evaluation. *Naval Research Logistics, Quarterly, 1*(1), 48–54.

Svorencík, A., & Maas, H. (Eds.). (2016). *The making of experimental economics: Witness seminar on the emergence of a field*. Cham: Springer International Publishing.

Tisdell, J., Ward, J., & Capon, T. (2004). Impact of communication and information on a complex heterogeneous closed water catchment environment. *Water Resources Research, 40* (9). doi:10.1029/2003WR002868/epdf

Vencorp. (2007). *Guide to the Victorian gas wholesale market*. www.vencorp.com.au/index.php?action=filemanager&doc_form_name=download&folder_id=466&doc_id=3468. Accessed 5 Aug 2007.

Vickrey, W. (1961). Counterspeculation, auctions and competitive sealed tenders. *Journal of Finance, 16*(1), 8–37.

Chapter 5
Prerequisites for Implementation

Abstract This chapter describes six prerequisites for a water market. The first five of these prerequisites should be met even if the policymaker does not have the goal of an active market; they are simply good management. First, stakeholders should share trust. Second, the regulator must have a hydrological model that describes the water flows with a level of accuracy acceptable to the community. Third, the regulator must specify the required environmental flows. Fourth, users must meter their water use. Fifth, the regulator must record and manage water rights and use. Sixth, the regulator must grant relevant parties full legal ability to trade, with as few impediments as reasonably possible.

Keywords Water market • Regulator • Stakeholders • Prerequisites to trade

5.1 Simply Allowing a Market Does Not Make It Work

Many countries have believed that rewriting the laws simply to allow a water market would be enough to create a well-functioning water market. This belief is far from realistic. Water regulation usually needs a number of changes to allow for any water market, including a smart market. This chapter describes these extra-market considerations.

> One take-away point from this chapter should be that the market design we are examining is not a laissez-faire free market, but a managed regulatory framework that enables voluntary market operations.

By far, the country that has done the most to establish the prerequisites for good water markets is Australia. There, stakeholders have seen the extra-market changes as worth the effort, because of the benefit from functioning water markets. The Australian example demonstrates that many of these changes are part of modern water management, and make sense whether or not a market is established.

To help analyze who might develop and oversee the changes needed for a smart market for water, we distinguish between the water regulator, who is responsible for setting and enforcing limits (like any modern water regulator should be), and the

J.F. Raffensperger, M.W. Milke, *Smart Markets for Water Resources*, Global Issues in Water Policy 12, DOI 10.1007/978-3-319-55008-4_5

market manager, who is responsible for operating the market. The next chapter discusses the requirements for the market itself, such as the office of the market manager.

5.2 Stakeholders Should Share Trust

First, for government to implement a smart market for water, the community must be ready to participate in a water market. Community readiness will be gauged by the elected representatives, so elected representatives need to agree that a smart market is the right approach.

Managers in the governing water agency must understand the value of the market. Then the water users, especially those who take the most water, must be convinced that a well-managed water market is in their interests. Implementers should inform other people in the community. Not everyone needs to agree that the government should implement a smart market, but the implementers should explain to everyone the goals and operation.

Holders of water rights will be concerned about losing rights in the transition to the market, or in some process after its introduction. This concern is justified, in some ways, because the market manager can buy only allocations that are highly likely to be real water. This means that implementers must convert water rights at some point to a system based on real water availability, not on paper rights. The rules associated with water rights must recognize that that actual water availability is variable, and often less than the total paper rights.

Environmentalists should see that the smart market approach would protect the environment. Environmentalists will be concerned that the market does not protect environmental flows. Hence, stakeholders must agree on the environmental flow requirements in advance, and the implementers must explain the mechanisms for respecting those flows. Most likely, the new flows will meet environmental flow requirements with more robustness and more certainty than ever before. Section 5.4 explores this prerequisite measure further.

A challenge with instituting a smart market for water is explaining the change to communities of interest that, based on experience, might lack trust in water authorities. Over many workshops, we collected a wide range of comments about water markets from New Zealand community members. We describe and answer these comments in Chap. 13. Implementers will be able to rebut many arguments against change, but rebutting arguments will not alone build the trust required for the major changes we describe.

Trust comes from parties respecting the goals of the other parties, listening and discussing, and working together to resolve disputes. The severity of the water problem in many regions, and the recognition that water concerns are interlinked, help to develop trust; however, trust is built slowly, and can erode quickly when individual parties flout the rules.

5.3 The Regulator Must Have a Hydrological Model that Describes the Water Flows with a Level of Accuracy Acceptable to the Community

Implementers should first develop a good hydrology simulation and the rules for improving it. The hydrology simulation is a critical prerequisite to implementing the market. The purpose of the hydrology simulation is to determine the effects of users' takes of water ("What if?"), immediately and over time, and thereby to regulate the users' takes based on objective science. The manager will be stress-testing this model every day, because people will be putting money on its results, and because the environmental flows will depend on its accuracy.

Fortunately, simulations of hydrology and hydrogeology are now ordinary tools of water management, apart from any potential use within a smart market. Hydrologists can build these simulations, using commercial or open-source software, with relative ease. We describe these simulations in Chap. 3. However, developing such a model to the standard expected by regulators and water users for a smart water market will be challenging. All parties will want a market that will be fair in terms of the distribution of money between water users, and reliable in terms of respecting the environment. The greater the fairness and reliability that a community expects, the higher the standard that the model must have. The model must be developed with a precision and care well beyond that required for the typical academic study, and must include detailed data about every user in the catchment—data that in pre-market days may have been recorded in messy formats, if at all. Because of high expectations, the model will likely require sophisticated data on forecasted inflows, in detail that is relevant to the structure of the catchment, e.g., by every reach of every stream.

Over time, the regulator will need to improve the hydrology model as the regulator better understands the hydrology or makes the model more precise. The consulting hydrologist must estimate a long list of factors, and these estimates are subject to error. Implementers may simplify the initial model, because some estimates are expensive to obtain, and the estimates could later be refined. When the regulator changes the model, the regulator may need to change users' allocations or the environmental requirements; implementers should include market rules for such changes.

Interestingly, the market itself can provide useful direction as to which aspects of the model the regulator should improve. The competent hydrologist should record the relative uncertainty associated with many of the simulation's parameters. For example, while concluding that the aquifer head near a coast must be at least *X* meters from the coast to prevent saltwater intrusion, the hydrologist will record that the estimate of *X* has some variance. The overestimate is different to the underestimate—one results in giving users a bit less water, while the other results in serious contamination of the aquifer. Further, reducing this variance requires the hydrologist to get more data at some cost. Fortunately, the market-clearing optimization model will provide a set of prices associated with the aquifer head constraint.

These prices convey the value to users if the regulator were to allow the aquifer head limit to fall. Should these values be low relative to the cost of more investigation, the hydrologist can leave the value of X as is. Should these values be high relative to the cost of more investigation, the hydrologist should see that market participants would likely benefit from a refined model with better data. The hydrologist can make a decision about whether to conduct the study using what the decision analysis literature calls "the expected value of perfect information" (Clemen and Reilly 2001).

Users are likely to question the hydrology model. They are likely to lobby the regulator to adjust various factors in their favor. Changes to a single factor can benefit some users and injure others. Most relevant here are the factors associated with water movement within the model, e.g., the response coefficients, or values sometimes called "trading ratios." Rules must be in place for a user to contest the model; perhaps the user can pay any of a list of approved consultancies for the improved hydrology, but then must agree to adhere to the result. Even without the market, the legal framework must accept the hydrology model as authoritative, with an agreed process for review of the model and changes to it.

We recognize that in a worst case, some hydrological effects and impacts could be highly non-linear and uncertain, and may not be easily modeled with the linear inequality constraints. If the price is not sensitive in this area, the market may be able to ignore the non-linearity (see Sect. 13.10).

Although we would expect that a sophisticated hydrological model would be a prerequisite for instituting a smart market for water, it is not necessary. Desperate times can lead to desperate measures, and extreme droughts could lead communities to agree on implementing smart markets for water with only crude models and data, and agreed on protocols for quickly adjusting tradable allocations as problems arise. In these unusual situations, as confidence grows that the smart market will resolve water allocation better for everyone, the desire will grow to improve and refine the model. In all situations where a smart market for water has been working for a number of years, a sophisticated hydrology model should underlie the market.

5.4 The Regulator Must Specify the Environmental Flow Requirements

When users take water previously allocated to the environment, ducks, salmon, and water lilies have no feeling of compensation and do not file suit in court. No one can compensate them for lost environmental flows. For this reason, we treat environmental flows not as an allocation that can be surrendered for compensation, but rather as constraints, as requirements that must be satisfied.

Management of environmental flows is symmetric to management of users' licenses, with specification of the required quantity, with measurement, and with enforcement. Just as the market design specifies a complete regulatory framework for users, the design also specifies a complete regulatory framework for the environment. Just as implementation of the smart market enables later continuous improvement for users, implementation also enables continuous improvement for the environment.

In some places, governments have not yet established good environmental standards. Although the process of setting environmental constraints is a key component of a smart market for water, it is a prerequisite for any system of good water management. After a systematic process involving community input, the regulator must explicitly specify non-commercial water requirements. The purpose of these requirements is to prevent the externalities of trades on the environment. Users can bid to buy or sell without regard to these requirements, because the responsibility for satisfying these requirements is on the market manager. The optimization treats the environmental requirements as constraints that must be satisfied in every time period, in every auction.

Just as the users' water allocations are precise and time-based, the non-commercial requirements must also be precise and time-based; e.g., *F* megaliters flow per day in a particular river at a particular location during the month of July, and *H* meters of head in the aquifer at a particular place in the month of July.

This precision in quantity, time, and space would bring an almost unheard-of robustness to the environmental flows, replacing vague 10-year "plans" for catchments with operational schedules of water. For example, the regulator can specify a surge in a river that would be otherwise difficult to arrange.

Mechanically, the regulator could impose almost any plausible environmental requirement. For example, the Australian government requires that trades do not cause channel erosion from sudden surges, nor overflow stream banks, nor water wetlands out of season (Murray–Darling Basin Authority n.d. 2, 6). While the Australian trading systems do not use a smart market and so do not convert these environmental requirements into mathematical constraints, making such a conversion is straightforward. The smart market optimization would then satisfy the constraints as a matter of normal operation.

Business people often feel ambivalent about precise environmental standards. However, clear standards can't be fudged and business people can't pretend they didn't know the standards. This precision gives the community confidence in the market's protection of the environment. Clear standards improve business certainty and ensure fair competition, which are two important aspects that are currently lacking when business people want water.

5.5 Users Must Meter Their Water Use

5.5.1 Metering Is Part of Good Water Management

Water metering is an essential element of modern water management. Implementers should make metering a requirement of each user's entry in the market, but it should be done anyway. If the water regulator does not require users to meter their water use, enforcing limits is nearly impossible, and so managing the catchment will be impossible. Unmetered is unmanaged.

Users may complain that water meters are expensive. We would envision that any smart market for water would start with the larger users and over time take in the more numerous smaller users. The regulator could implement water metering in stages, in decreasing order by quantity of water right, as new users enter the market. If users are reluctant to meter their water because of the initial cost of the meter, the existence of financial benefits for users will provide options for the market manager to negotiate with users to pay the initial cost, possibly recouped through trading fees.

But the extra cost of water metering for modern farmers could be much less than one might expect. Water meters can be valuable in detecting leaks within users' water distribution systems, and so they can help pay for themselves. Many farmers are already implementing sophisticated water measurement technology, with wireless water moisture sensors, Bluetooth communication of the logged measurement data to ruggedized computers in farm vehicles, and GPS for site selection of water application. The aggressive move of agriculture into smart irrigation technology demonstrates conclusively that metering is good for farmers, even without the advantages of improving allocations (See, for example, Davids Engineering (2015)).

5.5.2 The Market Can Trade Only What Is Metered, in Quantity and in Time

The regulator must verify allocations; the market manager's requirements are even more stringent, because the ability to meter directly affects the identity of the trader and the terms of an allocation. Participants can trade over quantities and time periods that users reliably meter. The quantity aspect should be obvious. When someone sells water, the market manager must be able to verify their agreement to take less water.

To see how metering affects the identity of the trader, consider a catchment with several water user groups. Each group has only one meter, which measures the water taken by two or more users. The market manager should be willing to clear trades only between the groups, not between individuals, because the manager cannot verify individual use. The users of a single meter will have to find some means to verify their individual use of water. Most likely, the users of a single meter will often argue, especially when the water is valuable, and the users suffer some problem such as a leak. To best align the market incentives with water use, water use should be metered down to the paying user.

The time aspect may be less obvious. If water use is measured and recorded only once per year, then the shortest time period associated with an allocation can be at shortest a year, because that is the time period that can be verified. Such a long period would prevent incentivizing users to change crop types to take advantage of different irrigation schedules. A long trading period also prevents the regulator from managing environmental flows to any finer level than annual. A preferred water meter would transmit data to the regulator hourly or even half-hourly. The regulator would need a communications and information system to accept these data. In some cases, users' meters automatically transfer water use data to water district servers. With this technology, users could trade water by hour if they wished, and the regulator could specify environmental requirements in near real-time.

Water users might be reluctant to install water meters out of a distrust of allowing government to have the data, or from suspicion that they may be charged for more water than they use or that their neighbors are undercharged. But lack of metering is lack of measuring, and that is poor management of the commons. Many other markets expect individuals to track consumption, so it is reasonable that a water market should as well. If water users have these strong reservations, then the issue is not one of meters but of trust. Water users who say they are ready for a water market but are not willing to have their water metered and their use tracked are not ready for a proper water market.

Perhaps new technology will reduce the need for individual water meters. Space technology already allows detailed remote sensing of water (Bastiaanssen et al. 2000), and farmers can profit from it (Tenkorang and Lowenberg-DoBoer 2008). Firms such as HydroBio (hydrobioars.com) are rushing to make money from selling the technology. To use for trading, the regulator and the users must agree that the remote sensing technology is satisfactory, that the sensor measures water abstraction sufficiently accurately in time and space and in the identity of the user.

5.5.3 Meter Error Is Important and Asymmetric

The regulator must routinely check users' meters to ensure that they measure all the water and are sufficiently accurate. For thousands of years markets have had regulators who check the accuracy of scales. Effective water markets will need regulated metering of water use. Regulators should have a rule to address the reliability of water meters.

For example, they could correlate users' electric power bills with their metered water quantities to check the reasonableness of the water meter readings because a light user of electricity is unlikely to be a heavy user of water. Because a hydrological model is running alongside the allocation model, the regulator has the opportunity to conduct forensic hydrological analyses to evaluate when a water meter might be working correctly. If, given the user's water use profile (as provided by the meter), the hydrology model cannot reproduce the water levels measured, then either the meter is misreading or the hydrology model needs to be refined.

Meter error is asymmetric in its incentives and effects. Users may wish to implement cheaper meters with lower metering accuracy. The regulator might be able to manage a smart market for water with some users using more accurate meters and others using less accurate meters, but the effects of errors need careful consideration because over-counting and under-counting have different consequences. To the regulator and market manager, meter accuracy specifications are irrelevant for over-counting (when the meter says the user has consumed more than actually was), but must be tight for under-counting to prevent damage to the water resource (e.g., by saltwater intrusion). Of course, the incentives are the opposite for the user—the user has an incentive to under-count (when the meter says that the user consumed less than actually was). Over-counting tends to affect primarily the user, while under-counting tends to affect primarily the commons. Thus, they are systematically different, and the regulations should treat them differently.

A given percentage error, say, 5%, has effects that vary drastically, depending on the location and size of the water use. Because users' abstractions can vary by six orders of magnitude (e.g., the difference between a well for a city and a well for one household), the meter error for a large user can be larger than the total rights for a larger number of small users put together. Metering errors matter greatly for large water users.

Figure 5.1 summarizes example data (dated roughly 2006) from Marlborough District Council, New Zealand.

- 146 permits have allocations of 99 ML per day or less. Their individual and total volumes are insignificant.
- 387 permits can take 100–999 ML per day, to a total of 110,000 ML per day. Each license alone is insignificant. Combined, they take a modest volume.
- 82 permits take 1000–9999 ML per day, to a total of 330,000 ML per day. These are medium-sized licenses. Their total volume can significantly affect their neighbors and the environment.
- 8 permits 10,000 or more ML per day, to a total of about 2,500,000 ML per day. These eight wells are the major takes in the catchment, taking more water than all other licenses combined.

The regulator in this example might wish to include smaller water users in the smart market while allowing them to have less accurate meters. After all, a 5% error on the eight largest wells in this example would be as much water as consumed by the smallest 500 users. Thus, maximum tolerable error rates in water meters could decline with increasing allocation, aiming for a maximum volume error, not a maximum percent error. For example, allowable error rates could be set to the following:

- 5% for licenses of less than 1000 ML, with a maximum volume error of 50 ML per day per user.
- 0.5% for licenses of 1000–9999 ML per day, with a maximum volume error of 50 ML per day per user. This would affect about 13% (82 of 623) of users.

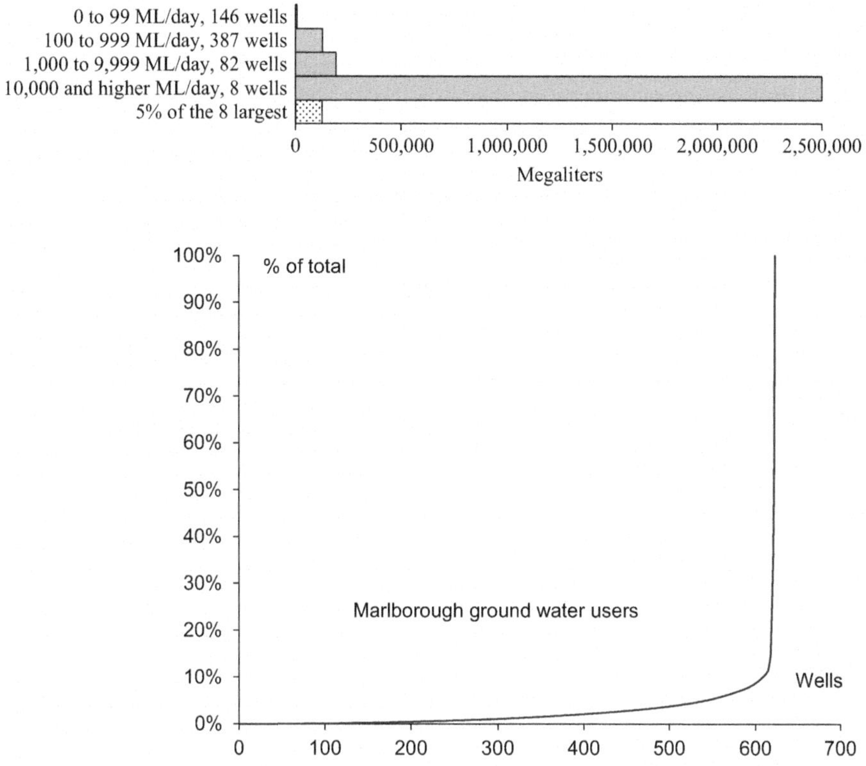

Fig. 5.1 Histogram of total takes of water in Marlborough, New Zealand, by size of permit

- 0.1% or lower for licenses of 10,000 ML and over, with a maximum volume error of say 10–500 ML per day per user. This will affect only the largest takes, a few users. This level of accuracy is possible with current technology.

Between users taking similar quantities of water, differences can arise in the importance of metering depending on the location of the water take. A groundwater user near a river with strict controls on flow (as in Fig. 1.6) will need an accurate meter because any under-reporting of consumption could have major effects on the environment. The ideal error tolerance rule would be asymmetric. Such a rule would require very small undercounting, say 0.05%, while tolerating large over-counting, say 5%. The asymmetry in the rule would incentivize users to have more-accurate meters.

For a smart market for water to function, water use must be measured, and the market manager and users should measure valuable water more precisely than cheap water.

5.5.4 The Environment Needs Metering

The enormous transaction cost of allocating water between users is dwarfed by the typical transaction cost associated with allocating water to the environment. Indeed, in many regions of the world, government has still not established firm environmental flows. When they are set, too often the regulator manages environmental flows on average annual flows, not for the detailed needs of the ecology. Farmers profit from closely managing the soil moisture (Tenkorang and Lowenberg-DoBoer 2008), which obviously cannot be managed by an annual average. Surely the environment would also benefit from more-detailed management.

To do such detailed management, the regulator must measure the environmental flows. At a minimum, the regulator should measure all flows associated with the required environmental flows, and the regulator should measure on a time scale that matches the auction schedule.

As the smart market develops over time, the regulatory hydrologist will be able to conduct value-of-information studies on the components of uncertainty in the measurement and modeling. Such studies will enable the regulator to determine when the market can cheaply ignore meter error and when more precise metering will result in a benefit. These value-of-information studies apply to both meters of users' takes and meters for the environmental flows. Where the environmental flow constraints are usually loose, the associated dual prices will usually be zero. Where the environmental flow constraints are often binding, the associated dual prices may be large. The regulatory hydrologist could compare the dual prices for environmental flow constraints over time, and the uncertainty of the parameters, to calculate the expected value of uncertainty reduction in the parameters. For example, Fig. 9.5 shows that well 12 faces the highest price, since it is adjacent to a location associated with a binding environmental flow constraint. Error reduction in measurement is likely to be more valuable at that location than near well 15, which faces a lower price.

5.6 The Regulator Must Record and Manage Water Rights and Use

5.6.1 The Regulator Must Maintain a Modern Water Rights Database

When economists complain that "water rights are poorly specified," part of that complaint can include the government's record-keeping. Implementers will have to digitize and parse scraps of old parchment for storage in modern relational databases. The careful review of old records could result in a finding that water rights would need to change, which could lead to contention. Despite the difficulty of the change to modern record-keeping, good water management requires that the regulatory institution must be able to record changes of ownership in water licenses with modern business practices.

But a state-of-the-art licensing system, stating who has rights to how much of a certain water resource, is insufficient for the market. Transactions involving licenses, where a new user buys into the market or users increase or decrease their rights permanently, will be relatively rare; instead, the transactions in the spot market will dominate the data management needs. We discuss this market database in the next chapter. The rights database should include a history of each user's actual takes as recorded from the telemetry on the water meters. These histories will be useful for the user, for the regulator and market manager, for the managing hydrologist, and for researchers.

5.6.2 The Regulator Should Redefine Initial Rights

We discuss initial rights in full detail in Chap. 8, "Converting Users' Rights to Tradable Allocations." Before we can explain how implementers must define and adjust initial rights, we must explain market operations in more detail, because we will define initial rights in terms of the smart market model. We outline the requirements for implementation here.

Water rights usually have two components. First, government gives permission to users to conduct activities related to taking water without their actually taking it, such as digging wells in a particular place. This permission can also specify the identity of the users, access rights, and the allowed uses for which users can extract the water. We will call all of these conditions the *terms of the license*. The second component regulates the extent of water use in time and quantity. Existing water rights would have some time length—5 years, 10 years, or perpetuity—and conditions regarding the quantity that the user can take per year. These conditions may depend on water availability, the date that the user obtained the permit relative to the date that other users obtained their permits, and effects on environmental flows.

The smart market design requires the water regulator to convert the second component of existing rights into new rights. Critically, implementers must convert vague "annual" quantities into quantities associated with much shorter time periods, e.g., a month or a week. January water is a different commodity than July water, because inflows of water naturally vary over the year, and because demand for water also varies over the year. By explicitly differentiating supply and demand of water by time, the market can price these differently. By pricing differently, users have an incentive to use water only when they need it, and to store it appropriately. Each period can have a different quantity limit, or it can be done as simply as taking an annual allocation and dividing it into a set of equal weekly allotments over a growing season. We will call this quantity an **allocation**: the tradable amount that the regulator assigns to each water user for a given time period in the auction schedule.

This conversion allows more precision in adjusting of rights. During drought, instead of the regulator telling the user, "Your rights have been cut by $X\%$," the regulator says, "Your allocation for July has been cut by $Y\%$," which is finer-grained control. This is a significant transaction cost to setting up the market, but the regulator should do this conversion anyway as part of good management of the water.

These allocations are the traded quantities in the market, not the terms of the long-term license, nor the quantity associated with the long license. A user trades an allocated quantity as a *lease* for its associated time period. When users sell to the market manager, they do not surrender their long-term license. They simply rent out to the market manager some of the water quantity associated with that license for a specific period. The lease is specific in quantity, specific in start and end times, specific to a location, and specific to whether the water use is consumptive or non-consumptive (or something in between). The hydrologist can use information about the intended use of the water to estimate the water that returns to the environment, so the model has a more accurate value for the hydrological impact of each user's abstraction. The market manager could run the market in a simplified fashion, giving no credit for return of water to the environment, so then the market manager does not need to know the use for the water.

More precisely, and perhaps more subtly, users do not trade water itself, but, rather, they trade permission to use a quantity of water at a given location. Users trade this permission directly with the market manager. When a seller gives up permission to take water at his or her location, the relinquished quantity is unlikely to match one-for-one elsewhere. Buyers will take up additional water with different relative impacts on the environment, for different levels of consumptiveness, and with different impacts on the water resource for nearby users. Thus, the role of the market manager is to match all the different types of allocations via the smart market optimization model.

With a smart market for water in place, the market manager's allocations of water must be definite and firm. Not only are the users' rights firm, but everyone can assume that the user is likely to actually use the allocation in each time period, because every user would be incentivized to sell allocation that they did not intend to use. This incentive to sell unused water eliminates significant man-made uncertainty for the environment; with ordinary water rights, the regulator has no knowledge of how much water any user will take, and therefore is uncertain as to the degree that the environmental flow requirements will be satisfied. With the smart market, all users, even those who do not wish to trade, have allocations that they are actually likely to take, so the market manager has on hand a detailed operational catchment plan, which extends out for a sufficiently long planning horizon.

Still, the regulator must permit users to *not* use their water, while still keeping the right. With the market, they have an incentive to sell unused water, partly because the allocation is time-bound. Without the market, users have an incentive to hold allocated water in reserve in case they need it, and this reserve may appear to others to be waste.

5.6.3 The Regulator Must Have Authority to Adjust Users' Allocations

Water managers typically adjust water rights to water availability only in the long term, because they do not know users' short-term abstractions, and because managing those short-term abstractions has an impossibly high transaction cost. In strong contrast, within a smart market for water, allocations will be time-bound for short periods, location-specific, and aligned with actual water availability. The consequence is the regulator will need to adjust users' allocations routinely. With a smart market for water, the regulator continually adjusts users' allocations to ensure that demand meets supply, so that the regulator never over-allocates water. Water could be under-allocated in wet periods.

The regulator must include every user in these adjustments, except perhaps for the very smallest users. The regulator's (and the market manager's) hydrology model should account for all water. This includes users who do not wish to trade in the market. While non-trading users are not part of the market, they are still part of the catchment, and still subject to regulation. Their allocations are subject to the same allocation adjustments as those users who trade. Non-traders' adjusted allocations will need to be included in the smart market optimization model as non-optimizable constants.

We envision several scenarios where short-term or localized over-allocation of water would require adjustment of water allocations under a smart market. The scenarios can have different regulatory and market consequences. Here are a few important scenarios:

a. Predicted Inflows and Hydrological Changes. As described in the previous section, and as described in more detail in Chap. 8, the regulator must adjust users' allocations to reflect the expected inflows. If the catchment is over-allocated, so users have more total allocation than the catchment has water (relative to the environmental requirements), and if the users offer those rights for sale, the market manager would have to buy enough of those rights to make them match the actual water. The result is that the market manager's accounts will come out in the red. Consider the realistic case in which users have rights totaling 100 units of water, when only 20 units of actual water are available. If no one wishes to sell, the market-clearing model will fail without a feasible solution. If users are willing to sell at least 80 units of allocation, then the market manager will have to pay out, and possibly dearly.

In some locales this adjustment is called "truing up" or "adjudication" of rights. In the smart market, this adjustment will be dynamic and precise, a regular and routine operation of the market manager as part of every optimization of offers to buy and sell.

b. Unpredicted Inflows and Hydrological Changes. Forecasted inflows may be lower than expected, and the regulator may have set users' allocations based on those estimates. Forecast errors would necessitate changes in users' allocations.

The smart market cannot eliminate over-allocation (though it certainly reduces the problem drastically) on its own. The market manager must use a forecast to adjust allocations, and these forecasts will often be wrong. Unless the government is willing to routinely buy over-allocated rights from users, the market manager must have authority to adjust water allocations. The point is that the regulator cannot be responsible for the state of weather forecasting science, and users must accept that the regulator can change allocations as better information unfolds over time.

Similar to inflow forecast error, the catchment hydrology can change in some systematic way, affecting users and the environmental flows. Such changes can occur from land movement, flood, and long-term temperature changes. A change could also occur because of improved understanding of the water resources; for example, hydrologists could find a confining layer that is limiting flow to a nearby stream, affecting the impact of water use on the stream. Since neither the regulator nor the market manager should promise something that does not exist, the cost and risk of systemic changes in hydrology should fall on the commercial water users themselves as an ordinary business risk.

c. Society's Choice to Change Environmental Requirements. Society can choose to tighten environment requirements. Such choices could come from better understanding of environmental flows, or simply from higher societal value for environmental flows. Business people may view these changes as a taking of their rights. Government must decide whether to compensate the commercial water users for their losses. Australia chose to compensate its farmers for the reductions, but its process is not the only solution.

The initial changes from the status quo to the precision allocations of the implemented market need not be instantaneous, and the market provides new tools to ease the process. Policymakers may find that commercial users can tolerate incremental changes from the status quo toward the desired end states. This incremental path may include a few years where the government buys allocations to satisfy the tightened environmental requirements, followed by a permanent reduction with no further compensation. This compensatory period would give commercial users some payment to make the changes, and get them accustomed to using less water. However implementers decide to adjust users' rights, implementers should specify the method for change before the first auction. Users need to know how their rights will change before they begin to trade, so they know what they are trading.

5.6.4 The Regulator Must Enforce Users' Allocation Limits

If the regulator will not enforce users' limits, the regulator is mismanaging the water. Enforcement of the rules is essential to basic water management, and all the more so for market operations. Basic management requires that people adhere to their requirements and commitments. The regulator's enforcement must be active, responsive, consistent, predictable, and firm, and it must follow the law.

With trustworthy metering and with the smart market, enforcement of allocations limits within the market can be surprisingly light. The market manager should be able to specify modest penalties for users who take small amounts above their allocations. The market manager could base these penalties on the current market price, perhaps with some modest premium. The market itself conveys information about the correct penalty for taking too much. The penalty rate for taking more than the allocation should increase with quantity, because the market manager has scheduled all water in the catchment (assuming dry conditions), and over-takings impose real shortages elsewhere. Implementers will have to set up the penalties in the context of the regulatory framework.

The attention to allocation enforcement can work in users' favor. In the context of the market, the market manager should pay users for taking less than their allocations. Again, the market manager should base the payment price on the current market price, perhaps at some discount. This would incentivize users to conserve water, to turn off the tap when they no longer need water. The discount incentivizes users to plan better in the first place.

5.7 The Regulator Must Grant Relevant Parties Full Legal Ability to Trade, with as Few Impediments as Reasonably Possible

People often ask us how the smart market can be made to work with existing water law, which for various reasons often impedes trading. Groundwater law in particular is inconsistent over legal boundaries. Implementers in some places may find that the law allows steps forward to start a market. In other places, implementers will have to seek legal changes to remove impediments, before the smart market can operate fully.

Other impediments to trade can include trade-distorting price-fixing, subsidies, and fees. Implementers must end such policies, or at least put them on a path to expiration. Guaranteeing prices for users' water supplies will distort the market, if not outright break it. Implementers must replace fixed fees for unlimited quantity with quantity-dependent market-based fees.

Historically, and out of simplicity, regulators associated water rights with land rights. Determining an approximate availability of water as a function of land area is relatively straightforward, if crude. Increasing water rights thus required people to buy more land, at enormous cost. With a smart market, however, the market manager will routinely calculate water availability based on the far more accurate hydrological simulation. This allows water rights to be separate from land rights.

Geography would still play a critical role in water regulation and the water market. The government must choose the geographical scope of the market: the degree to which, and conditions where, water can legally be shifted between regions, especially catchments and legal districts. The government need not concern itself whether water could *physically* be moved, particularly, but only whether the water could *legally* be moved. For example, a user with land in two adjoining catchments might find that water in one catchment is more expensive to purchase than in the other. That user might then construct a water pipe to transport water from one catchment to another. The regulator should allow this type of project, when done with a smart market, to proceed with no need for oversight specifically about trading per se (though still with concerns apart from trading, for example related to construction of a large pipeline or discharge of contaminated water). Most likely, users will first consider changing land use, or buying and selling land (and associated water rights) when confronted with variable water prices. Profit-motivated people will think of options that market implementers had not considered. Implementers should develop the market to allow creative approaches.

Before implementation, indeed even without the smart market, the regulator should work through all impediments to trade and remove them as much as possible. For example, trade is impeded by "ring-fencing" water, e.g., retaining specific quantities for a particular sector such as agriculture, or taxes on the water that leaves a water district. These changes may not be easy. McCann and Garrick (in Easter and Huang 2014, 17–18) pointed out the political backlash from proposed reallocation in the Murray–Darling basin, because the reallocation could have led to bankrupting of some irrigation districts.

Most pernicious, perhaps, are the "use it or lose it" rules, which incentivize people to waste, and put the user on hostile footing with the regulator and other users. The operator of a rural store may be worried—correctly—that allowing flexible water markets would put them out of business by leading to changes in local land uses. But this may be the right course of action for society as a whole, because society should make the best use of a scarce resource. Society may wish to help people harmed by actions that benefit society as a whole, but this help should not prevent those changes.

The Australian government appears to set the standard for removing legal restrictions to trade. Here is some text from the Murray–Darling trading rules (Murray–Darling Basin Authority n.d. 1, 2–3):

> The rules provide that a person may trade a water access right free of any condition as to the holding, buying, selling, obtaining, accepting, or terminating of a separate water delivery right, works approval or water use approval (section. 12.06).
>
> The rules also provide that a water access right can be traded free of any restriction that:
>
> - relates to the person being, or not being, a member of a particular class of persons (section 12.07)
> - relates to the purpose for which the water relating to that right has been, or will be, used (with limited exceptions) (section 12.08)
> - arises from the fact that water might be transported or used outside the Murray–Darling Basin (section 12.10)
> - is based on the fact that a water resource is over-allocated (for trade within that water resource) (section 12.13)

- is based on the historical or anticipated level of use of the water access right (except restrictions necessary to account for differences in reliability or water availability between places) (section 12.14).

  Under the rules, a person may:

- take and use water free of any restriction arising from the fact that the person acquired the water access right by trade (section 12.09)
- participate in a carryover arrangement free of a restriction arising from the fact that the water access right was acquired by way of trade (section 12.12)
- trade water allocation made available under a carryover arrangement in the same way that other water allocations are able to be traded (except where the carried over allocation is not yet available to be taken) (section 12.11)
- trade a water access right or an irrigation right free of any condition requiring them to hold, buy, sell, obtain, accept, terminate or vary the volume/share of a water delivery right (section 12.15).

The third paragraph of the above quote refers to "a particular class of persons." Part of the referenced section 12.07 points out that trading should not be restricted based on race, of course, but also, perhaps surprisingly, on whether the user is a particular user type.

> A person may trade a water access right free of any restriction that relates to the person being, or not being, a member of a particular "class of persons," as provided by section 12.07. Examples of "classes of persons" would include (but are not limited to):

- environmental water users or environmental water holders
- hydroelectricity generators
- landowners or non-landowners
- urban water authorities
- rural water authorities
- catchment management authorities
- irrigation infrastructure operators
- cultural groups.

We think the Australian market could operate better, because, among other reasons, the transaction cost remains high. But the Australian government has made great strides toward removing the impediments to trading.

References

Bastiaanssen, W. G. M., Molden, D. J., & Makin, I. W. (2000). Remote sensing for irrigated agriculture: Examples from research and possible applications. *Agricultural Water Management, 46*(2), 137–155.

Clemen, R. T., & Reilly, T. (2001). *Making hard decisions with decision tools* (2nd ed.). Belmont: Duxbury/Thomson Learning.

Davids Engineering, Inc. (2015, October). *Presentation regarding measurement technology implemented in California's Reclamation District no. 108*. http://www.rd108.org/wp-content/uploads/2015/10/ RD108_2015_Measurement_Program_Board_Update_151015_v6bab.pdf. Accessed 24 July 2016.

Easter, K. W., & Huang, Q. (Eds.). (2014). *Water markets for the 21st century: What have we learned? Global issues in water policy 11*. Dordrecht: Springer Science+Business Media.

Murray–Darling Basin Authority. (n.d. 2). *Guidelines for water trading rules, managing surface water trade restrictions*. http://www.mdba.gov.au/sites/default/files/pubs/03_WTG-FREE-TRADE-OF-SURFACE-WATER_final.pdf. Accessed 30 Apr 2016.

Tenkorang, F., & Lowenberg-DoBoer, J. (2008). On-farm profitability of remote sensing in agriculture. *The Journal of Terrestrial Observation, 1*(1), 6.

Chapter 6
Participants in and Components of the Smart Market Institution

Abstract This chapter describes the roles of market participants, starting with the market manager. The chapter assumes that the prerequisites, or extra-market considerations, for market implementation are already in place. It discusses the participants' bids and the bidding process, the market manager's transaction and allocation database, and the market-clearing model. A section describes the steps in routine operations, and gives additional rules for operations. Later sections describe how hydrological recharge affects the market design, and give an example from Australia of salinity discharge. The final sections discuss the problems that would arise if the private sector were to operate the market, various market structures, and likely questions that implementers would face.

Keywords Market participants • Regulator • Institution • Auction

6.1 The Water Regulator

We have described the role of the water regulator in the previous chapter. With a smart market, the water regulator's role will stay functionally the same as without a smart market, though the day-do-day chores should be less contentious with a smart market. The regulator's role in the market is to specify hydrological data collection and environmental requirements. The water regulator and market manager could be in the same office, but this is not necessary, so we will continue to describe them separately.

> The purpose of this chapter is to describe the components of the smart market institution, and the tasks that must be done in setting up the market.

Without a smart market for water, the role of a water regulator may be viewed as, "Allocate water when someone requests it, as long as no one else or the environment is too badly hurt." The water regulator uses the political process to identify water demands for non-commercial users and the environment. This process will involve working with water resource modelers to describe the relationship between water flows and environmental requirements. The water regulator uses an approval

J.F. Raffensperger, M.W. Milke, *Smart Markets for Water Resources*, Global Issues in Water Policy 12, DOI 10.1007/978-3-319-55008-4_6

process for commercial users. Again, the regulator needs some degree of water resource modeling to evaluate the impacts of individual water users on the environment and on other users. This can include some procedure for re-evaluation or renewal of water use agreements with users.

When water demands are less than the availability of water these processes can operate smoothly. As the total demand grows, however, the water regulator's role becomes more difficult. The main role of the water regulator is to try to get users to abstract less water. As catchments become fully allocated and then over-allocated, and requests are denied with increasing frequency, the water regulator is increasingly in conflict, even litigation, with users and water use applicants. As Sects. 1.2 and 2.1 explained, the approval process imposes a transaction cost on users and the regulator.

With the smart market regulatory framework, the transaction cost falls for both users and the regulator. The regulator's job will change from mainly denying and litigating demands for water, to enforcing the specific time-based contracts to which users have committed, with a focus on active monitoring of users' meters and the environmental flows. So the regulator's experience should improve considerably. With or without a water market, a user can take water without approval and overexploit the resource, and, in either case, the regulator should intervene. Regulation and enforcement of a smart market for water need not be more onerous than for a non-market water allocation system. But when people cheat in the context of the market, stakeholders will blame the market structure itself for allowing or enabling the cheating. To inspire trust in the market, a water market needs enforcement of rules, just as a Bureau of Weights and Measures has allowed trust in other markets.

A system of fines based on the market price could make penalties more effective under a spot market system, though fines alone will not be enough to ensure compliance. However, the water regulator should generally stay removed from market operations. The regulator should not be allowed to meddle with trading, such as to prevent politically unpopular trades, to limit the ability of any party to access the market, or to provide favors for any individual or group.

Should problems of competitiveness arise, where one party begins to exert market power, intervention should be by commerce regulatory authorities that normally enforce legislation to promote competition and prosecute restrictive trade practices.

6.2 The Market Manager

With the start of a smart market for water, the market will need an organization to operate it. We call this organization the **market manager**. Under the authority and regulation of the water regulator, a market manager acts as a broker, buying water from sellers and selling water to buyers, while ensuring that physical flows are feasible and that those flows satisfy the environmental requirements. From this single key role of broker, we can derive all other aspects of the market manager's job.

Operating a water market will require a number of specialized services. Some of these, such as water resource modeling, will likely already exist within a water regulator's office. Other services would not, such as an online commercial trading platform. The market manager will need access to computer programs and optimization experts; again, people with these skills are not likely to work currently in a water regulator's office.

The commerce regulator must recognize the market manager's power to run the market essentially as a monopoly. The legal authorization for the market manager to operate the water market would be similar to the legal authorization of the system operator for a regional power grid.

The market manager maintains the hydrological optimization. We distinguish the hydrological simulation—e.g., MODFLOW model—from the optimization. The water regulator can be responsible for the simulation. The market manager derives the market-clearing optimization from the hydrological simulation. The simulation specifies the water transfer coefficients, response coefficients, the percentage run-off from different types of water use, etc.

To support transparency, and to level the market between more and less sophisticated bidders, the hydrology and optimization software should be available to users. They can then conduct their own sensitivity analyses and check on the reasonableness of model parameters.

The market manager must specify users' prudential requirements, i.e., their ability to pay any money due and their agreement to give up water they have sold. Anyone wishing to buy must post sufficient financial information that the market manager can be sure of obtaining payments. Sellers must prove that they have forgone the water, i.e., they must have acceptable water meters. Participants must agree to follow market rules.

The market manager must ensure that market operations are efficient, open, and fair. Such assurance includes publishing market data, especially prices. While bids are private, the market-clearing prices should be public. The public should be able to view these prices in tabular and map form, to see how prices vary over time and space. The market manager records and maintains auditable transaction and allocation records. The market manager calculates users' initial rights based on the water regulator's rights database, but then users' trading of leases modifies those rights.

The market manager would be responsible for accepting users' bids for water. Most likely, the market manager will take bids through a website, but could allow bidding in almost any form, whether in person, by mail or email, by phone or text message, or through some other electronic message. Users themselves would be responsible for conveying to the market manager their wishes. Bids can be private. The market manager would ensure the security of the information received.

The market manager may require a landowner to inform a farming tenant of a bid, and the manager may require a tenant to inform the landowner of a bid. The market manager may require a title search before accepting a bid, especially in the case of a bid for an allocation far in the future; the market manager may require the bidder to notify lien-holders of a bid. In the case of a trader who has recently violated rules, the market manager may require extra prudential requirements, e.g., notarization of bids and allocation agreements, and a deposit for future purchases.

The market manager must have procedures for settling disagreements between users, between a user and the manager, and between a member of the public and the manager. The process can be a combination of an appeal process, arbitration, fines, and more-severe enforcement processes. We recommend administrative rather than court procedures where possible. Personnel working for the market manager should be technically trained to understand the issues.

The implementation process should specify rules for criminal penalties as well. Just as alleged violations in other markets need an enforcement and justice process, proper operation of the smart market for water requires a similar method of enforcement of use agreements. An unscrupulous water user could abuse a market for profit, yet not suffer financial penalties, for example, by declaring bankruptcy. The injured party will wish to know all avenues of redress.

Here are several types of potential conflicts:

- A member of the public may believe that a user is over-consuming water. To investigate alleged over-use, the market manager will physically check the well levels or flow levels. The market manager could also run the hydrology model to estimate the amount of water use that best matches the observed well levels and flow rates, and in that way check the reasonableness of the reported water use.
- A user may complain that the regulator has allocated too much water to the environment, or a public organization may complain that the regulator allocated too little to the environment. The allocation of water to the environment will be controversial, and stakeholders must settle this problem democratically. Some arguments are prone to abuse, e.g., that users can borrow water from the environment this year and pay it back in future years. Implementers may be able to resolve some disputes about environmental flows by running the hydrological model under various scenarios.
- Users may complain that the market manager is misallocating water over time. For example, suppose Period 1 is wet and the market agent allocates a large amount of water, and suppose Period 2 is surprisingly dry and so prices suddenly rise. Users may complain that the market agent should have allocated less water in Period 1 to have more water in Period 2. The market agent should try to avoid such complaints by providing forecasts of prices, and by setting values for storage carefully. Users must understand that spot prices will vary. Besides, if the user can forecast better than the market manager can, the user should be able to make money by speculating.
- A user may believe that the market manager's hydrological model is incorrect. The law should charge the market agent with keeping a responsibly accurate hydrology model, and should have a complaints and approval process to respond to users' concerns about the accuracy of the hydrology model. Certifying the hydrological model as sufficiently correct may require an outside auditor. The

market manager should publish the model on the market website. Market rules could require the user to pay for data collection and model changes in some circumstances; in others, the market manager may pay for the changes. Users should have the opportunity to present their evidence in support of a modification of the model. The regulator could specify the requirements for such evidence, such as the credentials required of the consulting hydrologist.

Any dispute resolution system will need a method for evaluating the suitability of the hydrological model and improving it as appropriate. Developing accurate hydrology models can sometimes be difficult, but an approximate model will allow a market to begin. Even without a smart market for water, the water regulator must have a sufficiently accurate model to set environmental limits that will withstand challenge. In many places around the world, regulators have worked to make hydrological models more accurate because of the sensitivity of environmental flows to identified model uncertainties. Water regulators identify which uncertainties have the greatest impact on environmental flows, and invest in improving the estimate of parameters that control those uncertainties. We anticipate that this process of focused model development would continue under a smart market regime. In fact, this process would likely become more efficient because the prices resulting from the market would identify which uncertain parameters have the greatest impact on the value of water, and the market would identify the relative value of water at various environmental assessment points. This additional information would allow for cost-effective model development.

6.3 Who Participates in the Market?

6.3.1 Participation Is Voluntary; Regulation Is Mandatory

With some caveats, we would envision that the smart market is completely voluntary. No one except the market manager would be required to participate.

Everyone would be subject to the regulation of the smart market framework. Every relevant user in the catchment would be subject to adjustment of their allocations in every period (day, week or month). Non-trading users would still be subject to regulation, and the market manager would include their adjusted allocations in the optimization model as unchangeable values.

> A wide range of people can participate in the market. The market manager should endeavor to reduce barriers to entry in the market as much as possible, making all reasonable attempts to allow everyone to participate.
>
> The most active participants will be landowners in the catchment who need water for production, and people who wish to buy water on behalf of the environment.

Implementers can take advantage of variations in geography and user type to decide who to include and who not to include. The implementers could decide to start small by experimenting in a sub-catchment where all significant water users are included in the market, including small and moderate-sized users. As the market matures in that sub-catchment, users in adjacent sub-catchments may wish to join, and they can be accommodated assuming hydrological connectivity.

Alternatively, implementers could start with a few large users in a large catchment. In time the next smaller tier of users may wish to join. Attracting new participants is as simple as publicizing the market availability and sending them an invitation with instructions for joining. This latter approach may be easier than starting with small users in a small catchment, because large users will have the technical skills to participate, and will have more to gain from trading.

Both of these approaches show how the set of participants might change over time. The participants might also change quickly after the smart market begins operation. At the first auction, many users, especially small users, might wait to see what happens.

Users may participate in the smart market in different ways. Some users might participate in the market in a simple manner, where they choose their allocations of water for the **hydrological year**, bid on those once, receive their adjusted allocations, and feel no need to revisit those allocations for the rest of the hydrological year. Other users may actively manage their allocations in every period.

Parties outside the catchment can participate, including overseas hedge funds and financial speculators (as an extreme). Of course, distant participants would not be able to take physical delivery of the water, and thus they risk holding allocation they cannot sell or use. The allocations they can buy must be associated with a physical location and time period, and would need the permission of the owner of that local allocation to obtain rights to the same water intake. Though trading with financial speculators and risk managers would be technically feasible, it is unlikely.

6.3.2 The Environment Plays a Role

Environmental groups will wish to buy water allocations on behalf of the environment. The market manager will need to make appropriate bidding arrangements for these parties, specifying their prudential requirements, but also proving that those buyers can see that the environment's allocations have increased. Thus, the actual flows associated with the environmental requirements must have good metering and be publically displayed.

The government could appoint an ecological agent for the region. The agent would be responsible for buying water for the environment above the agreed-on minimum environmental requirements. Since the smart market specifies allocations in detailed space and time, the ecological agent would have a rather difficult planning job. Most likely, an army of non-profits and academics would want to help. But the results of these purchases should produce a great improvement for the environment at low cost. For example, the ecological agent could purchase water to allow a brief river surge that the market model might otherwise smooth out. Through the market mechanism, the market manager would seamlessly compensate users: the market manager would pay them to reduce their allocations to allow the river surge. Further, the ecological agent's responsibilities could cross market boundaries; the agent could choose to purchase in one area in a given period, and a different area in another period.

Ideally, this agent would run a separate ecological simulation-optimization that assesses how much improvement the ecology could get for a given unit of water. Crudely put, the output from this model would be in units of (improvement in ecology)/megaliter. As the market produces prices in units of $/megaliter, the ecological agent would know exactly where to put money to improve the environment the most, by simply ranking water purchase opportunities as [(improvement in ecology)/megaliter]/$/megaliter = (improvement in ecology)/$. Murphy et al. (2006) studied such a mechanism. They found that taxation was more efficient economically than their market design was, because their market design allowed free riding. The free riding arose because the environmental value in that paper was environmental flows, not consumptive use. A buyer could "receive benefits from aggregate consumption at his node regardless of third party contributions."

6.3.3 Private Infrastructure Could Be In or Out of the Market

Market implementers should consider with care how privately operated reservoirs interact with the market.

In smart markets for electricity, the market manager usually directs the operation of private generators. Either the electricity generator is "in the market" and participates, or it is "out of the market". To say that the generator is "in the market" means that the market model contains explicit constraints about the generator operation, such as its capacities and the rates at which the generator can ramp up and down. In some cases, the generator owner would be responsible for physical feasibility, by always specifying bids that are consistent with the physical characteristics of the generator. If the generator owner's bid were in error and the generator were broken, the generator owner would be at fault. For these cases, the generator operation is "out of the market," because the constraints associated with physical operation of the generator do not appear in the market model. Of course, the generator owner would probably prefer the safety of having the explicit constraints in the market model, to ensure that the required generation was feasible, because an error in planning generator operation could be expensive.

In an analogous way, the owner of a privately operated reservoir could submit bids to supply water, but the owner would be responsible for ensuring that the reservoir operated feasibly. For example, the owner would be responsible for ensuring that the reservoir did not run dry, or for ensuring that the outflow channels could handle the water flow anticipated by water users downstream. The reservoir would be out of the market.

Alternatively, the market manager could include the reservoir as part of the market-clearing model with constraints on reservoir capacity, and the maximum outflow rate, for example. In this case, the market manager is responsible for ensuring that the reservoir operates feasibly. The owner still submits bids to supply. The reservoir would be in the market.

We expect that the market manager may not want to bother with very small reservoirs. Further, large reservoirs are likely to be publicly owned, in which case the regulator might stand in for the reservoir owner. These factors are important for stochastic operations (see Chap. 11).

6.4 Participants' Bids for Water

6.4.1 The Auction Schedule

The market manager must run auctions following a well-defined schedule, in which users know when auctions will start and end for the hydrological year, and when each auction will occur during the year. Further, users will want to plan for water in future periods for the current year. The schedule for the auction has multiple components, and must be set with some care:

- How often does the auction run, e.g., once per day, week, or season?
- When the auction runs, what is the length of time associated with use of a given quantity of water? For example, the auction could run once every 4 weeks, allowing trade of water by week at each auction. The length of time associated with an allocation quantity depends heavily on the regulator's ability to meter the water over this length of time; it can be no shorter than the shortest time between meter readings.
- How far into the future can users trade water in each auction? For example, the auction could run once each season, while allowing trade for the future six seasons. The answer affects how storage will be valued.
- How long should a period be for environmental constraints? For example, the market could trade weekly water, but the model could still have environmental constraints for each day. In this case, water users would need limits on how much they could take daily; the market manager could tie such limits to pump capacities.

From the user's point of view, submitting bids would not be quite as simple as in a normal auction, such as eBay, because water supply varies over time, and users have variable demand over time. These variations in supply and demand require the user to bid separately for different periods. Before farmers plant a crop, for example, they need to be reasonably certain of receiving enough water over several months, so they may need to bid for water for several future periods in a single auction.

Consider two extreme cases: a catchment where all users use fast-flowing river water, and a catchment that either has no storage or has one large reservoir. The river catchment without storage could likely allow a market that ran only one short period at a time: a day or a week. The optimization would essentially be a single-period model. While users could trade future allocations, the current allocations would be independent of future allocations. This situation has no drought protection, which comes from reservoirs and aquifers that, if managed properly, can provide water to its highest-value use during drought. In a catchment with a large reservoir, users could conceivably choose a single allocation for the year and take it at any time. The market has the potential to be so simple as to run once per season or once per year as a two-period model, with users bidding to reserve water for the following season or year.

In a smart market, the market manager ensures that storage is valued correctly. So users must give bids for each period remaining in the growing season. These bids give the future value of water, telling the market manager how much water to save for users in the future rather than allowing users to consume it immediately.

The auction schedule will naturally follow the local hydrological year, with the first auction of the year sometime before the first growing season begins, and the last auction of the year sometime near the last growing season's end. A hydrological year is unlikely to match a calendar year.

In dry times, the operator should operate the auction relatively frequently. A large rainstorm during a severe drought can change prices from day to day. In wet times, the operator can run the market less often. The market manager will have the responsibility of being responsive to users and of operating the market when an auction would improve welfare.

For a catchment with storage, we expect that the auction would run once per week, so users would have the opportunity to bid every week. In the first auction of the growing season, each user may wish to bid for every future week in the growing season. Thus, in every week, users have the opportunity to change their bids both for the current week and for future weeks in the remaining hydrological year. The market model and rules treat the current week as firm. Future weeks are treated as forecasts, where the allocations are subject to adjustment for the hydrological conditions.

The market manager must also manage storage beyond the end of the current hydrological year. In a catchment with some kind of natural hydrological reset, e.g., where reservoirs naturally refill with high certainty, or where reservoirs hold many years' supply of water, the market manager may not need to manage the storage aggressively. But in many settings for smart markets, the market manager must seek sufficient information from users to know how much water users will be able to take for the current year, and how much should be saved for future years. When users and the market manager expect the following year to be especially dry, users may want to use extra water in the current year to ensure a good crop, while planning to fallow fields in the next year. Eliciting this future value of water from users may be hard (see Chap. 11). Nevertheless, when the catchment has significant storage, the market manager will be responsible for assuming values for the water allocations at timescales beyond the current hydrological year.

The auction schedule may also depend on the environmental flows. As the market manager is responsible for them, the manager will need to decide how much man-made uncertainty is tolerable. For example, a monthly auction would allow users to vary their water use within the month, potentially leading to the environmental flows to be short in some weeks while over-satisfied in others. By changing to a weekly auction, users would commit to taking water over shorter periods, while reducing the within-period variability on the environment.

Optionally, the auction can run once every 2 or 4 weeks, with allocations set by week. Each user then effectively buys two or four items in each auction—the allocation for each week in the auction period of 2 or 4 weeks. This multi-week auction makes the structure of bids more complicated, but a multi-week auction would reduce the number of auctions.

6.4.2 The Structure and Nature of Bids

In the smart market for water, participants trade contracts and not water itself. Traders offer an agreement of some kind to the market manager, and the market manager chooses whether to accept the offer, using the market-clearing optimization. Because of the catchment complexities, one user's buy offer cannot be traded one to one with any single seller. In effect, the optimization model converts potential agreements from one trader into potential agreements that other traders would accept.

In the simplest possible bid, a user offers a price per megaliter and a quantity to buy or sell, with the availability to start and end at specific times and at a particular location. For example, Ann might ask to buy 0.1 ML at $20/ML, for the week starting 1 June. Should the market manager accept her bid, Ann would have either the option or the obligation to take the water allocated.

Whether the allocation is an option (that the user need not take) or an obligation (that the user must take) will depend on the hydrology and on the user's impact on others. For example, the market manager may view a reservoir operator's accepted bid to release water as an obligation, because the market manager is counting on both the downstream flows and the remaining storage capacity. An accepted bid from a farmer taking groundwater is likely to be an option, if other users are better off when the farmer takes less than the allocated amount.

The market allows trade in contractual agreements, not in water per se. This distinction is important legally and practically.

Legally, bids constitute requests to the regulator for changes in allocation. On behalf of the regulator, the market manager approves the changes, while accounting for all allocations simultaneously.

Practically, contractual agreements allow the market manager to distinguish consumptive from non-consumptive users, and allocate based on simplifications such as area and type of use. A user may sell a right to water an acre of almonds for a week, assuming little of the water becomes recharge. Another user may buy the right to wash tons of gravel, assuming much of the water returns downstream.

A bid is a firm offer from a user for changes to the user's existing allocation at a particular **water intake**. We define a water intake as a place where a user abstracts water from surface water or groundwater. A user could have several water intakes, and would usually need to submit separate bids for each. Although multiple users might use the same place to abstract water, each would need a separate bid and the market manager would consider each a separate intake with our terminology. To sell water, users can name a price at which they are willing to sell. If the market accepts the bid, the market manager gives the selling user the price asked or higher. If the user wants more water, then the user can name the price at which they are willing to buy. If the market manager accepts the bid, the buying user gets water in the current period at the bid price or lower.

However, traders' real value for water will be more complicated than a single value for a single quantity. Depending on the sophistication of the users and the maturity of the market, the market manager may ask users to bid in multiple parts, in which users offer to buy and sell different quantities at different prices. Each of these multiple bids is called a block. Bids in sets of blocks are harder for users to understand. But compared with single unit bids, multipart bidding enables users to convey more precisely their marginal values for different quantities of water. The first megaliter of water traded has the most value, and the last megaliter of water has the least. Let's take a farmer growing hay. If her land has been dry for several weeks, she could have reasons for wanting different allocations than she currently holds, and each increment would have a different value to her. Figure 14 shows an example set of values for water from the viewpoint of one water user, Ann. Assume for a moment that Ann has no initial allocation, so she is only buying; so ignore the dotted horizontal line for now. What does Fig. 14 say about Ann's value for water?

If the price were $9/ML or less, Ann would buy up to 4 ML. Because her fields are dry, some irrigation water is critical to avoid severe water stress, and so the first block of water is valuable. This is the right-most step in Fig. 6.1.

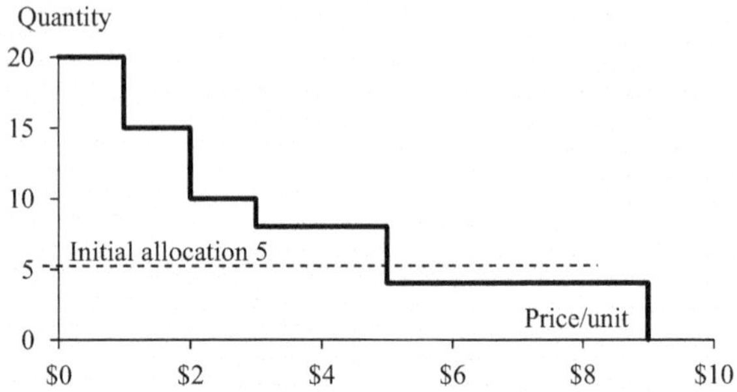

Fig. 6.1 Ann's approximate demand for water as a function of price

If the price were $5/ML or less, Ann would buy up to 8 ML. Once Ann gets that first block of water, the next one would add extra moisture to return the soil to normal conditions.

If the price were $3/ML or less, Ann would buy 10 ML. Additional water could help build up soil moisture in case of another dry spell, but otherwise is not that important.

If the price were $2/ML or less, Ann would buy up to 15 ML. More water would not help Ann's crops much. It might ensure that all the soil becomes thoroughly soaked, but much of this extra water would likely drain away.

If the price were $1/ML or less, Ann would buy up to 20 ML. By this point, extra water will almost certainly run off and not provide benefit. It might even increase risks. Maybe it could help wash out some salts or chemicals from the soils and be useful once a year. This is the left-most step in Fig. 6.1.

Now, let's assume Ann's initial allocation is actually 5 units of water, as shown with the dotted line. This same information conveys what she wants to do for any given price.

If the price were $2.75, Ann would buy 5 more units, bringing her allocation from 5 to 10. Why? She holds 5 units. If she bought 3 more units at $2.75, she would gain 3*($5−$2.75) = $6.75. If she bought 2 more units of water, she would gain, though a bit less, 2*($3−$2.75) = $0.50. If the price were $7, Ann would be willing to sell 1 unit, and (taking care with the signs) she would gain −1*($5 − $7) = $2. If the price were $11 per unit, Ann would have an incentive to sell all her water. A user will want to sell at a high price and buy at a low price. For any trade, users tend to have different values for different quantities of water. The bidding system is designed to capture these different values, while simplifying market operation.

Besides enabling users to specify their value for water more precisely, multipart bids avoid a "winner takes all" risk. For example, suppose Bob has 5 units, and Bob believes he needs 10 units of water, so he bids $2 per unit for the 5 units, but the optimization sets the price at $2.50. Then Bob gets nothing. After experiencing that once, Bob will see that he would be better off bidding in two blocks, perhaps $3 per unit for the first 3 units, and then $2 per unit for the next 2 units. Without multipart

Table 6.1 Multipart bids spread the water among users, lowering users' risk

	Ann	Bob	Cal	Don	Edna
Value for 1st ML:	$4.00	$5.00	$6.00	$7.00	$9.00
Value for 2nd ML:	$2.00	$2.50	$3.00	$3.50	$4.50
Allocation order:	6th	4th	3rd	2nd	1st
Allocation order:	10th	9th	8th	7th	5th

bids, Bob will tend to shade his bid higher than his true reservation price from risk aversion. This could lead to a less efficient allocation of water.

Similar economic inefficiency can arise when the auction has no trial or practice rounds before the final bid round. If the auction runs in several rounds, where users can change their bids before final clearing, Bob has multiple opportunities to change his bid. In the above example, he could see in the trial bidding round that the price at his location was $2.50 per unit. He could choose to bid higher in the next round.

Multipart bids, by contrast, allow a user to specify a range of prices, including a high price for the first quantity, and a low price for the last quantity, in some sense converting bids in several rounds of an auction into one multipart bid. When prices are high, many users are likely to get their first highest-value quantity, while only a few get their second value quantity. Table 6.1 shows an example. With only 6 units of water, everyone gets some.

We will assume that the demand function is a step function, a piece-wise linear relationship between price and the quantity that a user wishes to use at that price, and that this step function is a sufficiently good approximation of the user's demand. After each auction, buy and sell quantities are determined by comparing the allocation found by the optimization with the initial licensed water allocation. Based on those purchases and sales, the market manager then adjusts allocations for the next auction. Chapter 8 has more discussion about initial rights.

Thus, users can bid their value for total water, as if they had none. When they view their bidding web page, they simply post statements of the form, "If the price were $1/ML or less, I would take up to 20 ML . . . If the price were $2/ ML or less, I would use up to 15 ML." The emphasis here is on the word "use," rather than buy or sell. When users bid their value for total water rather than for a change in their allocation, the market manager simply allocates the water to the user with the highest value, subject to the various constraints. In the example above, Ann would buy at $1 per megaliter, and sell at anything over $5 per megaliter, but the market manager can do those calculations after the allocation calculation. If the market price were $7 and Ann's allocation 4 units, the market manager simply looks up in the allocation database to note that Ann's initial allocation was 5 units, and therefore registers a sale of 1 unit for Ann. When the market manager requires users to structure their bids as value for water, starting from zero allocation, the market is commonly termed a **gross pool market**. When the market manager requires users to structure their bids as a change from their initial allocations, with bids marked on their bidding forms clearly as "buy" or "sell," the market is commonly termed a **net pool market**. We discuss more implications of gross and net pool markets in Sect. 7.2.

Unfortunately, multipart bidding raises the complexity for users. In our tests with actual users, they were confused by multipart bids (though our training consisted of only a 15-min presentation). The concept of varying marginal demand for water is unnatural for most people. Just as importantly, they had to type in a lot of information into the web page, as Fig. 6.2 shows. Using this bid structure, in a catchment with storage, users would have to enter five bid quantities and five reserve prices for every remaining week in the auction schedule, up to 520 inputs. The web page shown in Fig. 6.2 had tools to ease this input, but it was still daunting for a user's first look at the market.

Users in our test group wanted to see their bids as clearly marked "buy" or "sell," because the gross pool interface implies to them that their water could be taken away from them. Indeed, the gross pool interface requires users to bid, in effect, for their own water. Most likely, users will prefer the net pool bidding structure.

In a market with many users and where the hydrology has storage, prices are likely to be relatively stable. Users may gain little or nothing from having multiple bid blocks, because their expectations about the market price are likely to be relatively accurate. They may be able to bid in single blocks without difficulty.

The solution to finding the right bid structure is to listen carefully to users, and assess how they want to submit bids, and then develop the model's data structure around that bidding structure. Different users will want different degrees of sophistication in their bidding structures. Large firms with full-time staff available may

Common Pool Trading Bid Page

Welcome pump 148 (easting 50XXXX, northing 24YYYY). These bids were previously saved.
The next auction (#2) will be on 08 Jan 2020. Click the auction date to see your trades in previous auctions.

Auction date	Take date	Adjusted start quota ML	Last clearing price £/ML	Bid quantity blocks, ML Really need .... Nice to have					Total ML on bid	Bid prices, £/ML Really need .... Nice to have				
01 Jan	01 Jan	7.75	£3	3.2	3.2	3.2	5	5	19.6	133	120	108	2	1
08 Jan	08 Jan	9.6	£4	2	2	2	2	2	10	133	120	108	2	1
08 Jan	15 Jan	9.6	£4	2	2	2	2	2	10	133	120	108	2	1
08 Jan	22 Jan	9.6	£4	2	2	2	2	2	10	133	120	108	2	1
08 Jan	29 Jan	9.6	£5	2	2	2	2	2	10	133	120	108	2	1
08 Jan	05 Feb	6	£5	1	2	2	2	2	9	133	120	108	2	1
08 Jan	12 Feb	6	£4	1	2	2	2	2	9	133	120	108	2	1
08 Jan	19 Feb	6	£4	1	2	2	2	2	9	133	120	108	2	1
08 Jan	26 Feb	6	£5	1	2	2	2	2	9	133	120	108	2	1
08 Jan	04 Mar	3	£5	1	2	2	2	2	9	133	120	108	2	1
08 Jan	11 Mar	3	£4	1	2	2	2	2	9	133	120	108	2	1
08 Jan	18 Mar	3	£4	1	2	2	2	2	9	133	120	108	2	1
08 Jan	25 Mar	3	£4	1	2	2	2	2	9	133	120	108	2	1
08 Jan	01 Apr	0	£5	1	2	2	2	2	9	5	4	3	2	1

Fig. 6.2 Example of part of a hypothetical multipart gross pool bidding web page, for an auction on 8 January

want the ability to enter complex bids. Small unsophisticated users will need to have simple bidding interfaces. Water users should not be disenfranchised by lack of technical skills, and the market manager should plan on training users. All users will need help, with formal training, online tutorials, online help pages, printed instructions, and user forums. As the market matures, users will become accustomed to interacting with the market mechanism, and should be able to enter bids that are more complex if they wish.

Implementers should extensively test the bidding procedure. Some unusual bids, say, a "Nice to have" bid for $0.000001, should be possible. The bid page should reject other unusual bids, say, for a quantity that is implausibly large.

As long as the market manager has the ability to scale initial rights, bidding need not be overcomplicated. Neither, though, should implementers over-simplify it.

6.4.3 Simplified Bidding by User Type and Automated Bidding

If the market manager has sufficient resources, the manager can develop web pages that simplify bidding for users, based on profiles of user types. During the registration process, the user could select their type of operation from a dropdown menu, e.g., "I run a dairy farm," or "I have an almond orchard." The manager can then construct the web pages to ask the dairy farmer about the number of cows, and the almond grower about the number and maturity of the trees. Using standard water requirements for dairy farms and almond trees, and based on the size of the user's operations, the web page could estimate the user's demand for water in each auction period, and enter initial bids, which the user could then adjust.

Users may wish to have automated bidding, in several forms, and for several reasons. With the gross market bidding structure, users will need standing bids in every period. They will want to be assured of an allocation even if they go on vacation or forget to input a new bid. With the gross market, users will want the ability to withdraw at least part of their allocation from the market. Standing bids would not be necessary with the net market bidding structure, because users would automatically have their initial allocation without bidding. Still, users may want standing bids in a net market, such as to buy if the price is especially low, or to sell if the price is especially high.

As an example of a simple rule for users who do not bid, the market manager could re-use previous bids, applying values for water from the previous auction to the next auction. Users need not be handcuffed to the bidding screen.

However, users should realize that they could improve their profits by planning their bids carefully. They will want to buy and sell the right amount of water in each auction period at the market price. Since the market price appears to them as an uncertain variable, they may want to use stochastic optimization methods to determine their own demand curves for water. Even with rule-of-thumb methods for their planning, users might benefit from data of their operations, such as farm soil moisture. Using technology now widely available, users can connect sensors to water control systems. They may wish to extend the connection to an automated bidding mechanism. The market manager will have to build and enable secure bidding interfaces, and develop all the related market rules, such as what happens when a user's sensors lead to robot bids that attempt to buy so much water as to deluge a field.

6.4.4 Pricing and Clearing

We recommend that the market manager use marginal cost pricing. This price is defined as the lowest price at which a buyer's bid is rejected, or the highest price at which a seller's bid is rejected. So the price is the increase in water value if one more unit of water were available. Users pay at a price equal to the marginal value by another user who would be unable to consume or unable to sell.

Paying at the marginal cost does not mean that the market price is the same everywhere, because the market price is adjusted for local constraints. A user close to an environmentally sensitive region is likely to face a higher price than a user far from any sensitive region.

Two different users taking water at the same water intake (e.g., the same well, the same section of a river) would face the same price, even if they bid differently; their water is from the same part of the hydrological system and has the same environmental impact. The users should receive a consistent message about society's value on each constraint at that time. For a simple hydrology, such as a single reservoir, each period would have a single price, which all users would face. Because users bidding for water at the same water intake face the same price, the user who was willing to buy water at a higher bid would be allocated the water for a price lower than the bid and would believe they had a good deal. The other user at that water intake who was willing to sell water for a low bid would enjoy the sale at the higher single price. A market based on marginal pricing allows two users at the same water intake to find a fair market price that allows them to effectively trade water allocation.

This single-price mechanism using marginal prices ensures that the smart market maximizes the value of water use for the participants, and a single price mechanism assures users that they can bid their true value for water. Assuming that the market is competitive—the subject of Chap. 10—then the market will incentivize people to behave in a way that maximizes the value of the tradable water to themselves and to society. We also believe that users and society will see marginal cost pricing as a fairer system than one where users face different prices for similar water use with similar impacts.

With marginal cost pricing, one bidder will typically be the "marginal user" who receives a partial block, depending on the complexity of the hydrology and various local conditions. Consider the example data in Table 6.1. The actual water in the catchment is unlikely to be exactly some integer number of megaliters. Suppose the market manager allocates 6.5 megaliters. Ann will get the sixth megaliter, and Don will get the remaining half megaliter. Since Don's value for this water is $3.50, that price should be the market price. As the lowest successful bidder, it is no surprise that Don got the last bit of water. Don is the marginal user, the one on the margin between users who got allocations and users who got none. His bid price is therefore the marginal bid, and serves as the market price. Since typically only one user will be the marginal user, most users will find that their bids tend to clear with full blocks.

For a variety of reasons, Don may not be able to use a partial quantity. Maybe his equipment may need a full charge, or maybe he doesn't want to water only part of his field. Don may wish to make an "all or nothing" bid, in which he gets either the whole megaliter or does not get the whole megaliter. This discrete bidding structure imposes potentially unreasonable computing difficulties on the market manager. Linear programming models are easy to solve when using only continuous variables. Linear programming models with discrete variables can be extremely difficult to solve, and do not naturally produce the convenient market-clearing price information. To continue our example, what price should the market manager use when rejecting Don's all-or-nothing bid? The market manager can leave the last half megaliter unused, and use Edna's value of $4.50 as the market-clearing price, or the manager can give the water to Cal for a $3 market-clearing price. Either solution will tend to cause confusion. One solution is simply to ask the user to refine their multipart bid. Don would have to raise his bid above Edna's to get the full block.

The all-or-nothing bid is one example of combinatorial bidding. Users may wish to bid in even more complex ways, e.g., they get water in either week 5 or week 6 but not both, or if they get water in week 5 then they must also get water in week 8. Combinatorial smart markets are well established, but do not generally use marginal cost pricing. Instead, they typically use "pricing as bid," which we believe the market implementers should avoid. Future implementers could extend the market design described in this book to allow combinatorial bidding, but we will assume the simpler bidding structure in this book. We think this is justified for two reasons. First, water is inherently a continuous commodity, unlike, say, airplane landing slots. A farmer who received only part of a bid could reduce the irrigated area accordingly. Second, the linear program tends to give most users full blocks anyway.

6.5 Transaction and Allocation Database

In collaboration with the regulator's long-term rights database, the market manager must operate a market transaction database. This transaction database records users' bids, and the accepted trades and market prices.

The market manager must have knowledge of users' license terms, and of their long-term licensed quantities, which come from the standard rights records. For a given auction, the market manager must be ready to record thousands of purchases and sales. Most likely, the first time that the auction is run, more trades will occur in the one auction than in the entire history of the region. The market manager must be prepared to execute thousands of trades for the first auction.

The rights database for the market must maintain a transactional history of the leases for allocation, to know the allocations for the next period. The market manager will use the database to adjust those allocations in response to changes in the forecast for inflows, and those adjustments must be recorded and auditable. The market must manage the money, with the attendant requirements for historical records and auditing. In the Appendix: Outline of Market-Clearing Code, we outline the database that would be required for the market operations.

Here are some of the fields for the traders' bids and the transactions and their functions:

- BidID: To identify the bid
- BidDate: To specify the date that the trader submitted the bid
- TraderID: To identify the trader
- AuctionID: To identify the auction
- AdjustedInitialAllocation: To specify the trader's initial allocation
- Automatic: To indicate whether the trader submitted a manual bid, or whether the trader used some automatic bidding function, e.g., to use the most-recent bid rather than submit a new one
- Quantity1, Quantity2, Quantity3, Quantity4, Quantity5: To indicate quantity blocks in the trader's bid, assuming multipart bidding
- Price1, Price2, Price3, Price4, Price5: To provide price blocks associated with the trader's bid quantities
- TradedQuantity: To specify the final quantity traded
- TradedPrice: To specify the price at which the final quantity was traded
- AdjustedFinalAllocation: To specify the trader's new allocation
- ExpirationDate: To identify the date at which the lease for this quantity ends
- IsBidAutomatic: To flag how the user has entered the bid; *False* if the user has entered a new manual bid, or *True* if the user has used a default bid

The trader database may have more fields, depending on the richness of the particular market characteristics.

6.6 The Market-Clearing Optimization Model

The market-clearing model is the heart of the smart market. Its output specifies both the final allocations and the market prices. We describe it briefly here and in full detail in Chap. 7.

After all market prerequisites are in place (Chap. 5), and after the market manager has an office and an adequate information system to record users' bids and trades, the market manager can prepare the market-clearing model. The market-clearing model is a linear program customized for the users' catchment and for some characteristics of the users themselves.

- The model has variables, which are the allocations that the market manager must choose. Dual variables (Lagrange multipliers on the constraints) specify the market prices. When the manager solves the model, these variables are the outputs.
- The model has constraints. Some constraints ensure the balance of water at relevant places, corresponding to the local hydrology. Some constraints require the flows to satisfy the environmental requirements. Other constraints specify the maximum quantity in each part of each user's bid.
- The model has an objective, which is to maximize the consumer surplus arising from the value of the water sold to buyers minus its cost to buyers.

Thus, the model captures all relevant information about the hydrology, the environmental requirements, and the users' bids. A correct market-clearing model serves as a checklist of factors that the market manager, and the market participants, must consider.

The general form of the net pool market-clearing optimization model is as follows, using lower case for variables (outputs) and sentence case for constants (inputs). We defined a water intake in Sect. 6.4.2 as a place where a user abstracts surface water or groundwater from the environment. For example, each bid could have $B = 5$ bid blocks and the catchment could have $W =$ a few hundred users' water intakes. This model formulation treats each combination of a user and place of water abstraction as a separate water intake. Assume the current period is period 1, and T is the last period, perhaps 52 or more weeks, depending on the hydrology. The current period advances, while T is constant (or advances in some agreed-on manner), until the full auction schedule completes.

The objective function maximizes the total consumer surplus (provided users bid truthfully) from water extraction over the planning horizon. The $BuyPrice_{b,w,t}$ and $SellPrice_{b,w,t}$ values come directly from each user's bidding page, which are the bid prices in Fig. 6.2.

$$\text{Maximize} \sum\nolimits_{b \text{ bids}} \sum\nolimits_{w \text{ water intakes}} \sum\nolimits_{t \text{ periods}} (BuyPrice_{b,w,t}\ buy_{b,w,t} - SellPrice_{b,w,t}\ sell_{b,w,t}).$$

(1) Maximum quantities that each user is willing to sell at his or her water intake w. These quantities come directly from each user's bidding page, the bid quantity blocks in Fig. 6.2.

$$0 \le buy_{b,w,t} \le BuyQty_{b,w,t},\ 0 \le sell_{b,w,t} \le SellQty_{b,w,t}, \text{ for bid blocks}$$
$$b = 1, \ldots, B, \text{ water intakes } w = 1, \ldots, W, \text{ and periods } t = 1, \ldots, T.$$

(2) Balance of each user's rights. The variable $q_{w,t}$ is the final quantity allocation for user intake w in period t. This allocation equals the user's initial allocation at

water intake w in period t, plus the quantity bought, minus the quantity sold. The dual variable $price_{w,t}$ is the price faced by the user with intake w in period t. We allow a user to offer both buy bids and sell bids in the same auction at each water intake. However, in an optimal solution, each user will either sell or buy at a given water intake, but not both.

$q_{w,t} = InitialAllocation_{w,t} + \sum_{b \text{ bids}} \sum_{w \text{ water intakes}} \sum_{t \text{ periods}} (buy_{b,w,t} - sell_{b,w,t})$
for water intakes $w = 1, \ldots, W$, and periods $t = 1, \ldots, T$. Dual variable $price_{w,t}$.

(3) Generic notation for hydrological balance of flows and environmental requirements. The structure of these constraints depends on the hydrology.

$\sum_{\text{intakes } w} \sum_{\text{periods } u=1}^{t} A_{i,w,u} q_{w,u} \leq L_{i,t}$ for each assessment point i and time period $t = 1, \ldots, T$.

(4) Buy and sell variables must be non-negative, and the quantity variable is unconstrained.

$buy_{b,w,t}, sell_{b,w,t} \geq 0$, for bid blocks $b = 1, \ldots, B$, water intakes $w = 1, \ldots, W$, and periods $t = 1, \ldots, T$.

$$q_{w,t} \text{ is a free variable.}$$

After building and solving the model, the market manager's software writes the solution to the allocation database. The output of the program is $q_{w,t}$, the allocation for water intake w in period t, and the price $price_{w,t}$ of water at water intake w in time period t. The market manager's server notifies each user of the results of the auction. The market manager charges user w: $\sum_{t \text{ periods}} price_{w,t} \sum_{b \text{ bids}} (buy_{b,w,t} - sell_{b,w,t})$, noting the signs carefully for the case where user w sells.

With marginal cost pricing, a user at water intake w buys or sells the cleared quantity at $price_{w,t}$, which corresponds to the marginal value to the market participants, as measured by all bids, given the environmental constraints. The $price_{w,t}$ depends on the water intake w and the time period t. This is because of the heterogeneity of the hydrology. Recall the example of Ann and Bob in Fig. 1.6. They had different hydrological impacts, and could not trade one-to-one. Therefore, they should face different prices for a unit of water. The user has firm rights to take the allocated water in the current period, and paper rights subject to scaling for future periods. Section 9.1 explores this issue more fully with a worked example.

When water is plentiful, the $price_{w,t}$ may be zero, in which case the market manager does not charge user w for water in period t. Thus, the price is the value of the highest rejected bid. Other auction designs clear the market with the lowest accepted bid, in which case $price_{w,t}$ could be zero only if someone entered a bid with a positive quantity and bid price of zero. (Implementers should consider this edge case.) Thus, with these other designs, the user with the lowest accepted bid uses up the last allocable water. To avoid this, the user could reduce the bid quantity in an attempt to pay a lower price. Hence, the latter price rule does not always incentivize the users to tell the truth (Schrage, personal communication, 16 Oct. 2016).

With the users notified of the results of the auction, that round of buying and selling is over—the market clears. Sellers close their web browser, understanding that they sold because they bid the same or lower than the market-clearing price. Buyers push away from their laptops, knowing they got water because they bid at least as high as the market-clearing price. The market price sends the correct signals to users whether to use or conserve water, where and when. Price information should be public; anyone could view a catchment map to see where water is priced high or low.

As with the use of any software, various things can go wrong. An external agency should develop optimization cases for testing purposes. The market manager would have responsibility to run these cases through the market-clearing software, and should report the solution back to the external agency. These test cases should cover all aspects of the market-clearing software, database, and model, and implementers may need to write market rules for edge cases. An example case is the problem of alternative optima, where multiple solutions have the same objective function value, or a given solution has multiple valid prices. Users will want the solution that is good for them individually. Linear programming software, such as LINDO, may have a facility to choose an average, or to find the lowest clearing prices, which can sometimes prevent prices from fluctuating. This behavior has appeared in operating smart markets, including those for electricity and radio spectra.

Aside: Steps in Routine Operation

Once the market is operating regularly, users will find themselves accustomed to its routine.

1. Users begin with a list of allocations for each period (e.g., a week) of the current hydrological year. These quantities are their initial market holdings, as granted by the water regulator in the previous market period, if not earlier.
2. Before the auction, the market manager estimates the amount of water available for allocation. The market manager must satisfy environmental requirements before allocating water to users. Dry conditions, at least drier than previously forecasted, require the manager to reduce users' initial holdings. This reduction matches the legal right to the physical quantity. We discuss initial rights at length in Chap. 8.

(continued)

3. In the auction, users can offer to sell water from their water allocations, and can bid to buy more water allocation, for the current week and for any future week. A user could bid a combination, offering to sell for a high price and to buy for a low price. The market manager can offer simple tools to help each user understand their general profile of water use over the calendar year, depending on type and size of operation, to assist them in bidding. By allowing trades for future periods at every auction, users can adjust their allocations for future periods. They can change their minds in the middle of a season, or have an assurance of water early in a season. As the given future period approaches, participants can know prices and allocations with increasing certainty.
4. The market manager constructs the optimization model, based on the hydrology of the catchment. The optimization specifies the allocations of water to users to maximize the sum of buyer and seller surplus, while ensuring that flows satisfy the environmental requirements.
5. Following any tentative rounds, the manager runs a final auction and gives each user a definite allocation for water for the current period and for any changes in each user's allocations for future periods. The manager pays net sellers and charges net buyers at each location at their local spot price. Users who do not trade get their scaled allocation without cost or payment. The manager notifies all users of the spot price of water at their location, and invites them to buy or sell water at future auctions. A user's allocation for the current week is a firm license to use any amount of water up to the specified value (or an obligation to take the allocation, depending on the contract), drawn from the user's location, for the current period.

6.7 Some More Rules for Auction Operation

Before each auction, the market system must update each user's set of future allocations based on any adjustments required by the market manager and based on purchases/sales by the users in the last auction. The market manager must also update the estimates of water availability, based on rainfall, actual water use in the last period, and projections for future use based on bids in the last auction. This regular update does not need to include an update of the response matrix, just an update of the status of the relevant water resources.

The auction could be multi-round, with a few indicative auctions before the final round, perhaps requiring diminishing changes to bids to limit gaming. Multiple rounds allow price discovery, and allow bidders to better anticipate outcomes (Cummings, Holt, and Laury 2004; Bourbeau et al. 2005). Most likely, the need for multiple tentative rounds will be most important at the start of each auction, when people are unsure of the market price. After many auctions, especially if the catchment has storage, users will anticipate the prices, and may not need the preliminary tentative rounds. Storage, even a modest storage, will smooth out prices considerably, as the storage facility would tend to add water when cheap and release water when dear, reducing or even eliminating price differentials between periods. If traders expect prices to stay within a range, they probably will not need multi-round auctions, especially with multipart bidding.

The regulator and market manager should develop a system of payments for over-use and under-use. Users should have some incentive to turn off the tap after the auction. Otherwise, users may view the cost of the purchased allocation rate as sunk, and use the water needlessly. So the market manager might wish to compensate users for their unused water, as it reduces the chance of later shortage. If many users came to the market manager asking for refunds, the market manager's cash may not come out even. One might also expect that some users may take somewhat more than they had been allocated, and then be required to pay the market manager. Any leeway allowed, over or under, would have to be regulated and managed by the market manager from auction to auction.

Some electricity markets clear these deviations using current prices. In our water context, the farmer taking excess at the end of the period would be charged prices calculated at the beginning of the period. Other electricity markets recalculate prices using actual quantities taken, once the market manager knows those quantities, and then the market manager charges users based on the recalculated prices. Recalculated prices measure the system impact of deviations from the agreed abstraction. If water turns out to be in shorter supply than expected, the recalculated prices will be higher, and the market manager will charge those who consumed extra, and pay those who conserved. But if water turns out to be in excess, recalculated prices will be low, and the manager will give little compensation to those who conserved and little penalty to those who did not. If the market manager does recalculate prices for deviations, the manager should also adjust future allocations to account for actual usage in the current period, and the market manager would reimburse users for unused water, or would charge them for excess taken.

6.8 Recharge and Smart Market Design

Consider two neighboring landowners who share one well and abstract the same amount of water per growing season. One landowner uses efficient drip irrigation for growing fruit, while the other uses flood irrigation for growing hay. Should both landowners pay the same price for their water? The surprising answer is that the hay grower may sometimes pay less.

In terms of their effect on the environment, they should both face the same marginal price for water use from that location. It does not matter if one is growing high-value fruit while the other is growing low-value grass. Of course, one of the two operations will be more economical when considering land value, the cost of irrigation, and the value of the product, but the markets for these (land, improvements, product value) exist to ensure the proper signal is given to the landowner in terms of land use and the consequent efficiency of water use.

They should pay different values for their water because they have different impacts on the water resource after use of the water. The efficient user of water allows little to reach the water table and so little water has an opportunity to be used again or to help meet environmental constraints elsewhere. On the other hand, the inefficient user of water may create more run-off into streams and more recharge of the water table. This means that the water can be used again. Because of this, the inefficient user might expect a credit in some form.

Finding an appropriate way to treat recharge within a smart market will be an important issue to resolve. Recharge can occur in groundwater systems, and in surface water systems where excess irrigation water enters shallow groundwater and then flows back into a stream. The extra linear programming required to institute a common-pool smart market including recharge does not appear to be onerous. The modeler has a number of ways to manage the issue, and water users will want credit for their recharge.

A proper estimation of recharge would require site-specific evaluation by hydrologists. The estimation depends on soil type, local weather, the surface conditions (smooth or rough), the type of crop, the time of year, and the rate of application. At one extreme, users might agree that the market should ignore recharge. A smart market can work in these situations. In this extreme, more water will be available in the future than the modeling predicts today. At first, users might over-value future water because the model would not assume that the recharge will happen. The market manager might recalibrate the hydrological model assuming some simple fraction of water reenters the system at a future place and time, and then adjust that fraction with experience. In this way, downstream water users would benefit from lower prices for their water while the up-catchment water users who provided the recharge would pay more than the true effect of their water use. Perhaps the amount involved would be so small that users would prefer to keep the market simple.

Any system for estimating recharge would need more information from water users. They may be reluctant to provide information on what crops they are growing, the method of irrigation, what type of soil they have, etc. Users may ask for an opportunity to vary this information from one auction to the next, and the market manager would need a way to validate this information. One could imagine a market designed so that water users who request a recharge credit would first agree to various methods for the regulator's hydrologists to estimate recharge on an ongoing basis. This would ensure a way to recover the extra transaction cost required to operate a spot market including recharge.

Recharge varies from place to place and is not a point function like water withdrawal. As a result, the market manager must develop calculations to account for where the water is applied and its areal extent, and then to estimate an equivalent recharge at the point of abstraction. We suggest that the market manager could reduce the abstraction price (calculated through the market-clearing model), based on agreed-upon estimates of recharge. Users could get a credit for a modest amount of unused water, perhaps at the price associated with the end of the period, but the market manager should not explicitly credit them for more-efficient use of water per se. Greater efficiency should benefit the user through reduced water purchases or greater sales. The benefit must accrue at the time and place of the extra water recharge. The efficiency of the water user's technology is not relevant in price-setting within a smart market framework.

If a number of water users in a catchment invest in a more efficient water use system, that will mean less recharge of water, leading indirectly to greater reductions (on average) to future water allocations. Although this might seem as if the smart market is not encouraging efficient use of water, we would disagree. The reason that users within a smart market would likely consider employing more-efficient methods of water use in the first place would be because of restricted supply due to, say, climate change. In the short run, a smart market could decrease investments in water-efficiency technology, because users can obtain the water they need through the market. The increased efficiency in water use comes from the reallocation of water between users. The introduction of a high-quality market, such as a smart market, would likely decrease the effective cost of water to users, counteracting to an extent any trend of a loss of supply from climate change or increased demand. One could imagine a situation where smart markets are introduced in a catchment that would otherwise need to switch to higher-efficiency (and more costly) irrigation methods. In this setting, the smart market would make water more available to those most interested in investing in increased efficiency, deferring on-farm investment in water efficiency for a period of years. Eventually, the regulator will need to tighten allocations, but the issue of tightening allocations would be more serious without smart markets.

Accounting for recharge will increase the importance of the water quantity's relationship to water quality. Recharge can positively or negatively affect water quality depending on the situation:

- More surface run-off could cause greater erosion and so greater sedimentation in streams and lakes.
- More recharge to groundwater could lead to greater leaching of pesticides, fertilizers, or salts.
- More recharge to groundwater could lead to dilution of the concentrations of toxins or salts in groundwater.

As you can see from this overview, if the market manager decides to include recharge as a credit, the issues can quickly mount. A common-pool smart market could start with ignoring recharge credits. After implementers have developed the simple market, with some trust and predictable prices, the users and water authorities could consider expanding the market to include recharge credits.

6.9 Avoiding Traps in Market Design

Implementing a smart market for water is a task for the strong-hearted. It is a radically new way of solving an old problem. The market can be badly designed and badly implemented in many different ways. The implementers should at least avoid some obvious traps.

Trap 1: Excluding Users

Some users may wish to develop a smart market that excludes certain types of users, perhaps from fear of losing water, perhaps from spite towards a different water sector, or perhaps from fear of competitiveness with another region. The implementers will need their greatest powers of persuasion to reject these wishes.

Implementers should try to include a variety of users, both those who would buy in abundance and those who would sell in drought. The market needs those who are desperate to avoid risk and those who have multiple avenues to manage risk. A market of fruit growers that excludes crop growers will miss the full potential of the market; a market of rural farmers that excludes urban or industrial users will have lower benefit. A market for only groundwater users will benefit more if it adds surface water users in the same catchment.

When a watershed spans regulatory jurisdictions, the market would work better if it crossed those jurisdictions. When water flows across a state boundary, so should the market. Crossing each regulatory or political boundary raises the difficulty of implementation of a smart market, but it also increases the benefits to all water users, due to increased market scope.

Trap 2: Indirect Regulation

Typically, environmental control means extra regulations on users. These regulations make sense without the catchment-wide optimization of the market-clearing model. But with the market-clearing model, implementers should eliminate all side regulations on users. Implementers should consolidate those regulations into direct constraints on the environmental flows in the model. Implementers should consider every regulatory constraint. Is this constraint redundant to what is really desired? The model's embedded physics allow direct constraints on the true endpoints of environmental control. The regulator can remove indirect requirements for "good engineering practice" or "typically proper conditions."

For example, the regulator may have restricted users' abstractions so that a river would have sufficient flow. The regulator should remove those side restrictions, and instead require the market manager to specify the river flows explicitly in the model. The model will then restrict users' abstractions implicitly, as a natural part of the model's solution. Here are a couple of examples.

The regulator may wish to protect a vulnerable wetland, and thus may restrict users' abstractions to ensure that the groundwater head is above a certain level in various locations. In the smart market, the regulator should eliminate these restrictions on users, and instead mandate that the market manager specifically constrain the water level in the wetland. With the constraint on the wetland water level hardwired into the market-clearing model, the regulator needs no separate constraints on users.

Consider a club of kayakers who have some rights to river flows for their recreation. The kayakers might say that a smart market for water in the watershed through which the river flows must have a series of flow constraints. They need a minimum flow to allow for a competition the last weekend of June. They also need a certain flow the Saturday before Easter each year for another event. They have scheduled a national event in August of next year, but are not sure which weekend, and so would like a minimum flow each weekend that month until the dates are set. If the club grows, they will want still other times with extra flows. The market implementers could burden the market with many of these constraints. However, these constraints are hard to change, so either users have extraneous constraints when the kayakers don't need water, or the kayakers miss out on their recreation. Instead, everyone may be better off by enabling the kayakers to trade for what they want. In a wet year, they might exchange allocations to allow for a long competition. In a dry year, they could trade for the weekend best for their competition.

Trap 3: Confusing Private Responsibilities with Public Externalities
Consider two neighboring wells: Ann's, which is 20 m deep, and Bob's, which is 30 m deep. Suppose the aquifer drops to 25 m, so Ann loses access. This case is similar to Fig. 3.2c, in which the house nearly loses access because of the factory's deep well. Should the regulator require the market manager to constrain water allocations to ensure that the water level at Ann's well stays above 20 m? Or should the regulator expect Ann to deepen her well to 30 m? This problem merits a subtle response from market implementers. Especially if the aquifer change occurs shortly after market implementation, Ann will argue, justifiably, that the market has imposed an externality on her through a loss of access to water.

The market rules should anticipate this situation, and specify reasonable well depths that the market manager would protect. If a user's well is shallower than market rules specify, the user bears the risk. If the user's well is deeper than the rules specify, the market manager bears some risk. For example, the market manager could pay half the costs for deepening the well, recouping the costs through higher fees on all users, since all users benefit from the looser constraint on the catchment allocation. In any case, well depth specifications should be set with care by a competent hydrologist, running a range of simulations.

Access to water is valuable. We anticipate that, before a smart market exists, and because of previous habits of overexploitation of water resources, many water users will have taken steps to ensure water availability (such as digging a deeper well). We expect that the smart market will tend to improve water availability, improve security of supply and reduce the need for these risk-reducing investments.

Trap 4: Including Water Quality in the Market, When the Physics Differ
Water quantity affects its quality, but attempting to trade them simultaneously complicates market design. As explained in Sect. 3.1, water use can lead to recharge that varies in quality. A downstream agricultural water user may tolerate water with extra nitrate, but a town may not tolerate extra nitrate for its water

supply. Both may object to having supplies with pesticides from users higher in the watershed, and so they may want water quality traded in the same market as water quantity.

We have studied smart markets for water quality (Prabodanie, Raffensperger, and Milke 2009, 2010; Prabodanie et al. 2011). Policy makers can apply much of this book to implementing a common-pool smart market for water quality. The practical challenges to implementing both at the same time are immense. If a particular watershed wants the market for quantity, but quality issues are so strong that implementers must consider quality, then implementers may wish to develop the market for quality first.

If the catchment is primarily groundwater, chemicals will move slowly, over the course of years or even decades. The appropriate auction schedule for water quality may have to encompass those decades, while the auction schedule for water quantity would be in days, weeks, or months. As an analogy, gasoline and automobile tires are closely related, but people buy them on different schedules. Their market operations therefore should operate independently, because the physics differ too much.

If the catchment is primarily surface water, the impacts of chemicals, such as pesticides, nitrate and phosphorus will affect assessment points at timescales similar to those for the effects of surface water use, and a joint market for quantity and quality makes sense and could work. If the physics are sufficiently similar, implementers could base a market for water quality on any quality factor that they can quantify. However, the regulator may find that measuring a user's effect on water quality is challenging, in contrast to measuring water quantity with meter reading or head level indicator.

We discuss a combined quantity and quality market further in Sect. 12.1.2.

Aside: The Australian Hunter River Market for Trading Salinity Discharge

The Australians have a widely hailed success story with the Hunter River Valley scheme to reduce river salinity (Environmental Protection Authority 2003). Salty water collects at the bottoms of coal mines, so the miners pump that water into the river. Salty water also collects in cooling towers of power plants, which pump that water into the river. Consequently, downstream irrigators found the water too salty for their crops. Fingers were pointed and voices were raised. So government stepped in, and gave the mines and power plants fixed licenses for total salt they could put in the river each year. The result was disastrous. River salinity changed day to day with the weather, while the mines and power plants put in the same amount of salt whether the river ran low or high. But the farmers needed water the most when the river was low. So they called in some hydrologists.

The Hunter River hydrologists noted that a rain shower first made the river saltier, as surface salt flushed in. As the rain continued to fall, the new water diluted the salt, and the river became less salty. So they suggested that dischargers should wait a little ways into a good shower before releasing

(continued)

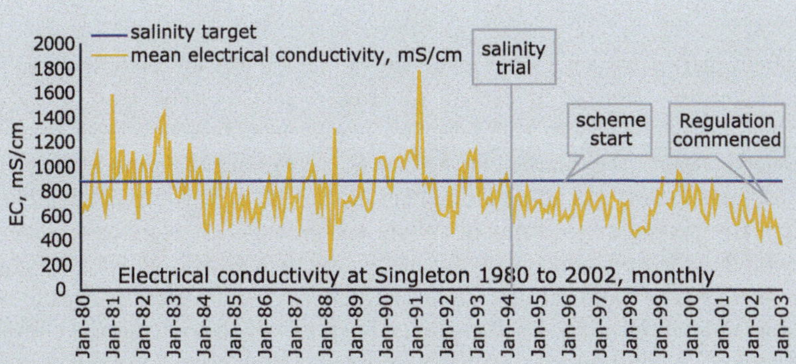

Fig. 6.3 Electrical conductivity in the Hunter River at Singleton over time. (Reproduced from www.epa.nsw.gov.au/licensing/hrsts/success.htm)

the salt. The government also hired some economists, who set up a trading scheme. The scheme gave licenses to the mines and power companies, but in a different manner to their old annual license scheme. The new licenses were shares that allowed users to buy and sell salt dumping rights between themselves, as long as river salinity was below a certain threshold. The government further set up a website that helps buyers and sellers find each other, which reduced the transaction cost. Now if companies want to dump more than their shares allow, they must buy more shares. If they dump less, they can sell shares. When faced with a *price* for dumping salt, some companies installed better equipment, so they have less salt to dump. And now the Hunter River has a lower salinity level. Fig. 6.3 shows the changes in electrical conductivity, which is a measure of salinity.

The Hunter River Valley salinity scheme had three key parts: good hydrological advice that informed the quantity and timing of release, a well-designed economic system of tradable shares, and reasonable initial allocations of shares. The website lowers the transaction cost. Really, the effect on the environment is the main point that the rest of society cares about. As long as the river is okay, no one cares whether the coal mines put in a bit more salt, as long as the power plants put in a bit less. So each user has a knob they can control, which is the number of shares they have. And the scheme has a knob for the environment, which is the level of salinity.

The Hunter River Valley scheme works because it has just one constraint—salinity. Because of the need for a hydrology model, the scheme is a step beyond an air emissions trading scheme, though it has similar simplicity of administration.

Because the Hunter River Valley scheme is relatively simple, and it depends on the water quantity, merging the market for salinity with water quantity would probably work.

6.10 Should the Private Sector Run the Smart Market?

If the smart market is such a good idea, why not let the private sector manage it? A smart market for water run by and for water users appears to be appropriate, even attractive philosophically to some, but we would expect such a smart market to work poorly. Many "free marketers" believe that government should leave markets to the business sector, with minimal regulation. This belief is widespread, including among many people in government. While our previous chapters have debunked the notion that the free market can allocate water efficiently, someone may suggest that a users' group could set up a smart market. The implication is that the water regulator could allow users to self-regulate. Such a mechanism would be consistent with Ostrom's (2009) approach to the commons. Why hasn't the private sector already developed and managed any smart markets for water resources?

Some might say that one private smart market is operating. As of October 2015, Mammoth Trading, Inc. (Mammoth Trading 2015) claims to be operating a smart market for groundwater, and to our knowledge is the only firm claiming to operate a smart market for water. They have not published their market design in detail, and it appears to lack some of the regulatory framework (e.g., metering) that we describe in this book. It also is not clear that the market protects users outside of the market from externalities imposed by users within the market. The market does not appear to generate publically available prices and so does not help society to assess the value of water.

While we hope that Mammoth Trading succeeds, this case highlights a weakness of private sector approaches to any kind of water market: the private operator will have an incentive to keep private their trading rules. This opacity inhibits trust among users, the regulator, and other stakeholders.

Before we go any further on the problems with a private operation, let's examine the reasons in favor of private operation of a smart market.

First, some users trust the private sector more than government. In corrupt places, some government employees demand bribes for resource allocation (UNDP 2011). In other places, users fear that their trades will result in government taxing the trades in money, or in water by surrendering some of the traded water to give the surrendered water to the environment (as is sometimes required in California). And in many places, people simply distrust government innately. Especially during market formation, users may trust private brokers more than they trust the bureaucrats running a centralized government system. On the other hand, many users distrust private enterprise, especially with water. They want and expect government to manage the public commons. In Chap. 13, we describe many of the concerns we have heard about water markets generally, and the smart market developed in this book specifically.

Second, users may think they can run the market more efficiently. Users may believe that they can more readily create a market system that meets their needs. They may be able to control costs of market operation more readily, perhaps with volunteer help or using staff already working for the users' group or broker. They might be able to arrange for transfers that do not require payment, and so avoid taxes. Proponents of a private market may even point out that private trading would relieve regulator staff time. Assuming a fee contingent on a successful trade, a private broker would have incentive to complete trades, and would therefore work to understand the trading process. By doing so, the broker would develop expertise in trade that most users would not care to accumulate themselves; they simply want to trade.

The main problem with a private market is that water use creates externalities not borne by water users, and that government is designed to manage these externalities. Private water markets can get around this major problem in rare situations when externalities are small or easily controlled. Consider a water reservoir and canal system that provides water to a number of users. These users could combine to form a cooperative that pays for the operation and maintenance of the canal system (and perhaps for its capital costs as well). Neglecting issues such as recharge, water quality, and discharge of excess canal water, the water regulator can ignore how much canal water each water user takes. Instead, the water regulator should attend to the water flowing into the reservoir, and should set various conditions on the rate and timing of water extraction. In this situation, the cooperative of water users could agree to institute their own smart market for adjusting their water allocations between themselves. If one water user takes too much water from the canal, depriving a down-canal user of water, the deprived user can take the issue to the cooperative for resolution. The community of those impacted by water misuse coincides with those who are in the water market.

Even this highly constrained scenario works only to the extent that the regulator and user group agree on the water extraction, have a system to measure it, and have agreed-on systems for resolving extra water use or unexpected externalities (say, erosion from discharging unused canal water into a river). The more common and general situation involving water users has many externalities, and trying to have a private company accountable for ensuring no negative effects raises great difficulties.

For the private market to work, it should meet specific conditions—conditions that we believe are unlikely for any real-world private sector market. First, the regulator must believe that the broker's hydrology and optimization models are correct. The broker or private operator will have an incentive to have incorrect models that protect against impacts of over-use of water. In the case that the regulator attempts to insist on thorough oversight of the broker, with audits of models and transactions, we have the problem that the private operator is not likely to accept the oversight because of impacts on intellectual property or invasion of users' privacy.

So consider the case in which the regulator wishes to maintain a light touch on a private operator. The water regulator would plausibly accept a user-operated market if the users' takings have only trivial externalities. In our reservoir example above, the regulator will monitor flows into and out of the reservoir. Should these flows be inadequate, the regulator would need to find a party to prosecute. This could not be the broker, because the broker does not take water. Instead, the regulator could make an agreement with an incorporated users' group. Considering the large potential negative effects of excessive water use (e.g., decimation of an endangered species), the regulator could require financial surety of this users' group. But users may prefer a government-managed market operator in place of contributing to large performance bonds within a private users' group.

Other problems with a private smart market for water are more subtle. The motivation for establishing a private operation rather than a public one could be solely to avoid government involvement. The private manager might wish to avoid all but only a cursory review of trades. Indeed, the private manager might prefer to have no regulation at all of pricing or other market behaviors, and might lobby the regulator to allow the market to operate without even metering users. Private firms will likely view their trading platform as proprietary, making the market harder to audit and opaque to users and society.

A private operator would have an incentive to sidestep or interfere with the regulatory approval process, to lobby for weakening regulations, and to violate regulations when those regulations inhibit trade. A private market will have a strong incentive to skirt obligations to both the environment and to water users outside the private club, because cheating on those obligations lowers the cost of water to inside users. Such behavior raises conflict between the private market manager and the regulator, because the regulator is responsible for ensuring that users' takings do not violate the environmental requirements.

The market manager—private or public—should hold a monopoly over a given catchment, because the manager should account for all water in the catchment at once, in order to know whether a given auction satisfies the spatial and temporal externalities. A private manager would very much enjoy a monopoly and would have incentive to take advantage of the monopoly power. Society has regulated monopolies for electricity and urban water supply. Those monopolies, where they are now private, evolved from government departments through painful political processes, and for good reason: a monopolist should be regulated.

A private market would likely fail for yet another simple reason: the private broker's view of the hydrology would be incomplete, in contrast with the full hydrological view (at least in principle) of the regulator. A private broker would not have the right to read the meters of users outside the market. Therefore, the private broker could not know what every user would do, and so could not write a complete smart market optimization model and could not forecast the environmental flows. In contrast, the regulator would have the right to read all users' meters, even of those who did not wish to trade, and so the regulator would have the full information required to manage the catchment and therefore could write a complete model and could forecast the flows.

Consider a catchment where only two users will bid for water in a given time period. The users participate in a closed-bid auction run by a commercial broker. The maximum aquifer drawdown allowed is 3 m. Ann bids \$2 per megaliter, and draws down the aquifer by 0.5 m for each megaliter she abstracts. Bob bids \$1 per megaliter, and draws down the aquifer by 0.333 m for each megaliter that he abstracts. The broker can model this with a linear program, using variable a for Ann's quantity and b for Bob's quantity:

maximize $2a + b$, subject to $0.5a + 0.333b \leq 3$, with $a, b \geq 0$.

The optimal solution is $a = 6$ ML, $b = 0$ ML, and the aquifer is drawn down by 3 m. (This example involves an auction and not a trade, but bidders could easily be added to the example.)

The assumption is that both users want to bid in the current auction. If only Bob joins the market, and Ann is not part of the private market, then Ann's behavior is unobservable—either because metering is not implemented, or because, for privacy reasons, the meter data are not available to the commercial firm running the market. Now the broker has no way to determine how much water can go to Bob. The broker needs to solve the following model:

maximize b, subject to $0.5a + 0.333b \leq 3$, with $a, b \geq 0$, and a is unknown, and not controllable.

The broker could guess that Ann will take her full existing right, perhaps 3 units as in the previous auction. Because users in the private market will wish to make profit, and because the broker's job depends on keeping the bidders in the system happy, the broker may be tempted to give something to Bob, even if the water were fully allocated.

The problem becomes harder when inflows vary. If inflows are less than expected, and Bob still has no allocation, Ann's license might require her to reduce her take. Who is responsible for enforcing Ann's reduction? That is clearly the regulator, not the broker. Are either the regulator or Ann required to inform the broker of Ann's new allocation? Probably neither is. The broker can demand information from Ann only if she wishes to bid in the auction.

To run a proper market, the broker requires knowledge of all users' abstractions to ensure that the environmental flow requirements are satisfied. Because a broker cannot see Ann's meter, the broker cannot ensure that the environmental requirements are satisfied. Trading regulations today (as in Australia) may not require water to balance, but a sustainable successful water market must do so in the long run, and preferably in the short run. Private water market operators that do not include all users will not be able to balance water use with water availability, and, in any case, they have an incentive not to do so. The hopeful middle ground of delegating regulation to the private firm appears to be infeasible.

The Australian national government relies on brokers to operate markets for shared natural resources, even when compelling evidence suggests it to be inappropriate (Nemes, Plott, and Stoneham 2008). One private Australian water-trading firm uses software to assess whether a trade is legal; the firm then transmits the proposed trade to the regulator, which uses similar software to check the trade. The regulator is the final arbiter of the trade, and is not relieved of regulatory duties. The result appears to be a sufficiently trustworthy system. The trading firm has an incentive to ensure that trades are valid. This seems to work so far, but no one completely accounts for water uses and supply, so the potential for failure is high. Indeed, Australia is having continuing difficulty satisfying environmental flow requirements. Further, this private firm is not a users' group, but a third-party broker. When business people profit as brokers, they become an entrenched constituency that opposes any changes that could endanger their livelihoods, and this appears to be the case in Australia. The same would happen with a user-operated smart market.

These are not minor technical issues. These problems go directly to the degree that government can delegate management of a commons to a private firm. Government regulators might perceive that a smart market should be run not by government, but by private interests. Regulators cannot wash their hands of their responsibility by allowing private organizations to run a water market with only a light hand by government. Government must regulate private water markets so closely that we believe that a government-managed market structure would be most successful.

Most of this book follows what we believe to be the simplest and most logical structure: that the market manager is a government agency that is part of or closely supervised by the water regulator.

6.11 Workable Market Structures

Discounting the option of a private organization running the smart market, implementers could structure the smart market in various ways, each with particular challenges.

The smallest logical organization for a smart market is the catchment, because catchments are relatively independent hydrologically. Of course, catchments are often sub-catchments of a larger catchment, which raises the question of the extent to which implementers should divide a region into separate markets. Each catchment could have its own market manager, because the catchments are relatively independent hydrologically, or a single manager could operate multiple independent markets.

A similar question relates to political divisions. A single institution may be difficult to start because of the transaction cost of negotiating across political boundaries. Unfortunately, political and hydrological boundaries rarely match. Indeed, a political boundary is often a river, which splits a catchment in two!

The decision over who operates the market will depend on how quickly abuse of water can lead to serious environmental problems. The more quickly that abuse can lead to problems, the more reasonable it will be that the water regulator operates the market. The more slowly that problems might arise, the less need for the regulator to operate the market.

A single market manager could operate a state-wide or larger market, with responsibility for all market operations within that area, under the authority of the state water regulator or under the joint authority of multiple state water regulatory agencies. A few variants of this ideal structure are worth mentioning.

The Water Regulator Could Operate the Market. In effect, under this structure the market manager is an immediate subordinate of the water regulator. This makes sense, because the market manager will have many quasi-regulatory functions. Because the water regulator agency has responsibility to protect the environment, and because adjustments in water allocations affect the environment, the regulator will be interested in controlling those adjustments. Further, the regulator would be more likely to provide sufficient resources to operate a market if it were directly under its management. And finally, because the water regulator's responsibility is to the public welfare, the regulator will want to be in a position to prevent financial impropriety and tax avoidance, and to publish prices to ensure that society receives a proper signal for the relative value of water in the catchment.

The Water Regulator Could Contract Out the Market Operation to a Private Firm. In effect, under this structure the market manager is a contract employee of the water regulator. Economically and legally, we see this as almost identical to the previous market structure.

The State Water Regulator Could Allow Local (e.g., County) Government Water Agencies to Run Their Own Markets. A local region may want its own market manager rather than allow a political body responsible for a larger area to run it. This is plausible if the hydrology is similar enough to the political boundaries. Because negotiating across political divides can have a high transaction cost, local market operation may be easier to implement than a single state-wide structure. For example, some California counties currently prohibit or limit water sales to buyers outside their county; a county market would not require changing those limitations. If two neighboring counties each has separate smart markets, this will naturally lead to a desire to join the markets. Eventually, local users will demand the right to trade across the political boundaries, and society will see the benefits of doing so.

Unfortunately, the local market managers trying to operate local smart markets will have difficulty managing the mismatch between political and hydrological boundaries. For example, suppose two separate markets share only a single river, and users wish to trade between the markets. The market managers will have to coordinate separate optimizations, with bids between the market managers themselves. A single market clearing would be simpler to manage, with less overhead. If the hydrological connection is complicated, the bids between the market managers will be complicated, too. Ultimately, the market managers will recognize the need for a single market and will cooperate. So this approach is probably viable, and certainly easier politically, but such a pair of smart markets will eventually evolve into one of the above mechanisms.

6.12 Likely Questions for Implementers

1. This Proposed System Is Just for a Spot Market, So What Happens When a User's Water Right Is Up for Renewal?
This question is a problem of initial rights, not about the smart market. It applies to any water allocation policy. The regulator has many options:

At one extreme, rights could expire when renewal is considered, and the user would then purchase all water from the spot market, which gradually evolves to a full user-pays system. At the other extreme, rights could be permanent.

In between, the water regulator could renew rights based on specified criteria. Part of the water right could extend for free, e.g., based on the quantity actually used, and part could extend for a fee, possibly based on a related mechanism for trading long-term rights. In this way, new water users would have a means to gain long-term water rights rather than always purchasing through the spot market. The criterion based on the user's historical use is still the "use it or lose it" policy with its perverse incentives, but the negative effects would appear only in the long run and in a weakened form.

Interestingly, a spot market informs the water user of the value of the user's long-term rights. If the value is high, the user has improved ability to borrow money to buy such rights or to obtain a loan with the water rights as collateral.

2. If a User Does Not Take the Full Water Allocation Between Auctions, or Finds No Buyers for the Allocation on the Spot Market, What Happens to It?
The user's buyer is the market manager. Finding a buyer is easy, but the buyer—the market manager—may not be willing to pay the seller's asking price if no one is buying on the other side of the trade.

If the user received no compensation for water not taken, they would be tempted to waste excess water, so the market manager should pay attention to this case. What happens to allocation not taken depends on the market rules. The market manager may require the user to take the water, such as in a case where downstream users rely on run-off, but generally the market manager will not require users to take their allocation.

The market manager could run an end-of-period market, in which the market manager reimburses users for allocation not taken at the prices in the following auction for water not taken. We discuss this in Sect. 6.7. This would treat the allocation as an exchange-traded European-style options contract, in which the user is not required to take the water, and users may exercise the option only on specified dates.

3. Could the Market Manager Scale Users' Water Allocations to Zero?
Yes, theoretically. It would occur only when the catchment had no water to allocate. (We discuss scaling in Chap. 8.) Users should have quite a bit of warning about such scaling by simply following the weather forecasts.

It could also happen when allocable water is available. The market manager might scale down users' initial allocations to zero in order to force all users to be buyers. In this case, the manager would gain all the revenue from the sales. But the market should have rules regarding the manager's revenue.

4. What if Users Have All the Water They Need? Wouldn't That Mean Thin Trading?
Yes, that's right. Users will have no interest in trading allocation during a season of floods. The market manager should run the market when users ask for it.

However, some people observe thin trading, and then believe that users have all the water they need. This is false. Current water markets can have thin trading because of these markets' high transaction costs. The transaction cost of the smart market is much less than that of other water trading systems, so trading will expand under such a market, and new users will join.

One lesson from Australia (Young and McColl 2005) is that trading can be "thickened" by reducing water allocations by some modest fraction e.g., 5% or 10%, to encourage users to enter the market. This would also help cover water regulator's costs of operating the market. Section 7.2.5 addresses this issue further.

5. What Would Happen During Flooding? Could Users Be Paid to Take away Excess Water?
We have assumed general conditions of scarcity, but, with care, the market manager could adjust the market model to provide negative prices. In this case, a user could be liable for *not* using water, or a user may be paid to store water, or to export it safely from the catchment. We think the market should be implemented first for scarcity.

6. How Would Hydroelectricity Generators Behave in this Market?
Hydroelectricity generators would bid into the regular auction just as any other user. The market-clearing model would encapsulate their non-consumptive hydrological effects.

Hydroelectricity generators usually have different concerns than do consumptive users. Hydroelectricity generators are concerned with timing of releases on a roughly half-hourly basis; the electricity market trades much more frequently than the water market probably would. In contrast, a water market is likely to trade on a monthly or weekly basis. Thus, hydroelectricity generators would factor in the weekly price of water into their half-hourly management. The market model may need to constrain the generator's releases. For example, a release schedule of "low releases all week except for a 1-h surge" may have different downstream effects compared with uniform releases all week.

To incorporate this kind of transient use, the market manager could adjust the smart market optimization model and develop associated market rules. The manager would have to expand the bid structure to allow different uses that have different hydrological and environmental consequences over shorter periods. A hydroelectricity generator could bid to draw water from one location and period, and return the water to another location and period.

7. What Is the Smallest Amount of Water that Users Are Likely to Trade?
The smallest tradable quantity depends on the length of time between auctions and the transaction cost. We expect that some amount of water use will be so small as to be "noise"—too difficult to measure relative to the typical variability in water flows. This is an advantage, in that a large regional wholesale water market can exclude the many users who take only tiny quantities.

However, a small steady use over a long period can become substantial. The market manager may demand that some small-volume users pay for their water, but the transaction cost may motivate these users to contract for water supplies over a period longer than the auction period. The private sector could provide such contracts, along with a range of futures and options contracts. City water departments already make contracts on behalf of city homeowners.

8. Could Water Users Agree to Small Trades Through the Market Agent, but Without Going Through the Auction?
The basic problem of trading water still applies. To trade outside of the smart market, users would have to have similar, if not identical, third party effects. This is less common than it might appear. In some circumstances, the regulator and market manager may allow small trades without an auction. If the market manager were unsure whether the trade was appropriate, the manager could run the optimization model to check whether the trade satisfied the environmental requirements.

9. What Happens if a Buyer Bids Very High?
Then the price may go up. A high bid indicates a high value for water. Other users would have an incentive to sell.

10. Will the Market Penalize a User for Being Close to an Environmentally Sensitive Region?
Prices will tend to be higher near environmentally sensitive places. This is reasonable—society should want heavy users to move away from environmentally sensitive places. The high prices near sensitive locations during water shortages should not come as a surprise for these users. Before these users got permission to take water for that period, the market manager would simulate the environmental flows for at least a full year into the future, using the best available hydrology data. The objective is not to punish these users; it is to ensure much more reliable flows for the environment. Higher prices are better than banning abstractions.

11. Will a Spot Market Mean That People Overconsume Now, Increasing the Risk of Future Shortage?
Water use now can reduce availability in the future if the catchment has storage. Prices should be higher now if shortages are forecasted. Fortunately, the market model can readily incorporate any information regarding future water availability.

If set up correctly, the market-clearing model will consider future water supplies and demands. The market manager should establish a bid structure that seeks users' future value of water, incentivizing users themselves to disclose whether they would rather have water now or later.

12. Can the Smart Market Incorporate the Downside Risk of Dry Periods?
"Downside risk" technically means the risk of returns below a target. To explicitly manage users' risk within the market-clearing mechanism, the market-clearing model would have to be a stochastic optimization (Chap. 11). Implementers should select the ratio of risk to return in consultation with the water users, since the risks and the returns fall on them, probably through the bidding structure.

To take this question in a more informal way: yes, even with a deterministic market-clearing model, a smart market lowers risk for dry periods. Prices in dry periods will naturally rise, incentivizing some users to sell and release from storage. Of course, people may blame the market for the high prices, when in fact the market is helping to alleviate the problem. Sellers alleviate shortage to some extent, except when the market manager can allocate no water at all. The use of a common-pool smart market would identify strategies to reduce risk. The environment would likely face low risk in the smart market, though this depends on the structure of the environmental requirements.

13. How Does the Smart Market Change the Role of the Water Regulator?
While the water regulator's job should get much easier, the role should not change much. The water regulator still grants initial rights and approves type of use. The water regulator retains responsibility for short-term water allocation and trading, but the regulator no longer has to approve pair-wise trades. Instead, the regulator controls such trades by setting the environmental requirements, e.g., required flows in surface water and aquifer levels, and, second, by determining that the market-clearing model's calculated allocations would actually satisfy those environmental requirements. The water regulator no longer needs to impose restrictions in drought, per se, because the market manager automatically scales water allocations to fit the available water.

14. If the Water Resource Moves from Fully Allocated to Over-Allocated, Will the Market Manager Generate Surplus Revenue?
No, just the opposite. In over-allocation, by definition users' rights to water exceed the available water. If the market manager attempted to clear the market in this condition, at worst the market-clearing model could be infeasible, leaving the catchment without a plan. At best, the market manager will have to buy rights from users to bring the rights into balance with the water. The smart market institution is likely to fail if the market manager loses money over a long period. To resolve this, we think the market manager should have a method to adjust allocations downwards, and the legal means to enforce this method. Chapter 8 treats this issue in more detail.

If water is under-allocated, by definition users' rights to water are less than the available water. This could result from two cases. First, the regulator could intentionally under-allocate. In this case, the market manager has allocation to sell, and will make money. The government could use these funds for other purposes. Second, water in the catchment may be in excess, perhaps even to the point of flood, in which case prices are likely to be zero, and the market manager may choose to skip the auction in that period.

15. What Are the Implications of Errors and Uncertainty in the Hydrological Model?

We think that the largest errors associated with existing water allocation are due to the lack of metering and of knowledge about what users plan to take, and the inability or unwillingness of water regulators to enforce basic sustainability. Water regulators often allow users to draw down aquifers and reservoirs much further than should be allowed. Even when regulators are willing to enforce sustainability criteria, the regulators have poor ability to manage catchment hydrology because they do not know what users will do, and wouldn't know even if their hydrological models were perfect, because they have only limited knowledge of users' plans.

Contrast this existing situation with a fully instrumented catchment in which the market manager specifies the allocations and has testable simulations every week or even every day. The smart market would provide the ability to specify flows that are physically possible and satisfy the environmental requirements. Managers would have much improved ability to manage hydrological flows, while also using the model to enable a market.

The implications of uncertainty depend on the seriousness of the errors, the magnitude of the uncertainty, and where those problems appear. Here is a partial list:

- Errors in structure of the hydrology (e.g., the market manager incorrectly modeled an aquifer's connection to other water bodies) or errors in response coefficients (e.g., the market manager modeled the impact of a water use as smaller than it is).
- Errors in environmental requirements. For example, the market manager set the aquifer head too low to avoid coastal saltwater intrusion.
- Errors related to users. For example, in scaling allocations, in recording bids and allocations, in operation of the market.
- Errors of metering, in recording users' abstractions, or in measuring environmental flows.

Each of these errors will have different implications. Implementers should address these by taking reasonable steps to get it right the first time, and by specifying market rules to manage the cases where things go wrong.

The hydrology model will need ongoing verification and perfecting. Furthermore, users must have access to a complaints process. Users need to appreciate that, when the regulator sets environmental flow requirements and adjusts allocations, the regulator's models have errors and uncertainty. The water prices that will come from a market will give clear signals to users and water regulators about how much errors and uncertainties are costing water users, and allow the market manager and regulator to use cost–benefit analysis to evaluate the need for better hydrological models.

16. Can You Give a Simple Recipe for Setting Environmental Flow Requirements? Stakeholders must choose allocation of water to the environment through political processes, with wide community involvement. Every resource must be treated differently, though the tools—hydrology simulations and optimization models—will often be similar.

Stakeholders should find that some environmental allocations are easy to agree on. For example, the aquifer should have enough fresh water to prevent incursions of coastal salt water that will ruin the resource for productive use. Other allocations will be more troublesome, e.g., how much water should be allocated for river kayakers?

The social contention in setting environmental flow requirements will never go away. In dry times, some users will want government to lower their costs by reducing environmental flows. The government may decide that some reductions may be appropriate. But the smart market approach would reduce the crisis thinking of droughts, where short-term economic motivations cause long-term environmental damage.

The recipe could be multi-step: First, immediately set requirements on which society already agrees sufficiently. Second, as society can agree to them, enable the market manager to impose new requirements for some period, perhaps 5 or 10 years. During this time, some users face regulatory uncertainty in at least some of their rights, because the market manager may impose new requirements. In the third step, the regulator implements a standard process for changing the environmental requirements. This process could be one of buy-backs, condemnation via eminent domain, or a mix.

References

Bourbeau, B., Crainic, T. G., Gendreau, M., & Robert, J. (2005). Design for optimized multilateral multi-commodity markets. *European Journal of Operational Research, 163*(2), 503–529.

Cummings, R. G., Holt, C. A., & Laury, S. K. (2004). Using laboratory experiments for policymaking: An example from the Georgia irrigation reduction auction. *Journal of Policy Analysis and Management, 23*(2), 341–363.

Environmental Protection Authority. (2003). *Hunter River salinity trading scheme: Working together to protect river quality and sustain economic development.* EPA (Environmental Protection Authority), Department of Environment and Conservation (NSW).

Murphy, J. J., Dinar, A., Howitt, R. E., Mastrangelo, E., Rassenti, S., & Smith, V. L. (2006). Mechanisms for Addressing third-party impacts resulting from voluntary water transfers. In J. A. List (Ed.), *Using experimental methods in environmental and resource economics.* Cheltenham: Edward Elgar Publishing Ltd.

Nemes, N., Plott, C. R., & Stoneham, G.. 2008. *Electronic BushBroker exchange: Designing a combinatorial double auction for native vegetation offsets.* https://excen.gsu.edu/plott.pdf

Ostrom, E. (2009). Social-ecological systems: A general framework for analyzing sustainability of social-ecological systems. *Science, 325*(41), 9.

Prabodanie, R. Ranga, A., Raffensperger, J. F., & Milke, M. W. (2009). *Simulation-optimization approach for trading point and non-point source nutrient permits*. 18th World IMACS/MODSIM congress, Cairns, July 13–17.

Prabodanie, R., Ranga, A., Raffensperger, J. F., & Milke, M. W. (2010). A pollution offset system for trading non-point source water pollution permits. *Journal of Environmental and Resource Economics, 45*(4), 499–515.

Prabodanie, R. A., Ranga, John F., Raffensperger, E., Read, G., & Milke, Mark W. (2011). LP models for pricing diffuse nitrate discharge permits. *Annals of Operations Research*. doi: 10.1007/s10479-011-0941-0

UNDP. (2011). *Fighting corruption in the water sector: Methods, tools and good practices*. United Nations Development Programme, Bureau for Development Policy.

Young, M., & McColl, J. (2005). Defining tradable water entitlements and allocations: A robust system. *Canadian Water Resources Journal, 30*(1), 76–72.

Chapter 7
The Smart Market Optimization Model

Abstract This chapter focuses on the optimization model that is at the heart of the smart market for water, expanding on the model introduced in Sect. 6.6. We first contrast net pool markets with gross pool markets, and the implications of those market designs. The following sections provide market model formulations for specific hydrological cases: a basic reservoir, surface water systems without storage, surface water systems with reservoirs, groundwater systems without surface water intakes, and conjunctive use with groundwater and surface water. We show how the hydrology affects the model, and discuss the market issues specific to each hydrological situation.

Keywords Optimization • Market clearing • Gross pool • Net pool

7.1 The Heart of the Market

After all market prerequisites are in place (Chap. 5), and when the market manager has an information system to record users' bids and trades, the market manager can prepare the smart market optimization model for use. The model is a linear program, customized for the users' catchment, and for some characteristics of the users themselves.

A written algebraic formulation for the model serves as a checklist of factors that the market manager, and the market participants, must consider. Its output specifies both the final allocations and the market prices. It even specifies the precise rights that users actual traded. The model has to capture all relevant information about the hydrology, the environmental requirements, and the users' bids. Implementers must therefore develop the model with care. Different hydrological settings (e.g., surface water use or groundwater use) will lead to different model formulations, and different hydrologies lead to different modeling and market design issues that implementers should address.

J.F. Raffensperger, M.W. Milke, *Smart Markets for Water Resources*, Global Issues in Water Policy 12, DOI 10.1007/978-3-319-55008-4_7

7.2 Net Pool Versus Gross Pool Market

7.2.1 An Important Market Design Decision

As we mentioned in Sect. 6.4.2, the market could operate as a net pool market or as a gross pool market. This decision has important implications for the expectations of users when they bid, for how the market manager defines the model and handles market data, and for incentives of potential sellers.

In a net pool market, users bid for changes from their existing initial allocations. The optimization model's decision variables correspond to changes in initial allocations, i.e., "buy" and "sell". Thus, the market considers only net changes to users' allocations. As we will see in the example, every change in initial rights requires extensive modification of the net pool model, including the bid variables, objective coefficients, and bounds.

In a gross pool market, users bid for the total quantity of water they wish to have, regardless of their initial allocations. All initial allocations are on the table for trade. See Fig. 6.2 for an example of a gross market bid structure, noting the absence of clear "buy" and "sell" blocks. If a user does not want to sell, the user simply enters a high bid on their own water; if it is high enough, they will get their own water, effectively buying from and selling to themselves, with zero cost. (They could pre-bid and have their desired result automatically, as described in Zetland 2013.) Each decision variable corresponds to a user's total water allocated after trades, not changes from initial allocations. Thus, the market considers the gross quantity allocated to each user.

With a gross pool auction, the formulation is always feasible mathematically, so the market manager never has to deal with a broken model. As part of the financial settlement process, simple accounting code can handle initial rights, purchases, and sales, after solution of the model. Consequently, the programmer can ignore initial rights during model formulation, and the programmer need not rebuild the model to change the initial right. The market manager may find that the gross pool formulation will be easier to manage.

Let's consider an example and algebraic formulations, to see the implications of the two formulations.

Aside: Notation for the Market-Clearing Models

We will use the following notation in the linear programming models.

Indices

b: bid block, $b = 1,\ldots, B$. Typically $B = 1$ or 2, at most 5. See Fig. 6.2 for example bid blocks.

w: water intake, with $w = 1,\ldots, W$. Typically W is between a dozen and a few thousand. More than one user could use a single place to take water, in

(continued)

which case they each would bid separately for their own allocations, and a user could have rights at more than one place to take water. W is the total of all user/intake combinations.

k, l: assessment points, with k, $l = 1, \ldots, K$. Typically K is between 1 and several hundred. These are points where the regulator has specified environmental requirements.

t, u: time period, $t = 1, \ldots, T$. Depending on the auction schedule, a time period could be a day, a week, a month, or a season. Period 1 is the current period, and period T is the final period in the auction schedule.

Parameters, Title Case

$BidPrice_{b,w,t}$, $/m^3, the price that a user submits for water use at that user's water intake w for bid block b in period t.

$BidQty_{b,w,t}$, m^3, the quantity that a user submits for water use at that user's water intake w for bid block b in period t.

Decision Variables, Lower Case

$q_{w,t}$ = quantity, m^3, that the market manager allocates to a user at a water intake w for period t. This is the total amount of water that user can take at w in period t.

7.2.2 Example Data

Consider a simple catchment with two users, Ann and Bob, each with different water intakes. Let's use the following variables:

Aq and *Bq* for the final allocations, in cubic meters of water per day, that Ann and Bob will take.
Abuy and *Bbuy* for the cubic meters per day that Ann and Bob will buy.
Asell and *Bsell* for the cubic meters per day that Ann and Bob will sell.

To determine the allocation of water between Ann and Bob, the market manager needs to know the environmental requirements, the initial rights, and Ann and Bob's values for water.

Regarding the environmental requirement, Ann is next to the river, and Bob is further away. When Ann takes a cubic meter of water in a day, it comes out of the river almost directly, one for one. When Bob takes a cubic meter of water in a day, much of it comes from a strong aquifer, so his impact on the river is only 0.1 m^3. The maximum water extraction allowed is 9 m^3 per day. Thus, the market-clearing model has constraint $Aq + 0.1Bq \leq 9$.

As for the initial rights, suppose each user has an initial right to 10 m^3 per day. The catchment is over-allocated, because $10 + 0.1 * 10 = 11 > 9$.

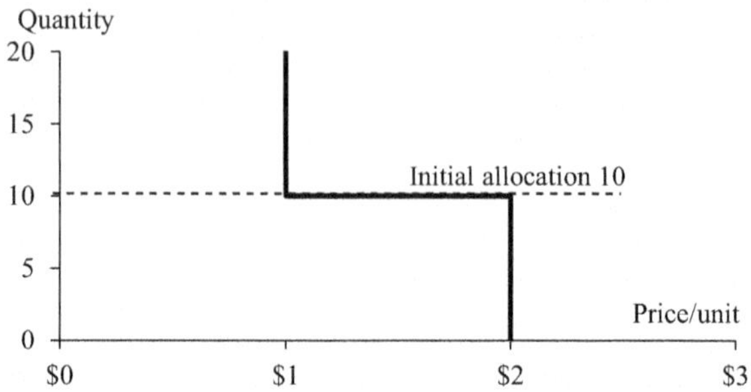

Fig. 7.1 Example value for water, for both Ann and Bob, showing the initial right for 10 m^3

Finally, for their values for water, suppose each user has a value of $2 per unit for 0 to 10 units, and a value of $1 per unit for any amount more than 10 units. Figure 7.1 shows this value function and the initial right. We next examine the two different formulations with this example data.

7.2.3 Net Pool Formulation

Algebraically, the net pool formulation (ignoring time periods for simplicity) can be given as follows:

$$\text{Maximize} \sum\nolimits_{\text{users } w} \sum\nolimits_{\text{buy bids } b} BuyPrice_{b,w}\, buy_{b,w} - \sum\nolimits_{\text{users } w} \sum\nolimits_{\text{sell bids } b} SellPrice_{b,w}\, sell_{b,w},$$

$$q_w - \sum\nolimits_{\text{buy bids } b} buy_{b,w} + \sum\nolimits_{\text{sell bids } b} sell_{b,w} = InitialAllocation_w \text{ for each user } w,$$

$$buy_{b,w} \le Buylimit_{b,w},\ sell_{b,w} \le Selllimit_{b,w} \text{ for each user } w, \text{ and each block } b,$$

$$Aq \le d, \text{ some arbitrary hydrology constraints in generic matrix form.}$$

Note that if users refuse to sell, then $sell_{b,w}$ variables are all 0, and q_w could be too big, and the model could be infeasible. So implementers must establish some market rules to ensure that the market manager can operate the auction realistically, as discussed in Sect. 5.6.3.

Here is the net pool model for the two-user example:

$$
\begin{aligned}
&\text{Maximize } Abuy - 2Asell + Bbuy - 2Bsell, \text{ subject to}\\
&\text{ATotal}: \ Aq - Abuy + Asell = 10,\\
&\text{BTotal}: \ Bq - Bbuy + Bsell = 10,\\
&\text{Environment}: \ Aq + 0.1Bq \le 9,\\
&0 \le Abuy, \ 0 \le Asell \le 10,\\
&0 \le Bbuy, \ 0 \le Bsell \le 10.
\end{aligned}
$$

Observe that if Ann and Bob were only willing to buy, and not sell, the model would be infeasible.

As a technical point, variables Aq and Bq should not be further constrained with non-negativity constraints, to avoid degeneracy. To see this, observe that the buy and sell variables are non-negative, and the sell variables are bounded by the initial right. Therefore, the q variables are already guaranteed to be non-negative. Introducing $Aq \geq 0$ or $Bq \geq 0$ would then be redundant to the ATotal and BTotal constraints, so both could be binding, and the dual will have alternative optima, so the dual prices (which the market manager needs to clear the market) will be unreliable.

The objective coefficients can be interpreted as the marginal profit to the user for a change from the user's initial allocation. Buying is worth \$1/m^3, and selling is worth \$2/m^3, with different signs. Solving the model above, the solution is:

	Sell	*Buy*	*q*	Price
A	10	0	0	–\$10/m^3
B	0	80	90	–\$1/m^3

Since Ann sold 10 m^3 at \$10 each, and Bob bought 80 m^3 at \$1 each, the market manager has paid out a net of \$20. The catchment was over-allocated, and the environmental constraint required the market manager to bring down the rights to match the available water.

The maximum objective value in this case is \$60. This is the total market surplus for Ann and Bob. Bob's price is exactly his willingness to pay, so he breaks even on all the water he bought. Ann was paid more than her sell bid, so she gains 10 * (\$10 – \$2) = \$80. The remaining \$20 came from the market manager.

Since the market manager is not keen to lose money, and wants to be sure the model is feasible, the manager may want to adjust users' rights downwards. Let's see what happens when the initial allocations arbitrarily change from 10 to 8. Figure 7.2 shows the same value-for-water function, with the new initial right.

To adjust the above model for the new initial right, maybe the 10s could just be changed to 8s? No, that is not enough. Units 8–10 have flipped from *sell* quantities to *buy* quantities. The objective would give a value of only \$1 per unit, when they

Fig. 7.2 Example value for water, for both Ann and Bob, showing the initial right for 8 m^3

are actually worth \$2 per unit to each user. So the model requires two more *buy* variables, worth \$2 per unit. This is annoying and difficult to code correctly. The new model is:

$$\text{Maximize}: \ 2Abuy1 + Abuy2 - 2Asell + 2Bbuy1 + Bbuy2 - 2Bsell,$$
$$\text{subject to}$$
$$\text{ATotal}: \ Aq - Abuy1 - Abuy2 + Asell = 8,$$
$$\text{BTotal}: \ Bq - Bbuy1 - Bbuy2 + Bsell = 8,$$
$$\text{Environment}: \ Aq + 0.1Bq \leq 9,$$
$$0 \leq Abuy1 \leq 2, \ 0 \leq Abuy2, \ 0 \leq Asell \leq 8,$$
$$0 \leq Bbuy1 \leq 2, \ 0 \leq Bbuy2, \ 0 \leq Bsell \leq 8.$$

The result in this case is:

	Sell	*Buy*	*q*	Price
A	8	0	0	$-\$10/m^3$
B	0	82	90	$-\$1/m^3$

The total users' welfare is 68. User A gains 8(10−2), while user B gains 2(\$2 − \$1) on the first 2 units purchased and breaks even on the last 80 units purchased, with \$2 to the market manager. Net to the market manager = \$2.

Note two key points. First, the final prices and quantities are the same in the two examples. That is the Coase Theorem, which implies the users would not change their bids because of the initial allocations. (Though this may not be true, as discussed later in the chapter.) Second, adjusting the model for different initial rights was annoying. A net pool market will require the market manager to adjust the model in complicated ways whenever initial rights change.

7.2.4 The Gross Pool Formulation

Algebraically, the gross pool formulation can be given (ignoring time periods for simplicity) as follows:

$$\text{Maximize} \sum_{\text{users } w} \sum_{\text{bids } b} BidPrice_{b,w}\, bid_{b,w},$$

$$q_w = \sum_{\text{bids } b} bid_{b,w} \text{ for each user } w,$$

$$bid_{b,w} \le BidQty_{b,w} \text{ the quantity in bid block } b \text{ for each user } w,$$

$$Aq \le d, \text{ arbitrary hydrology constraints in generic form.}$$

The initial allocation does not appear in the linear programming (LP) formulation. After solution of the model, the market manager can use the initial allocation to calculate the net sales.

$$buy_w = \max(0, q_w - InitialAllocation_w) \text{ for each user } w.$$

$$sell_w = \max(0, InitialAllocation_w - q_w) \text{ for each user } w.$$

For matrix $A \ge 0$, as it is usually for groundwater, the gross pool formulation is always feasible. To find a feasible solution, then, the market manager can just set the $bid_{b,w}$ variables to zero. Users who do not want to sell would want to set their relevant $BidPrice_{b,w}$ coefficients to some huge value. If the catchment is over-allocated, the model is feasible, but the market manager could have to pay out an enormous sum. As with the net pool formulation, implementers must establish some market rules to ensure that the market manager can operate the auction realistically.

Here is the gross pool market model for our example:

$$\text{Maximize } 2 \cdot Abid1 + Abid2 + 2 \cdot Bbid1 + Bbid2, \text{ subject to}$$

$$\text{ATotal}: \ Aq = Abid1 + Abid2,$$

$$\text{BTotal}: \ Bq = Bbid1 + Bbid2,$$

$$\text{Environment}: \ Aq + 0.1Bq \le 9,$$

$$0 \le Abid1 \le 10, \ 0 \le Abid2,$$

$$0 \le Bbid1 \le 10, \ 0 \le Bbid2.$$

This model has the following solution:

	bid1	*bid2*	*q*	Price
A	10	0	0	−$10
B	0	80	90	−$1

The objective function's value is now \$100. So \$100 is the users' total value for water, ignoring their initial allocations. Ann gains nothing, as she used no water. Bob gains \$2 per unit on the first 10 units and \$1 per unit on the remaining 80. Of course, money changed hands during settlement, but the linear program was ignorant of the initial allocations.

In the gross pool market, the market manager can easily calculate buys and sells *after* solution. Ann's total quantity is 0, and her initial allocation was 10, so she sold 10. Bob's total quantity is 90, and his initial allocation was 10, so he bought 80. Net to the market manager is $-10 * \$10 + 80 * \$1 = -\$20$, just as before.

As in the net pool case, let's see what happens if the initial allocations change from 10 to 8. The model does not need to be touched at all! The settlement, how much each user has bought and sold, can be calculated at the end. Using the same gross pool model as above, we calculate the buys and sells, again after solution. The objective value is unchanged at \$100 because the market manager has not solved a new model. The reader should be able to calculate the trades now in your head, again finding that the market manager gained \$2.

In our examples above, the market manager either lost money or gained money. What initial rights would guarantee that the manager breaks even? Let us assume—while acknowledging that some people may not accept this—that users will behave the same in the gross and net pool markets. This means that their initial rights do not affect their values for water. Using the solution to the gross pool market, we can determine a break-even initial allocation.

To break even financially, the market manager needs $\$0 = \$10 * (0 - \mathit{AInitial}) + \$1 * (90 - \mathit{BInitial})$. In the name of fairness, let's scale Ann and Bob's initial rights equally, by a fraction α. Then we have a single equation to solve:

$$\$0 = \$10*(0 - 10*\alpha) + \$1*(90 - 10*\alpha) = 90 - 110*\alpha, \text{ so } \alpha = 90/110 = 0.818182.$$

The catchment is about 18.2% over-allocated. By reducing users' rights from 100% to 81.8%, the market manager and the users can feel confident that the market manager will break even.

7.2.5 Further Implications of the Gross Versus Net Decision

Our examples above indicate that the two formulations are the same, and that the market manager will prefer the gross pool formulation for its simplicity. The gross and net pool formulations are the same mathematically. This is no surprise, because the Coase Theorem says that initial rights do not affect the optimal allocation. As long as the market manager has the authority to scale users' rights (to ensure feasibility of the net pool model), the net and gross pool models will give identical results, given the same bids and the same initial rights. But would users want the gross pool design, and would they bid the same in the two market designs?

Users may not want a gross pool market for two important reasons. First, the display used with gross market bidding is likely to confuse users. If users see the raw results of the optimization, they may be confused about how much water they can use and what they paid for; the market manager will have to translate the solution into a display that bidders can understand. Of course, this is a relatively simple problem, to be solved with a good user interface and training.

Second, more seriously, users will often want to hold back initial rights from sale. Forcing them to put all their allocation up for sale is likely to cause outright fear. Users may argue that the gross pool market does not really give them all their rights, because it allows all water for sale, as in an "all-in auction," as described by Zetland (2013). Zetland's all-in auction is a gross pool market. That market is compulsory, because the market manager forces users to allow their water to be allocated elsewhere. To quote Zetland:

> All-in-Auctions (AiAs) maximize the reallocation of water or water rights among water users to direct water to its highest economic and social uses. AiAs force water rights holders to put their water up for sale, a weakening of property rights that does not have a material impact on owners due to their ability to buy back their own rights without concern for price or liquidity—meaning, in other words, that the AiA nests the status quo (no trade) within its potential set of outcomes, less trivial transaction costs. The benefits from reallocation and value discovery in the AiA more than offset the small costs of operating the AiA or inconvenience to participants who merely want to buy back their water.

In general terms, a gross pool smart market need not be an "all-in auction," because some users could choose not to participate. The gross pool market can guarantee that trading will be active. However, users may argue that the market would incentivize all beneficial trading, without an "all-in" auction that requires all users to bid for allocation in each round.

Market designers' choice between a gross or net pool design should also consider the likelihood that water users will not be "rational" economic optimizers. For several decades, economists have used laboratory experiments to test for "endowment effects," where a seller demands a higher price than the "rational" selling price. Endowment effects may be stronger with water than with other commodities, such as fertilizer, because water rights are often associated with land that was inherited, and because a water right might have been received as a gift from agents in government or as an allocation by a community organization with cultural ties (Zetland 2013).

The question is whether users will bid differently when they know they start from a known water allocation (the endowment) versus when they receive no water if they do not bid. Kahneman and Tversky (1979) found compelling evidence that people demand more to sell an initial allocation than they are willing to buy the same good. Other researchers seemed to find the same thing, until Plott and Zeiler (2005, 2007). These authors ran careful experiments finding no significant endowment effect, contradicting the previous literature. They concluded that previous findings of an endowment effect might have been because of the experimental procedures, such as the subject's familiarity with the rules of the game. Sprenger (2015), recognizing the cautions raised by Plott and Zeiler, conducted experiments finding endowment effects associated with risk. Zetland's 2013 paper concludes that increasing market participation is important when efficiency results in positive externalities, as it does with water: market prices incentivize users to use water more efficiently, which frees up water for others and the environment, and the more market participation there is, the greater this effect is. Perhaps more research is needed to test whether the endowment effects are likely to be large in this specific context, in case the market needs a mechanism to address those effects.

Implementers could choose a middle ground where users who take volumes above a certain amount are required to be all in, and smaller users are not. This will cause the market manager some grief with the modeling, though.

We discuss initial rights some more in Chap. 8. For simplicity, we usually use the gross market formulation in this book, while not assuming an all-in design.

7.3 Model Formulation for a Basic Reservoir

Reservoir planning has a long history (Read 1979; Read and Hindsberger 2010; Lund and Ferreira 1996).The smart market optimization model for the case of a basic reservoir is relatively simple, though it would have specific features that merit explanation.

In this case, all users take directly out of a single reservoir. Figure 7.3 is a diagram of the associated catchment. The reservoir has a single inflow and a single outflow. The environmental requirements in this case are typically for the outflow, to satisfy minimum river flows downstream. The reservoir also has physical limits: upper and lower bounds on storage. Usually, to model the net water availability as starting from zero, the operations researcher can subtract the lower bound from the upper bound, so the model needs only the adjusted upper bound. The upper bound on storage imposes a requirement on the market manager to ask users about the future value of water. Thus, the bid structure contains bids for the current period and bids for some set of future periods. The auction schedule depends on how often people wish to trade and the size of the reservoir. Users will need to make choices relatively more often with a small reservoir (relative to total demand) than with a large one.

In general, reservoirs provide value to users in two ways. First, reservoirs allow users to transfer water from the inflows in one period to meet demand in a later period. This value does not depend on an inability to perfectly forecast future water input—users know that they need to hold water from the wet season to the dry season. Second, reservoirs allow users to obtain water in situations where the forecast is wrong. This looks like the same thing as a transfer from a wet time to a dry time, which it is, but the value in this second case is because the dry period was drier than expected. Unfortunately, a deterministic model can sometimes underestimate the value of storage, because, with perfect foresight, people would

Fig. 7.3 Diagram of a catchment with a single reservoir

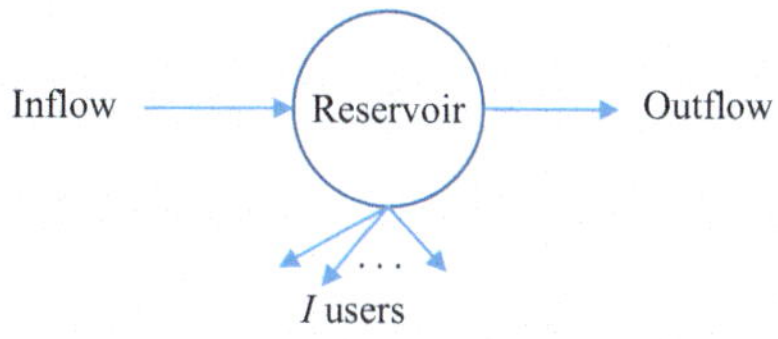

select crops that ideally match the anticipated inflows, but under uncertainty, people need storage to protect their crops against the risk of an incorrect forecast. Chapter 11 explores this further.

In addition to the terms defined in Chap. 3, some more terms are needed for the case of the reservoir.

Parameters, Title Case

$Inflow_t$, m^3, the estimated natural inflow in period t. The market manager assumes this inflow is a constant, unaffected by decisions within the market. This parameter includes inflows from the catchment above the reservoir and from expected rainfall on the reservoir, minus evaporation that occurs in the reservoir. In dry periods, $Inflow_t$ could be negative due to low inflows and high evaporation. The inflow must occur at a rate commensurate with users' abstractions. If the inflow arrives very late in the market period, after users have taken their allocations on the assumption that the inflow would have already occurred, the reservoir could run short. A simple solution to this potential problem is to run the auction more often, with smaller allocations and more frequently updated values of actual and predicted inflows.

$InitialVol$, m^3, quantity of water in the reservoir at the beginning of the current period.

$Capacity$, m^3, maximum quantity of available water in the reservoir.

$MinOutFlow_t$, m^3, the minimum required outflow in week t, typically to ensure flows in the downstream river.

$MaxOutFlow_t$, m^3, the maximum allowed outflow in week t, such as to help avoid downstream flood.

Decision Variables, Lower Case

$bid_{b,w} \geq 0$, m^3, the amount of bid block b accepted from each user's water intake w in period t.

$outflow_t$, m^3, the outflow in period t.

$price_t$, \$/m^3, the market-clearing price in week t. This is the dual variable on the reservoir balance constraint for period t in the model below. The market manager could just as well use the price from the user allocation constraints; these prices should be equal.

$volume_t$, m^3, the quantity of available water in the reservoir at the start of week t.

Market-Clearing Model for a Simple Reservoir

(1) The objective maximizes the sum of each user's bid price times the amount of accepted bid over all users at all water intakes and all periods in the auction horizon.

$$\text{Maximize} \sum\nolimits_{\text{bids } b} \sum\nolimits_{\text{water intakes } w} \sum\nolimits_{\text{periods } t} BidPrice_{b,w,t}\ bid_{b,w,t} \ \text{ subject to}$$

(2) Each user has a set of constraints, associated with each bid block and each time period. The accepted bid quantity is less than the total quantity in the user's bid block.

$$0 \le bid_{b,w,t} \le BidQty_{b,w,t} \text{ for each block } b, \text{ well } w, \text{ and period } t.$$

(3) The total allocation is the sum of the accepted bids.

$$q_{w,t} = \sum\nolimits_{b=1}^{B} bid_{b,w,t} \text{ for each well and period } t.$$

(4) Reservoir flows must balance.

$$volume_0 = InitialVol,$$

$$volume_{t-1} - \sum\nolimits_{\text{water intakes } w} q_{w,t} - outflow_t + Inflow_t, = volume_t \text{ for each period } t$$
$$= 1, \ldots, T. \text{Dual price } price_t.$$

(5) Reservoir volume cannot exceed capacity.

$$0 \le volume_t \le Capacity \text{ for each period } t.$$

(6) The reservoir outflow cannot flow in reverse. Market implementers should double-check whether flows always move in one direction. For example, pumping can sometimes intentionally cause backward flows, and these backwards flows should be included in a model formulation.

$$outflow_t \ge 0 \text{ for each period } t.$$

(7) The outflow must be at least the required minimum.

$$MinOutFlow_t \le outflow_t \le MaxOutFlow_t \text{ for each period } t.$$

Some Observations

To construct this linear program, the market manager needs only few data compared with what would be required in other hydrological cases. The optimization model is trivial. Forecasting the inflows will be the most difficult part of planning the model.

The market price is the dual price, $price_t$ on constraint 4. The market manager can take prices from either constraint 3 or constraint 4. In each period t, all users using the reservoir face this same $price_t$.

The market manager can easily scale users' allocations—simply apportion the inflow to users following some agreed-on formula, such as by proportion or by seniority. However, after rescaling, the price may change because of the influence of the reservoir capacity constraint and the outflow requirement.

This market model allows bids for non-consumptive use, in particular a hydroelectricity generator that uses the release to generate revenue (see Sect. 6.12). To recapitulate, a generator would bid in the auction as any other user. However, it would be concerned with timing of releases on a roughly half-hourly basis, as it interacts with the electricity market. In contrast, the water market is likely to trade on a monthly or weekly basis. Thus, a generator would factor in the weekly price of water into its half-hourly management. The market manager may need to include some constraints on the generator's releases. For example, a release schedule of "low releases all week except for a 1 h surge" may have different downstream effects compared with uniform releases all week. Agreements by hydroelectricity generators to release in the future would likely have to be construed as commitments, not options, at least to the extent that those releases were required for downstream environmental flows.

The market does not explicitly value the storage after period T, so the price for water will be wrong to the extent that the value of water in the final period is wrong. If users over-value the water allocated after 12 months from now, then today they will take less than the most efficient amount. If users under-value the water allocated 12 months from now, they will take more than the most efficient amount.

A partial solution is to have a rolling horizon. If the market runs monthly, then next month the uncertain future price is now 11 months away; the market manager adjusts the price based on the previous price for that block of future water as well as new information on water flows. As the auction date advances, the future price comes closer, so the prices should become more properly valued.

Still, current bidding could under-value future water to the point that the market manager plans for the reservoir to run dry 12 months ahead, losing the ability to allocate water to more productive uses. The market is likely to benefit if the manager limits the risk of a future dry reservoir.

The market manager could add a constraint requiring some minimum final quantity in the reservoir, or the manager could enter a bid in period T implying a future (and less risky) value for water. The manager could use a combination, e.g., a fixed minimum, with a bid for additional water. In either case, the manager will need some separate analysis, or market rules, to guide the decision. The market manager could base the final quantity or value on the quantity and prices at the end of the previous year, or on a more complex analysis that estimates users' values for the following year.

This market model does not account for the fixed costs of storage. Users would have to pay for the maintenance of the reservoir through some other payment mechanism. Nevertheless, the ability to trade changes the value of infrastructure. Implementers might start the market more quickly and more cheaply than infrastructure can be built. Existing infrastructure may be found either to be highly valuable because it enables trade, as when a reservoir allows excess water from a wet period to be used in a dry period, or the infrastructure may be found nearly worthless when the new water allocation system results in the infrastructure becoming unused.

The shadow price on the capacity constraint 5 indicates the value of additional capacity. This shadow price would convey the value to the market participants for an increase in the reservoir capacity. This may be useful for planning. The price for water will signal when the reservoir should release water. The marginal value for reservoir capacity signals whether the storage capacity should be increased or decreased. Most of the time, the reservoir will be below its capacity, and the shadow price of capacity will be zero. For example, when a reservoir has capacity of 5 ML, and currently holds 2 ML, increasing the capacity in the current periods would gain nothing, since the current level is already below capacity. When it does reach its capacity, the shadow price on the capacity will become positive—an increase in capacity would improve the objective. Further, when this happens, the market manager's money will no longer balance. Market rules should specify how the market manager disburses excess money. The rules could require that the money be used for maintaining the reservoir, for market operations, for the government's general purse, or that the manager returns it to users. We discuss capacity rights further in Sect. 10.5.

These questions about infrastructure raise a higher-level market design issue: who owns the reservoir? The market model above assumes that the market manager has full control over the reservoir, so the reservoir is publicly owned, or at least the private owner is willing to let the market manager control it. Consider the three possibilities: public ownership with public market operation, private ownership with public market operation, and private ownership with private market operation.

Given public ownership, the market manager is the logical controller of the reservoir. All constraints associated with the reservoir should appear in the market model. The reader can refer back to Sects. 6.10 and 6.11 for further discussion on the issue of private versus public operation of a smart market for water.

With private ownership of a reservoir and private market operation, the private owner cedes management of the reservoir to the market manager. All constraints associated with the reservoir will occur in the market model. This is the case when hydroelectric generators operate in modern electricity smart markets. With a privately owned reservoir, the market could plausibly be operated privately, without government, with the reservoir operator running the market-clearing model. The regulator then need only check that outflows met the environmental requirements, and perhaps respond to any users' complaints that the reservoir owner might be exercising monopoly power. This private–private combination can work, but it implies a private system of water infrastructure and provision that has no externalities outside the market—an unlikely situation even with a reservoir because of the likely effect of discharges of water by users or the reservoir operator.

A private reservoir owner may not wish to give the market manager control over the reservoir's operation, particularly in the case with a private owner of the reservoir and a public owner of the smart market. In this case, where a simple reservoir is supplying multiple users, the market model becomes almost vacuous, because the reservoir variables (such as release rates) and constraints are not under the market manager's purview, and so they should not be part of the market-clearing optimization model. The inter-temporal constraints (for example, constraints that require release when water would exceed reservoir capacity) would be entirely the purview of the private reservoir owner. The market would not be able solve for allocations of future water that met constraints on water availability when the constraints are not certain from the viewpoint of the market operator. As a result, users will want to make contracts for water with the reservoir owner to ensure their future water.

The single reservoir market-clearing model is especially suitable for stochastic programming. We discuss this further in Chap. 11. Briefly, the market manager could use a variation of the above formulation to account for users' risk. For a simple catchment with a single reservoir, we recommend that the implementers use a stochastic model instead of the above deterministic model. Interestingly, for a large complex catchment where the market manager wishes to have a stochastic market-clearing model, having a private reservoir owner control their own reservoir outside of the market-clearing model would simplify the market model (see Sect. 6.3).

7.4 Model Formulation for Surface Water Systems Without Storage

In this case, all users take directly out of a catchment that has no storage. Figure 7.4 is a diagram of the catchment. The catchment has a tree-like network of fast-moving streams that converge into larger and larger streams, and finally to a single river with one outlet. The outlet could flow to a larger river or to a lake or ocean. The stream network can be described as a graph, with arcs for river reaches, and nodes for confluences and points of interest. This tree graph structure is reflected in the market-clearing model. In each period, each reach has an inflow that depends on local land area, land use, weather, aquifers, and other factors. Users take water at points on the network. Environmental requirements are typically to satisfy minimum flows on selected reaches of the stream network. The regulator may define these flows as average flows at the middle of a reach, or as average flows at nodes.

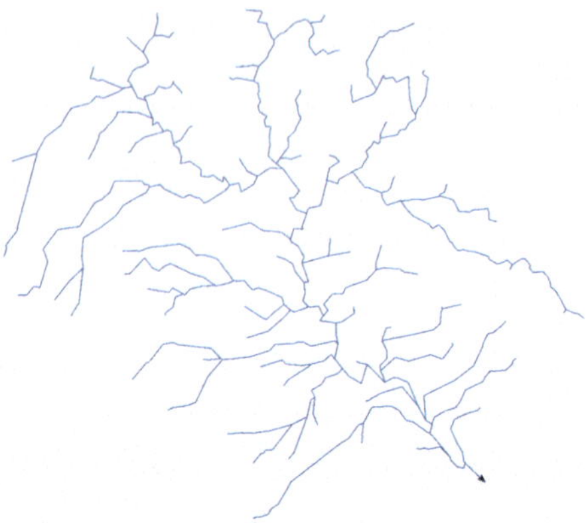

Fig. 7.4 Diagram of a river catchment with no storage. The outlet is the arrow at bottom right

In a small stream network, the bid structure contains bids only for the current period. Variables and parameters in the market-clearing model therefore need no subscript for time, and the market manager can update these outside of the model. In a large river network, a big storm could take days to flow downstream, so the bid structure may contain bids for the next few days. The auction schedule depends on how often people wish to trade and the rate of change in inflows over time. Users will need to make choices relatively more often with highly variable inflows than with less variable inflows. Of all hydrological cases, this one is most like an electricity network.

The following formulation is similar to the CALVIN model (see Sect. 3.4.3), with bid information added.

Indices and Sets

b, bid block, $b = 1, \ldots, B$.

Set *Nodes*: river junctions, nodes in the network, with indices i, j, k.

Set *AssessPoints*: within *Nodes*, the set of nodes at which environmental flows will be measured and constrained.

Set *Headwater*: within *Nodes* = {i in *Nodes* such that degree[i] = 1 and i not in Mouth}. These nodes are the most distant headwater source in the river network.

Set *Mouth*: within *Nodes*, the set of nodes that drain naturally to the outlet. Usually, a catchment has only one.

Set *Reaches*: (i, j) within *Nodes* cross *Nodes*, arcs in the river network.

$w = 1$ to W, water intakes.

Set *WaterIntakeArcs*: (i, w) within {*Nodes* cross *WaterIntakes*}. Arc (i, w) goes from river node i to water intake w. Each WaterIntakeArc (i, w) has exactly one water intake w. However, river node i could supply multiple water intakes.

Parameters, Title Case

$BidPrice_{b,w,t}$, \$/m^3, the price that the user at water intake w submits for bid block b, for each period t.

$BidQty_{b,w,t}$, m^3, the quantity that the user submits for water intake w and bid block b, for each period t.

$Capacity_{i,j}$, m^3, the maximum capacity on reach (i, j). We assume that water is generally in short supply, so a natural reach (i, j) is unlikely to reach its capacity in normal market operations. In near-flood conditions, downstream users might want to pay for upstream users to take water, but we will ignore this case. However, when reach (i, j) is a constructed waterway, users may wish to take especially large amounts of water along it, and the flow could come to the reach's capacity.

$Gain_{i,j,t}$, % gain on river reach (i,j) in period t. This parameter would be 1 for no loss or gain on the arc. It would be less than 1 for loss, such as from evaporation from the surface, or water use by streambank vegetation (e.g., willows).

$Inflow_{i,t}$, m^3, the estimated natural inflow to node i for each period t.

$MinimumFlow_{i,t}$, m^3, the minimum regulatory flow for assessment point $i \in AssessPoints$, in each period t. We have defined this constraint based on a node; requiring a minimum flow on a reach is simple.

$NaturalInflow_{i,t}$, m^3, the natural inflow for assessment point $i \in AssessPoints$, in each period t.

Decision Variables, Lower Case

$bid_{b,w,t} \geq 0$, m^3, the amount of bid block b accepted from water intake w in period t.

$flow_{i,j,t} \geq 0$, m^3, flow from node i to node j on reach (i, j) in period t.

$price_{w,t}$, \$/m^3, the price to charge user w in period t. This is the dual variable on constraint set 3 in the model below.

$q_{i,w,t}$, m^3, the allocated water flowing from river node i to a consumptive user at water intake w, in period t.

$q_{(i,j),w,t}$, m^3, the water allocated to a non-consumptive user at water intake w, flowing across reach (i,j), in period t.

Market-Clearing Model for a Fast Stream Network Without Storage

(1) The objective maximizes the sum of users' bid prices times the accepted bid quantities.

$$\text{Maximize} \sum_{\text{bids } b} \sum_{\text{water intakes } w} \sum_{\text{periods } t} BidPrice_{b,w,t}\, bid_{b,w,t} \text{ subject to}$$

(2) Each user's bid quantity must be less than the quantity in the user's bid block.

$$bid_{b,w,t} \leq BidQty_{b,w,t} \text{ for each block } b, \text{ water intake } w, \text{ for each period } t.$$

(3a) For consumptive users, allocation equals total bid.

$$q_{(j,w),t} = \sum\nolimits_{\text{bids } b} bid_{b,w,t} \text{ for each water intake } w, \text{ for each period } t, \text{ dual price } price_{w,t}.$$

(3b) For non-consumptive users, allocation equals total bid.

$$q_{(i,j),w,t} = \sum\nolimits_{\text{bids } b} bid_{b,w,t} \text{ for each water intake } w, \text{ in each period } t, \text{ dual price } price_{w,t}.$$

(4) Each river reach has some forecasted inflow, and the flow at every point must balance.

$$Inflow_{j,t} + \sum\nolimits_{\text{arcs } i,j:i \in Headwater} Gain_{i,j,t}\, flow_{i,j,t} - \sum\nolimits_{\text{water intakes } w} q_{(j,w),t}$$
$$= \sum\nolimits_{\text{arcs } j,k} flow_{j,k,t} \text{ for each node } j \text{ and period } t.$$

(5) Flow at each assessment point is above the required minimum. The model contains a constraint for each assessment point, for every period in the optimization horizon. These constraints ensure that all trades respect the environmental requirements. The min() function on the right-hand side ensures that the model remains feasible (possibly requiring all $q_{(i,w),t} = 0$ in the extreme) when the natural inflows are lower than the regulatory minimum required flows.

$$Inflow_{j,t} + \sum\nolimits_{\text{arcs } i,j:i \in Headwater} flow_{i,j,t} - \sum\nolimits_{\text{water intakes } w} q_{(j,w),t}$$
$$\geq \min(MinimumFlow_{j,t},\ NaturalInflow_{j,t}) \text{ for each assessment point } j,$$
$$\text{in each period } t.$$

(6) The flow on reach (i, j) must not exceed the capacity.

$$0 \leq flow_{i,j,t} \leq Capacity_{i,j,t} \text{ for all reaches}(i,j), \text{ in each period } t.$$

(7) For non-consumptive users, the flow along the user's reach must be at least the user's accepted bid.

$$flow_{i,j,t} \geq \sum\nolimits_{w \in (i,j)} q_{(i,j),w,t} \text{ for all reaches}(i,j), \text{ in each period } t.$$

(8) Bids and flows cannot be negative.

$$bid_{b,w,t} \geq 0, flow_{i,j,t} \geq 0.$$

In some cases, flows can be negative. Market implementers should double-check whether flows always move in one direction. For example, pumping can sometimes intentionally cause backward flows, and these backwards flows should be included in a model formulation.

Some Observations

To construct this linear program, the market manager should start with a good vector map. The structure of the map data will provide definitions for reaches and nodes. Setting up the optimization from this map data may require cleaning up the map data first, but is straightforward using a scripting language and algebraic modeling software, as the from–to structure of the map objects translates directly to from–to objects in the optimization model. The model is trivial to solve.

Forecasting the inflows will be the most difficult part of planning the model. A simple approach might take measured rainfall at various places in the catchment, measured stream flow at a limited number of places, and then apply a simple run-off model to estimate stream inflows at various points. With a smart market in place, the market manager could find that water users would welcome the use of more-sophisticated methods, including greater collection of data on slope, soil types, soil moisture, and temperature and more-detailed stream flow and rainfall data. With good meter data from users, along with basic hydrological data, this model should be easy to calibrate because of the quick response in catchments dominated by surface water.

For each user w and period t, market prices are set by $price_{w,t}$ from constraint 3. Every user w could face a different price in every period t. Since constraint 3 is an equality, and the variable $q_{(i,j),w,t}$ is free, $price_{w,t}$ is a weighted sum of the prices on the other constraints. Thus, this structure has nodal prices as in an electricity market.

Users may think they are trading inflow in constraint 4, but they are also trading the right-hand sides of constraints 5 and 6. Scaling of users' allocations is much more complicated than in the case of the reservoir. We take up this problem in Chap. 8.

Like the simple reservoir case, this market model allows bids for non-consumptive use, such as a hydroelectricity generator that uses the release to generate revenue, or a gravel washing facility that uses water and then returns it. We discuss a hydroelectricity generator's likely behavior in Sect. 6.12. To recapitulate, a generator would bid in the auction as any other user. A generator would factor in the daily or weekly price of water into its half-hourly management of its role in the electricity market. The market manager's agreements with non-consumptive users are commitments to provide flow across the reach, but the users would have no requirement to use the water, as long as the flow passed downstream with little hindrance.

Some "semi-consumptive" water users return significant fractions of water to the same catchment at a much later time. The market can address delayed returns in at least two ways. The user could simply buy water when abstracting and sell water when returning; the market manager would price the two separately. Or the market manager could introduce coefficients similar to the groundwater response matrix that reflect semi-consumptive use, and then the market manager plans and prices abstraction and return simultaneously. This issue is similar to the issue of providing credits for water use that results in groundwater recharge, and is discussed in more detail in Sect. 6.8.

Because this market fully clears all its water in every period, it does not have the end-of-horizon problem that the reservoir market has. The downside is that prices are likely to vary wildly.

Some reaches could be constructed waterways. As with all the markets proposed in this book, this market model does not account for the fixed costs of infrastructure. Users would have to pay for maintaining infrastructure through some other payment mechanism. This shadow price would convey the value to market participants for an increase in the capacity of a constructed waterway, e.g., by widening or relining an aqueduct, though the rapid changes in the shadow price that one might expect will complicate such an analysis a bit.

As with the reservoir market, the ability to trade changes the value of infrastructure in this case. Policy makers could implement trading more quickly and more cheaply than they could build infrastructure. On the other hand, stakeholders may find existing infrastructure to be highly valuable because it enables trade, as between periods for the reservoir, or nearly worthless, stranding assets and potentially disrupting communities. In these cases, the market would be telling society that water was misallocated in the first place, and the policymaker should protect the water trades, while mitigating community damage separately.

The capacity constraint associated with infrastructure capacity (say, the aqueduct capacity) can generate rents when it becomes binding, so the market manager's revenue will not balance. In theory, this excess revenue would go to the infrastructure owners. The market implementers will have to consider who owns the infrastructure, and develop a formula to disburse the extra money. Eventually, users may want to trade the right to use the infrastructure, as fractions of the available capacity. An infrastructure owner may wish to lease capacity to parties, giving them security of conveyance. We discuss capacity rights further in Sect. 10.5.

The issue of a shadow price on the water conveying capacity extends to the case where the catchment has no constructed infrastructure. Various stakeholders could increase the capacity of a river to convey water (particularly in a drought) by dredging it or by removing plants along the riverbanks that transpire large amounts of water. The shadow price on the capacity constraint 6 suggests the value of additional capacity and can help evaluate how much effort relevant stakeholders should expend on improvements in water availability. These cases will be more complicated, because the changes that can help increase capacity of a waterway might not always help society to achieve other goals with river management. Dredging the river could improve navigation, while creating seasonal wetlands alongside rivers could increase capacity and improve biodiversity (Hey and Philippi 1999). Removing native trees along a riverside might hinder biodiversity goals of society. A smart market would help society to decide whether water users should contribute to the costs of improving capacity or whether society should compensate users when initiatives decrease the capacity to convey water for productive use.

Most likely, developing a stochastic optimization of a smart market model for surface water use without a reservoir would have little value, because the market manager should know inflows in the current period with high confidence, and because the model has no recourse in the case of uncertainty. However, if the inflows vary together, as for a small catchment, developing the stochastic optimization should be relatively easy, if the market manager finds use for it.

7.5 Model Formulation for Surface Water Systems with Reservoirs

This case combines the above two cases, with a general stream network and any number of reservoirs, though without groundwater. Like the stream network, this catchment has a tree-like network of streams that converge into larger and larger streams, with reservoirs at various points on the network. Often, the reservoirs will be in series along a major river. Users take water at points on the network, both from streams and from reservoirs. Environmental requirements are typically to satisfy minimum flows on selected reaches of the stream network. Figure 7.5 shows an example schematic of a river network with four reservoirs.

Users high in the catchment without storage above them have only the current inflows. They can bid for future water, but the market manager does not guarantee that water. Users at or below a reservoir can bid for water from inflows or from the reservoir. Much of the discussion from the above two cases applies here.

Indices and Sets

b, bid block, $b = 1, \ldots, B$ (e.g., 5). To each user, this corresponds to 5 input text boxes of bid quantities and 5 input boxes of bid prices, for each time period.

Set *Nodes:* i, j, k, points of relevance in the stream network.

Set *AssessPoints*: within Nodes; the set of nodes at which the regulator will measure and constrain environmental flows.

Set *Headwater*: within Nodes = $\{i$ in Nodes such that degree$[i] = 1$ and i not in Mouth$\}$. These nodes are the most distant headwater source in the river network.

Set *Mouth*: within Nodes; the set of nodes that drain out of the catchment.

Set *Reaches*: within Nodes cross Nodes, (i, j); arcs in the river network.

Set *Reservoirs*: with Nodes, I; nodes that are reservoirs.

$t = 1, \ldots, T$, time periods.

$w = 1$ to W, water intakes, i.e., all abstraction points.

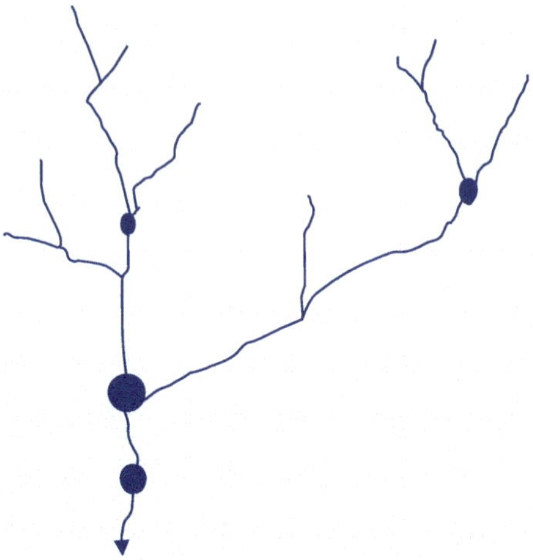

Fig. 7.5 Diagram of a catchment with four reservoirs represented by ovals. The outlet is at the bottom

set WellArcs within {Nodes cross Water intakes}: arc (i, w) goes from river node i to abstraction point w. Each arc (i, w) has exactly one abstractor w. However, river node i could supply many abstractors.

Parameters, Title Case

$BidPrice_{b,w,t}$, \$/m^3, the bid price that a user submits for water at the water intake w for bid block b in period t.
$BidQty_{b,w,t}$, m^3, the quantity that a user submits for water at the water intake w for bid block b in period t.
$Capacity$ = maximum inventory (capacity) in the reservoir at node i.
$Gain_{i,j,t}$ = % gain on river reach (i,j) in period t. This parameter would be 1 for no loss or gain on the arc. It would be less than 1 for loss, such as from evaporation or transpiration.
$Inflow_{i,t}$, m^3, the estimated natural inflow to node i on period t.
$InitialLevel_i$, m^3, initial inventory in the reservoir at node i.
$MinimumFlow_{i,t}$, m^3, the minimum regulatory flow for assessment point $i \in AssessPoints$ in period t.
$NaturalInflow_{i,t}$, m^3, the natural inflow for assessment point $i \in AssessPoints$ in period t.
T is the number of periods in the optimization horizon.

Decision Variables, Lower Case

$bid_{b,w,t} \geq 0$, m^3, the amount of bid block b accepted from water intake w in period t.
$flow_{i,j,t} \geq 0$, m^3, flow from node i to node j on Reach (i, j), on period t; m^3/period.
$volume_{i,t} \geq 0$, m^3, volume of water in reservoir i in period t.

$price_{w,t}$ = \$/m^3, the price to charge a user at water intake w in period t. This is the dual variable on constraint set 3 in the model below.

$q_{(i,w),t} \geq 0$, m^3, the allocated water flowing from river node i to abstraction node w in period t.

Market-Clearing Model for a Stream Network with Storage

(1) The objective maximizes, over all users and all periods in the auction horizon, the total of each user's bid price times the amount of bid to be accepted.

$$\text{Maximize} \sum\nolimits_{\text{bids } b} \sum\nolimits_{\text{water intakes } w} \sum\nolimits_{\text{periods } t} BidPrice_{b,w,t}\ bid_{b,w,t} \text{ subject to}$$

(2) Each user's accepted bid quantity is less than the quantity in each of the user's bid blocks.

$$bid_{b,w,t} \leq BidQty_{b,w,t} \text{ for each block } b, \text{water intake } w, \text{and period } t.$$

(3) The total allocation is the sum of the accepted bids for each user.

$$q_{(j,w),t} = \sum\nolimits_{b=1}^{5} bid_{b,w,t} \text{ for each water intake } w \text{ and period } t, \text{ dual price } price_{w,t}.$$

(4) Each river reach has some forecasted inflow, and the flow at every point must balance.

$$Inflow_{j,t} + \sum\nolimits_{\text{arcs } i,j:i \in Headwater} Gain_{i,j,t}\, flow_{i,j,t} - \sum\nolimits_{\text{water intakes } w} q_{(j,w),t}$$
$$= \sum\nolimits_{\text{arcs} j,k} flow_{j,k,t} \text{ for each node } j, \text{period } t.$$

(5) Regulatory limits on minimum flows. Flow at assessment points are above the required minima. The model contains a constraint for each assessment point, for every period in the optimization horizon. These constraints ensure that trades respect environmental requirements. The min() function on the right-hand side ensures that the model remains feasible (possibly requiring all $q_{(j,w),t} = 0$ in the extreme) when the natural inflows are lower than the regulatory minimum required flows.

$$Inflow_{j,t} + \sum\nolimits_{\text{arcs } i,j:i \in \text{Headwater}} flow_{i,j,t} - \sum\nolimits_{\text{water intakes } w} q_{(j,w),t}$$
$$\geq \min(MinimumFlow_{j,t}, NaturalInflow_{j,t})$$
$$\text{for each assessment point } j \text{ and period } t.$$

(6) Reservoir flows must balance.

$$volume_{i,1} = InitialLevel_i,$$
$$volume_{i,t-1} + Inflow_{i,t} + \sum\nolimits_{\text{nodes } j} flow_{(j,i),t} - \sum\nolimits_{\text{intakes } w} q_{(i,w),t} = volume_{i,t}$$
for each reservoir i, and each period $t = 2, \ldots, T$.

(7) Reservoir volume cannot exceed capacity.

$$volume_{i,t} \le Capacity_i \text{ for each period } t.$$

(8) For non-consumptive users, the flow along the user's reach must be at least the user's accepted bid.

$$flow_{i,j,t} \ge \sum\nolimits_{w \in (i,j)} q_{(i,j),w,t} \text{ for all reaches}(i,j), \text{in each period } t.$$

(9) Bids and flows cannot be negative.

$$bid_{b,w,t} \ge 0,\ flow_{i,j,t} \ge 0,\ volume_{i,t} \ge 0.$$

In some cases, flows can be negative. Market implementers should double-check whether flows always move in one direction, because pumping can sometimes intentionally cause backward flows, which the model may need to address.

Some Observations

As with the stream network, the market manager should use a good vector map to start constructing this model. With many users and many reaches in the river network, this model may take some minutes to compile from data into the optimization matrix, but solution should be fast. With good water meter data, this model should be easy to calibrate.

The market price for user w and period t is $price_{w,t}$ from constraint 3. Every user could face a different price from other users, and a user at their intake w could face a different price in every period.

What is traded? Users may think they are trading inflow in constraint 4, but they are also trading the right-hand sides of constraints 5, 6, and 7. Scaling of users' allocations is much more complicated than in the case of the reservoir. We take up this problem in Chap. 8.

This market model allows bids for non-consumptive use. We discuss a hydroelectricity generator's likely behavior in Sect. 6.12. The market manager's agreements with non-consumptive users are commitments to provide flow across the reach, but the users would have no requirement to use the water, as long as the flow passed downstream with little hindrance.

As with the single reservoir case, this market has an end-of-horizon problem, which the market manager will have to address in some way.

As with the stream network without storage, the market manager could have an ability to manage the infrastructure (or the water-conveying capacity for natural sections of the system). As before, this market model does not account for the fixed costs of infrastructure; users could expect to pay for maintaining infrastructure through some other mechanism, though the shadow price on constraint 7 suggests the marginal value of capacity.

As with the single reservoir case, the capacity constraints can generate rents when they become binding, so the market manager's revenue will not balance. In theory, this excess revenue would go to the infrastructure owners. The market implementers should consider who owns the infrastructure, and could develop a formula to disburse the extra money. Eventually, users may want to trade the right to use the infrastructure. The market manager could trade these rights as fractions of the available capacity. An infrastructure owner may wish to lease capacity to parties, giving them security of conveyance.

Developing a stochastic optimization of this general case would be quite difficult. We discuss this further in Chap. 11.

7.6 Model Formulation for Groundwater Systems Without Surface Water Intakes

In this case, developed further elsewhere (Raffensperger and Milke, 2005; Raffensperger, Milke and Read 2009), all users take water out of groundwater, from aquifers. Groundwater provides an opportunity and a challenge for common-pool smart markets. It is an opportunity because, even more than for surface water, groundwater is a common resource and the regulator should manage all abstractions simultaneously. A market for surface water flowing along a canal might have only six different water users, and users can trade one-to-one without considering where they take their water, the impact on other users, or the environment. In contrast, one user's groundwater abstraction affects other users; the impact varies with distance from the abstraction point. A one-to-one trade is not possible without a detailed assessment of the effects. A market for groundwater needs to be a common pool market and, because water flow is complicated, it needs to be a common-pool smart market.

Groundwater flows underground in complicated and hidden ways. An aquifer may be viewed crudely as a reservoir with upper and lower bounds on storage, but the quantity in storage is not readily measurable, and the flows can be more complicated than for an open body of water. Storage logically implies that the bid structure has bids for current and future periods. The bid structure should also account for recharge, as we discuss in Sect. 6.8.

The bid structure in this market design should also account for recharge, as discussed in 6.8. That is, while the market manager does not restrict trade based on user type, e.g., flood irrigator or drip irrigator, the market manager needs to know a user's type to correctly calculate the effects of the user's abstraction. The user should have registered the use type with the market manager. If the user wants to adjust allocations between types at every auction, the market manager should be able to make this adjustment via the bidding mechanism. The market manager could require the user to note the type of use on the bid page, or to register in advance as a certain type, with agreed recharge parameters for each type of use.

The auction schedule depends on how often people wish to trade and the rate of change in inflows over time, but also on characteristics of the groundwater and how quickly and predictably it responds to changes. Users will need to make choices relatively more often with a fast-reacting aquifer (relative to total demand) than with a slow-reacting one. The bid structure might also account for recharge (see Sect. 6.8).

An operations researcher without knowledge of hydrogeology may find the network-oriented models for a stream network and storage to be straightforward but the groundwater case more obscure. We will explain this case in a bit more detail than the previous cases, in particular how a hydrogeologist develops the optimization model, and the interpretation of prices. Rather than simply starting from a good vector map, the hydrogeologist develops the optimization model from a groundwater simulation. Briefly, the hydrogeologist uses the simulation to calculate how much each water intake affects each environmental assessment point. The effect of one water intake on an assessment point, for each future period, is the *response*. This response depends on the user's type, especially the amount of recharge, if considered. The complete list of effects, for each well, for each assessment point, for all periods into the future, is the *response matrix*. We will use parameter $R_{k,w,t}$ for this response. When water intake w takes one unit (e.g., a cubic meter) of water in the current period, the rate of drawdown in meters per period at assessment point k at $t-1$ periods in the future is $R_{k,w,t}$. (We define $R_{k,w,t}$ as $t-1$ periods in the future, so that $R_{k,w,1}$ is the effect in the current period.) The market-clearing model for groundwater requires these response coefficients $R_{k,w,t}$.

Hydrogeologists can calculate $R_{k,w,t}$ by solving the flow equations in various ways. The flow equations can be solved with a variety of analytical and simulation methods: for example, with the eigenmodel approach, or with finite difference modeling (e.g., MODFLOW).

Sahuquillo (1983) showed how to model unsteady flow of groundwater using a series of eigenvectors. Bidwell (2005) calls this the eigenmodel approach. The eigenvectors represent the impact on head by pumping at various locations. Researchers have refined this method, but many groundwater modelers overlook it. The eigenmodel approach provides a simple way to calculate the response coefficients $R_{k,w,t}$., and it has been used to develop a hypothetical smart market optimization model for groundwater (Plagmann and Raffensperger 2007).

Instead of the previous two methods, we discuss development of the response matrix R using MODFLOW. MODFLOW is public domain software in wide use, and hydrogeologists routinely maintain MODFLOW models for many regions in the U.S. and internationally. Many government agencies use MODFLOW to make decisions about water allocation and as evidence in court. MODFLOW is computationally robust and regularly updated by the USGS. While MODFLOW was originally a groundwater flow simulation, it now has modules for linking to surface water flows.

MODFLOW simulates changes in head over time. To simulate water flow, MODFLOW approximates the differential equations of flow using a finite-difference approximation. MODFLOW reads text files with inputs on hydrogeological boundary conditions, abstractions, natural recharge, initial conditions, aquifer porosity, storativity, and transmissivity. The hydrogeologist chooses a discretization of the subject catchment into three-dimensional rectangular cubes or cells stacked in layers, a bit like stacked ice cube trays (see Fig. 7.6). In each cell and time period, the relevant state variable is the head. The bottom of each well is at the bottom of a layer. MODFLOW treats all wells that connect a given pair of cells as one well. MODFLOW works best when the aquifers are flat layers beneath the ground. When layers are not flat, or when flow properties vary within a layer, or near certain critical areas (say, near wells) the modeler might choose more cells to give a finer resolution. So a MODFLOW model can be relatively slow computationally for complex systems.

Researchers have developed programs to get the response matrix $R_{k,w,t}$ from MODFLOW models. These programs allow easy interface between MODFLOW and linear and integer programming models, such as minimizing the cost of contaminant mitigation, maximizing the output of a set of pumps subject to environmental flow constraints, or minimizing ground subsidence from aquifer drawdown (Lefkoff and Gorelick 1987; Greenwald 1998; Ahlfeld and Mulligan 2000). We have used GWM (Ahlfeld et al. 2005). GWM is based on MODFLOW, and has self-contained LP software. As of this writing, GWM was last updated on 17 Sep 2015. GWM estimates $R_{k,w,t}$, the effect of pumping at each well on the head at each assessment point in the system.

To create the response matrix $\underline{R}$, GWM uses MODFLOW as a subroutine to estimate a baseline head at each assessment point, using well pumping rates close to the expected conditions. Next, for each well w, GWM simulates each well w being pumped at rate $q_{w,t}$ in period t, to obtain the change in head (from the baseline) at each assessment point k over each time period in the planning horizon t=1,. . .,T. This simulation provides a W by K by T response matrix $\underline{R}$, with components $R_{k,w,t}$. $R_{k,w,t}$, which approximate the first derivative of head at each assessment point and time period, with respect to abstraction from each well. The MODFLOW simulation can include assumptions about each user's recharge (e.g., some irrigation water returning to groundwater), so the response matrix therefore can do this too.

The response matrix $\underline{R}$ depends on the aquifer and the period length. Abstraction at a particular rate will have a larger effect on head if each model period is longer. At one extreme, if all wells and assessment points lie in an undifferentiated reservoir, $R_{k,w,t}$ is a constant for all w, k, t. At the other extreme, $R_{k,w,t} = 0$ for w and k in unconnected catchments. Generally, $R_{k,w,t}$ tends to zero for large t, and for increasing distance between w and k. Note that $\underline{R}$ does not depend on rainfall, assuming linearity of the hydrogeology.

Hydrogeologists have found that these effects tend to be linear. That is, when well w takes q units of water in the current period, assessment point k falls by $qR_{k,w,t}$ meters at $t - 1$ periods in the future. Total effect at one assessment point from abstraction at all wells is close to the sum of the effects from abstraction at each well. When the effects are linear, the hydrologist can compute the response matrix independently of the future use of water; the baseline head need not be precise.

The behavior is not always linear, especially if water level falls below a well bottom, or if a cell that is assumed to be saturated in the solution becomes dry, or if a cell assumed dry becomes saturated. Unconfined flow especially will sometimes exhibit a significant non-linear response, so the hydrogeological simulation may find that the solution does not satisfy the required environmental constraints. In these non-linear cases, as Greenwald (1998) and Ahlfeld and Mulligan (2000) explain, the hydrologist should use the optimal solution of future pumping rates, q_{it}^*, from the smart market optimization model as the baseline when finding $\underline{R}$. By doing this, the hydrogeologist can come much closer to the expected drawdowns. If the simulation indicates that the environmental constraints would be violated—or if they have much more slack than the hydrologist expected—then the hydrologist would simply update the response matrix $R_{k,w,t}$ and re-run the LP. The concern is whether the solution will change enough to have a significant effect on users' allocations, prices, or environmental flows. In rare cases, the hydrologist may need to recalculate the response matrix several times in order to ensure that a properly updated response matrix is used. We expect this to be an unusual occurrence because hydrogeologists have found through experience that linear equations often approximate the hydrological behavior accurately. Even when necessary, the extra cost and time for recalculating response matrices should not be onerous.

At this point, we will assume that the market manager has obtained the response coefficients R_{kwt}, and other relevant information about the groundwater catchment.

So how are the environmental flow requirements set for groundwater? The regulator will specify environmental flow requirements to ensure that the users' abstractions do not draw down aquifers too far. Roughly speaking, the environmental requirements are the aquifer levels: how far down one must dig to reach water. Aquifers feed surface water bodies, including streams, lakes, and wetlands, and sufficiently high aquifers keep coastal salt water at bay. To ensure that the aquifers are high enough to keep water bodies sufficiently wet, and to prevent coastal saltwater intrusion into the freshwater aquifer, the regulator will specify the environmental requirements as head, which is the aquifer pressure, usually defined in meters of height. Thus, the response matrix indicates users' impacts on the environmental requirements, which the regulator can now define quantitatively as limits on changes in aquifer head. Besides drawdown constraints, the regulator may impose head difference constraints to maintain flows in certain directions. Although market managers have multiple ways that they could incorporate environmental constraints, we reiterate the caution in 6.9 that market managers should be careful not to over-constrain the model.

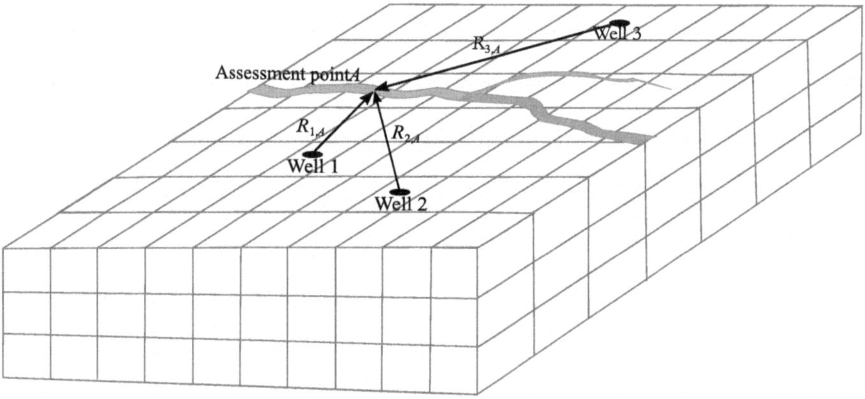

Fig. 7.6 Hydrologists divide land into cubes, and then simulate flow between cubes over time

Indices

b, bid block, b=1,. . ., B. Typically B = 1 or 2, at most 5.

k, l, assessment points, with k, l = 1, . . ., K. Typically K would be 10 to 100.

u, t, time periods, with u, t=1,. . ., T. Period 1 is the present period. Typically, the time periods would be weeks or months, so T would be 52 or 104 weeks, or 12–24 months.

w, water intakes, w =1,. . ., W. Typically W would be somewhere between a dozen and a thousand.

Parameters, Title Case

$BidPrice_{b,w,t}$, \$/m^3, the bid price that a user submits for abstraction from water intake w for bid block b in period t.

$BidQty_{b,w,t}$, m^3, the quantity that user submits for abstraction from water intake w for bid block b in period t.

DiffSet = the set of pairs of assessment points (k, l) between which head difference is constrained.

$N_{k,t}$ = "natural" head at assessment point k in period t, given the initial heads in all cells and projected natural inflows from periods 1,. . ., t, and assuming no further abstraction. This is a reference point.

$P_{b,w,t}$ = the bid price for demand quantity $Q_{b,w,t}$ at water intake w, period t.

$Q_{b,w,t}$ = the bound on demand block b for abstraction at water intake w, period t.

$R_{k,w,t}$ = meters drawdown after t periods at assessment point k, due to abstraction in the current period at water intake w, in meters of head. Note that $R_{k,w,1}$ is the effect in the current period from abstraction in the current period. The variables have units of volume of water per bidding period, but could be scaled in other ways.

$LD_{k,l,t}$ = the lower bound on head difference between assessment points k and l, period t, in meters, where assessment point k has higher head than l.
$UD_{k,l,t}$ = the upper bound on head difference between assessment points k and l, period t, in meters.
$L_{k,t}$ = the lower bound on head at assessment point k, period t, in meters.
$U_{k,t}$ = the upper bound on head at assessment point k, period t, in meters.

Decision Variables, Lower Case

$drawdown_{k,t}$ = the cumulative drawdown (in meters of head) at assessment point k after t periods.
$price_{w,t}$ = market price per unit of water at water intake w, period t. This is the dual price on constraint 3.
$q_{w,t}$ = total abstraction from water intake w during period t.

The regulator could define the formulation with variables for head, but these are not needed, because head is just initial head minus drawdown.

Market-Clearing Model for Groundwater

(1) The objective maximizes, over all users and all periods in the auction horizon, the total of each user's bid price times the amount of bid to be accepted.

$$\text{Maximize} \sum\nolimits_{\text{bids } b} \sum\nolimits_{\text{water intakes } w} \sum\nolimits_{\text{periods } t} BidPrice_{b,w,t}\, bid_{b,w,t} \quad \text{subject to}$$

(2) Bid block quantities limit abstraction. So long as each water intake has only one user, the regulator can omit constraints on water intake capacity, because of the assumption that rational traders would not purchase water beyond their capacity. In other words, the offer/bid limits imply the water intake capacity constraints.

$$bid_{b,w,t} \leq BidQty_{b,w,t} \text{ for each block } b, \text{ water intake } w, \text{ and period } t.$$

(3) The abstraction at the water intake equals the total of the water exchanged in each price block. The dual variable $price_{w,t}$ of this constraint is the marginal value to the market for another unit of water at water intake w, period t.

$$q_{w,t} = \sum\nolimits_{\text{bids } b} bid_{b,w,t} \text{ for each water intake } w \text{ and period } t, \text{ dual price } price_{w,t}.$$

(4) Drawdown at assessment point k in period t is a weighted sum of prior abstractions, from all water intakes.

$$drawdown_{k,t} = \sum\nolimits_{w=1}^{W} \sum\nolimits_{u=1}^{t} R_{k,w,t-u+1}\, q_{w,t}$$
$$\text{for each assessment point } k = 1, \ldots, K, \text{ and period } t = 1, \ldots, T.$$

(5) Constrain drawdown. Head at each assessment point must be between lower and upper bounds.

$$L_{k,t} \leq N_{k,t} - drawdown_{k,t} \leq U_{k,t}$$
$$\text{for each assessment point } k = 1, \ldots, K, \text{ and period } t = 1, \ldots, T.$$

(6) Constrain difference in head between two locations. Head difference constraints between assessment points ensure that water flows in a certain direction, and these constraints can limit the rate of flow.

$$LD_{k,l,t} \leq (N_{k,t} - drawdown_{k,t}) - (N_{l,t} - drawdown_{l,t}) \leq UD_{k,l,t}$$
$$\text{for assessment points} \{k, l\} \in DiffSet, \text{ periods } t = 1, \ldots, T.$$

(7) Bids are non-negative. The quantity and drawdown variables are unconstrained.

$$bid_{b,w,t} \geq 0 \text{ for each block } b, \text{ water intake } w, \text{and period } t.$$
$$drawdown_{k,t},\ q_{w,t} \text{ free.}$$

Discussion

The regulator can use constraints 5 and 6 to prevent injury to the environment and third parties in many ways. The regulator can view the upper and lower bounds on heads and head differences as boundary conditions, which can vary over time, such as by season. For example, constraints of type 5 could be used to prevent the market from taking groundwater beneath a lake and so drawing down a lake level too far, from preventing saltwater intrusion from a coast, and from maintaining minimum flows in streams and rivers, changing with time of year as appropriate. More-general constraints could employ weighted sums of head and abstraction rate variables. Some catchments would not need all such constraints; a hydrologist in consultation with the regulator would select and develop the constraints, based on community expectations and standards for the environment.

From an operations research point of view, the model above is an easy linear program. Following solution of the optimization, the market manager would tell the users how much water they got and at what price. User w faces price $price_{w,t}$, from constraint 3, for water in period t. Every user's water intake w could face a different price from other users, and a user's water intake w could face a different price in every period t.

The mathematics of the optimization guarantees that buyers never have to pay more than they bid and often will pay less. It also guarantees that sellers always get at least what they bid, and often will get more. Finally, the mathematics guarantees that every environmental constraint is satisfied. After the auction clears, the market manager credits and debits users' accounts for purchases and sales, in water allocations and money.

What is traded? Users are trading the right-hand sides of constraints 5, and 6. Scaling of users' allocations is not too hard (see Chap. 8). The $q_{w,t}$ variables will be usually be non-negative, because of constraints 2 and 7. However, the market manager may wish to allow bids for injection. For example, water managers in California coastal areas inject groundwater to replenish it, and to prevent saltwater intrusion. The mathematics will specify the payments for these injections. Hydrogeologists typically define extractions as negative flows, and injections as positive flows; we have done the reverse to more closely follow traditional definitions in optimization models.

As with the reservoir cases, this market has an end-of-horizon problem, which the market manager will have to address in some way. Implementers may think that the market manager need only require that the head at the end of the hydrological year be the same level as at end of the previous hydrological year, or that the market manager should target an average level over the year. Because of the delay in response, however, the model may need constraints for periods beyond the end of the hydrological year, even far beyond it, depending on the time lags in the response matrix. For example, suppose the auction schedule ran weekly from week 1 to week 52, and the groundwater response time lags were typically 26 weeks. Then when users take water in week 51, the model would not constrain their allocations unless it had constraints to at least week 78. A rolling horizon of at least 26 weeks could work in this case. Alternatively, the objective could include a value for stocks in storage at the end of the trading horizon. Ideally, the market manager would learn these values by allowing trading for usage in periods beyond that horizon. Alternatively, the objective could include a value for stocks in storage at the end of the trading horizon. Ideally, the market manager would learn these values by allowing trading for usage in periods beyond that horizon.

As before, this market model does not account for the fixed costs of infrastructure, in particular the wells in the groundwater market. Because wells are private, the regulator should expect users to pay for their own infrastructure. However, a user's well depth could become contentious if the aquifer levels fell below the well bottom. Section 6.9 considers this issue.

Implementers should carefully consider which aquifers will be within the market. Many groundwater systems have a series of aquifers with an unconfined aquifer near the surface, and then confined aquifers (with more or less connectivity between them) at greater depths. A community of groundwater users who are interested in developing with the water regulator a smart market for groundwater could find advantages in developing markets that jointly consider multiple aquifers. Deep water is more expensive to extract because users must pump the water from a lower head. As a result, users would likely prefer to extract shallow water, but might also be willing to pay the extra pumping costs during droughts when water use from the upper aquifer is limited because of environmental constraints. A smart market for water would work in this situation and would likely provide greater benefits than a market in only the shallower or only the deeper groundwater.

Joint consideration of water use from both deep and shallow aquifers within a smart market will benefit users and regulators. Layered aquifer systems interconnect to a greater or lesser extent. Over- extraction of a shallow aquifer could lead, in varying degree, to water flowing up from a deeper aquifer. But that supplemental water can create effects elsewhere in the water system, because the deeper water might otherwise have travelled downgradient for many years to nearer the coast, to emerge as a spring or to prevent saltwater intrusion. Environmental constraints may differ for deep and shallow aquifers. In many situations, deeper groundwater also has a different quality (better or worse for users) than shallow groundwater. A smart market that allows bids into two markets—one for the shallow aquifer and one for the deep aquifer—would provide the water regulator with the assurance needed for use of the deeper aquifer during certain times by certain users. Response coefficients enable a smart market to consider layered aquifers that influence each other.

Developing a stochastic optimization for groundwater smart markets would be quite difficult. We discuss this further in Chap. 11.

7.7 Model Formulation for Conjunctive Use: Ground and Surface Water

From Pulido-Velázquez (1997):

> Conjunctive use is the coordinated management of surface and groundwater resources, taking advantage of their complementary properties. Both surface and groundwater storage are used to redistribute water in time to match supply and demands. However, surface and groundwater storage differ in storage capacity, recharge and depletion rates, capital and operation costs and constraints. Jointly operating all manageable water resources in a region can increase the yield, efficiency, supply reliability and cost-effectiveness for a system.

This hydrological case is for a complex catchment with groundwater, stream flows, and reservoirs, all connected. A smart market optimization model can address the full complexity. Users may take water from all of these resources, and the regulator can specify environmental requirements for all of these resources, as well.

Surface water users high in the catchment without storage above them have only the current inflows. They can bid for future water, but the market manager cannot guarantee that water. Any surface water users at or below a reservoir can bid for water from inflows or from the reservoir. The upper bound on storage imposes a requirement on the market manager to ask users about the future value of water. Thus, the bid structure contains bids for the current period and bids for some set of future periods.

The auction schedule depends on how often people wish to trade and the size of the storage that affects them. A large catchment will more likely contain a wider variety of users, increasing the complexity of the market manager's task. A large number and variety of users does not make a smart market model more difficult to develop, but it does make it more challenging to develop the rules for the market.

As an example, Fig. 7.7 shows a schematic based on the Rangitata Diversion Race (RDR) irrigation scheme in Canterbury, New Zealand. The scheme has five rivers, a lake, a major canal, hydroelectricity generators, and irrigation offtakes to large areas of farms, some of which take groundwater, with more groundwater users downhill. Developing the market-clearing model would be a significant undertaking, but it is dwarfed by the work to gain cooperation among a variety of users and the community.

We will not develop a detailed market-clearing model. But here is the general case, with the mass balance and environmental flow constraints defined in generic matrix notation. Except for the generic matrix, we have previously defined these parameters and variables several times.

(1) Maximize $\sum_{\text{bids } b} \sum_{\text{water intakes } w} \sum_{\text{periods } t} BidPrice_{b,w,t}\ bid_{b,w,t}$ subject to
(2) Quantity limit on each bid block: $bid_{b,w,t} \leq BidQty_{b,w,t}$ for each block b, water intake w, and period t.
(3) Allocation equals total bid: $q_{w,t} = \sum_{\text{bids } b} bid_{b,w,t}$ for each water intake w and period t, dual price $price_{w,t}$.
(4) Mass balance and environmental flow requirements: $L \leq Aq \leq U$.
(5) $bid_{b,w,t} \geq 0$ for each block b, water intake w, and period t.

Discussion

To construct this linear program, the market manager should start with a detailed simulation of the water flows in the catchment. To help model these complicated cases, hydrologists have tools such as the Water Evaluation and Planning (WEAP) software produced by the Stockholm Environment Institute. This and similar software can help the hydrologist develop optimization models. Forni (2010) extended WEAP for use with economic optimization. The resulting linear program could be large (hundreds of thousands of variables), but with modern solvers and hardware would still not be difficult.

The eigenmodel approach could be an alternative to the development of complex software applications. Pulido-Velázquez et al. (2006) presented an economic optimization of surface and groundwater considering highly linked water resources. Their method will also allow a smart market that includes a combination of surface and groundwater.

User w faces $price_{w,t}$, from constraint 3, for water in period t. Every user's water intake w could face a different price in every period t. Since constraint 3 is an equality, and $q_{w,t}$ is free, the prices derive from weighted sums of the prices on the other constraints. As a result, the market community should understand that the rights being traded are primarily the bounds on constraint set 4. Scaling of users' allocations for variable inflows is complicated, as discussed in Chap. 8.

Like the reservoir cases, this market model could have bids for non-consumptive use. To recapitulate, a non-consumptive user would bid in the auction as any other user. The market manager's agreements with non-consumptive users are commitments to provide flow across the reach, but the users would have no requirement to use the water, as long as the flow passed downstream with little hindrance.

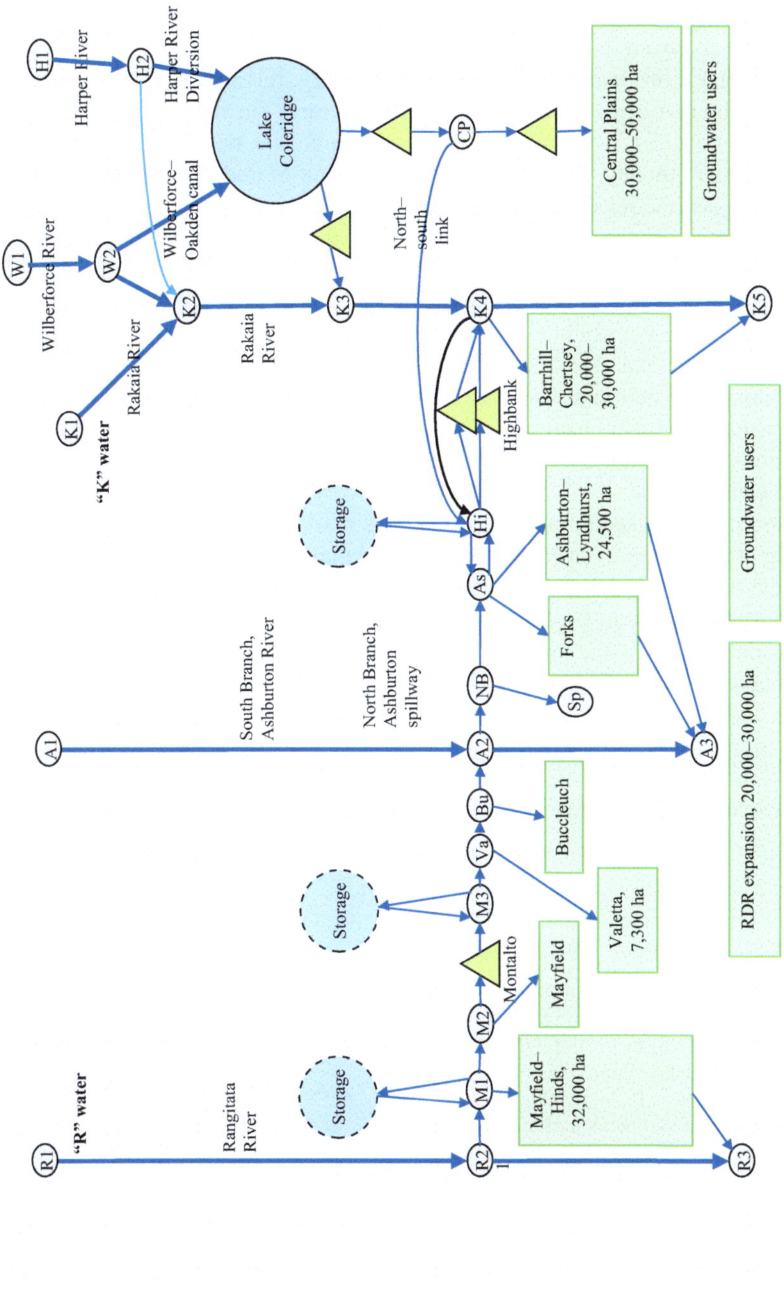

Fig. 7.7 Schematic based on the Rangitata Diversion Race area. The link from *K4* to *Hi* is pumped. Triangles are generators

As discussed before, this market model does not account for the fixed costs of infrastructure. Users would have to pay for maintaining infrastructure through some other payment mechanism. The model produces price information that would convey the value to participants for an increase in the water-conveying capacity. This may be useful for future planning.

The capacity constraint can generate rents when it becomes binding, so the market manager's revenue will not balance. In theory, this excess revenue would go to those who manage and maintain the infrastructure. The market implementers will have to consider who owns the infrastructure, and develop a formula to disburse the extra money. Eventually, users may want to trade the right to use the infrastructure, as fractions of the available capacity. An owner may wish to lease capacity to parties, giving them security of conveyance. We discuss capacity rights further in Sect. 10.5.

Just like with a smart market for surface water, the design of a conjunctive use smart market for water could consider the impacts on water-conveying capacity resulting from societal decisions on natural waterways. After a smart market properly prices the marginal value of a river's water capacity, society has a firm basis to evaluate whether changes to a natural waterway should come with financial contributions from water users or compensation to them for a reduction in their access to water.

Developing a stochastic optimization of this market-clearing model would be extremely challenging; even harder would be explaining the model to users, and gaining the users' confidence in it would be still harder again. We further discuss stochastic optimization in Chap. 11.

The complexities of the simpler hydrological cases now merge. As with the reservoir cases, this market has an end-of-horizon problem for each storage system, reservoir, and groundwater, which the market manager will have to address in some way. Ideally, the market manager would learn these values by allowing trading for use in periods beyond that horizon. As with the groundwater case, the regulator and market manager should study well depths, and consider making them part of the market.

References

Ahlfeld, D. P., & Mulligan, A. E. (2000). *Optimal management of flow in groundwater systems*. San Diego: Academic Press.

Ahlfeld, D. P., Barlow, P. M., Mulligan, A. E. (2005). *GWM – A ground-water management process for the U.S. Geological Survey modular ground–water model (MODFLOW–2000)*. (U.S. Geological Survey Open–File report 2005–1072). http://pubs.usgs.gov/of/2005/1072. Accessed 15 July 2006.

Forni, L. (2010). *Economic and hydrologic models integration new method: Sacramento Basin, California*. Master's thesis, Agricultural and Resource Economics, University of California, Davis.

Greenwald, R. 1998. *MODMAN: An optimization module for MODFLOW version 4.0*. Freehold: GeoTrans.
Hey, D. L., & Philippi, N. S. (1999). *A case for Wetland restoration*. New York: John Wiley & Sons.
Kahneman, D., & Tversky, A. (1979). Prospect theory: An analysis of decision under risk. *Econometrica, 47*(2), 263–291.
Lefkoff, L. J., & Gorelick, S. M. (1987). *AQMAN; linear and quadratic programming matrix generator using two-dimensional ground-water flow simulation for aquifer management modeling* (Water-resources investigations report 87-4061). Menlo Park: U.S. Geological Survey.
Lund, J. R., & Ferreira, I. (1996). Operating rule optimization for the Missouri River reservoir system. *Journal of Water Resources Planning and Management*, ASCE, *122*(4), 287–295.
Plagmann, M., & Raffensperger, J. F. (2007). *A smart market for ground water using the Eigenmodel approach*. MODSIM Conference, Christchurch.
Plott, C., & Zeiler, K. (2005). The willingness to pay–willingness to accept gap, the 'endowment effect,' subject misconceptions, and experimental procedures for eliciting valuations. *American Economic Review, 95*(3), 530–545.
Plott, C. R., & Zeiler, K. (2007). Exchange asymmetries incorrectly interpreted as evidence of endowment effect theory and prospect theory? *American Economic Review, 97*(4), 1449–1466.
Pulido-Velázquez, M. A. (1997). *Conjunctive use opportunities in Southern California*. Master's thesis, University of California, Davis.
Pulido-Velázquez, M., Andreu, J., & Sahuquillo, A. (2006). Economic optimization of conjunctive use of surface water and groundwater at the basic scale. *Journal of Water Resources Planning and Management, 132*(6), 454–467.
Raffensperger, J. F., & Milke, M. W. (2005). A design for a fresh water spot market. *Water Science and Technology: Water Supply, 5*(6), 217–224.
Raffensperger, J. F., Milke, M. W., & Read, E. G. (2009). A deterministic smart market model for ground water. *Operations Research, 57*(6), 1333–1346.
Read, E. G. (1979). Stochastic long-term scheduling models for a power system. In *Proceedings of 15th annual conference, Operational Research Society of New Zealand*. New Zealand Operational Research, special issue (pp. 41–52).
Read, E. G., & Hindsberger, M. (2010). Constructive dual DP for reservoir optimization. In P. M. Pardalos, S. Rebennack, M. V. F. Pereira, & N. A. Iliadis (Eds.), *Handbook of power systems I* (pp. 3–32). Heidelberg: Springer.
Sahuquillo, A. (1983). An eigenvalue numerical technique for solving unsteady groundwater continuously in time. *Water Resources Research, 19*, 87–93.
Sprenger, C. (2015). An endowment effect for risk: Experimental tests of stochastic reference points. *Journal of Political Economy, 123*(6), 1456–1499.
Zetland, D. (2013). All-in-Auctions for water. *Journal of Environmental Management, 115*, 78–86.

Chapter 8
Converting Users' Rights to Tradable Allocations

Abstract This chapter describes the process of converting users' rights to tradable allocations. The first section discusses the nature of the water rights in the smart market, why initial rights are important, and whether this conversion is privatization. The second section discusses the need for the new license system to be able to manage new entrants. The third and fourth sections discuss the degree of security for the new rights, and make a case against long-term and perpetual licenses within smart markets. The final two sections discuss trading of long-term licenses, and converting disparate existing water quantity rights to smart market allocations.

Keywords Water rights • Security • License conversion

8.1 The Nature of the Water Rights in the Smart Market

8.1.1 Overview

How can the regulator create initial tradable allocations from the legal medley, the statutory potpourri, the juridical jumble of real world water rights? The conversion process will be complicated and perhaps messy.

In most places around the world, government allocates rights through a licensing process. In too many places, this licensing system is open to political influence and in some places subject to corrupt practices. Easter and Huang (2014, 3) note that when government employees can extract rents for allocating water, those employees will resist implementation of water markets. Thus, a smart market for water provides an opportunity for society to help protect itself from corruption. Just as Adam Smith saw an open market as a way to remove the economic privileges of those with import and export licenses, an open market for water can be developed in a way that lessens the ability for influence peddling. Section 14.5 develops this matter more fully.

For some types of rights regimes, implementers will be able to convert the existing rights regime into a recognizably similar allocation mechanism. For those rights that can be converted, the conversion of each type (farm or hydroelectricity generator, appropriative or riparian, surface water or groundwater) may follow a different process. In each case, implementers must find some conversion

J.F. Raffensperger, M.W. Milke, *Smart Markets for Water Resources*, Global Issues in Water Policy 12, DOI 10.1007/978-3-319-55008-4_8

formula that users will accept. The regulator may be unable for legal reasons to convert some types of rights to a tradable allocation; for these cases, the government will have to change the law. In particular, the new rules must address the license duration and nominal quantity.

We will use the word **license** to mean government permission to users to conduct activities related to taking water. This permission includes access rights such as digging wells in a particular place, specification of the allowed uses for the extracted water, and the identity of the users. The license also specifies a term (e.g., 5 years, 10 years, or perpetuity), and a nominal quantity that the user can take or trade per period (week, month, season, or year). The nominal quantity is subject to regulatory change, especially in dry periods.

Whether the existing rights are riparian, appropriative, or in some other form, implementers must specify the appropriate duration of the water right after implementers convert the right into a smart market allocation. The tradable allocation is specific for a short period of time—an hour, a day, or a week—and the regulator must assure users of security in the term length of that right for the duration of the allocation. But security in the term length of a long-term license is associated with how the regulator allocates them. A given water right may be valid for only a short time, but the quantity associated with the water right may be nearly certain. A water right may be valid permanently, but the quantity associated with it can vary wildly.

These issues depend on whether implementers decide to convert long-term and perpetual licenses to short-term licenses, and, if not, how to enable trading of long-term and perpetual licenses.

8.1.2 Initial Rights Are Important

Implementers must make initial rights and tradable allocations clear before the smart market can begin, otherwise people don't know how much they have to buy or sell. Some people reject markets on the basis that some rights holders will enjoy windfall gains, or that some rights holders will suffer severe losses. While people have to know their initial rights and tradable allocations to know whether they need to buy or sell, real or imagined fears about initial rights can cause a good market design to fail at implementation.

Theoretically, initial rights do not matter, at least from the point of view of economic efficiency for society as a whole. In a properly designed market, people will buy up or sell down to what they want, based on the market price each person faces. So initial rights and tradable allocations are independent of the operation of a well-designed market. This is the Coase Theorem, repeated confidently by economists everywhere.

To be more precise, Coase claimed that the initial rights do not matter if rights are properly defined, if the transaction cost is zero, and if the parties have no budget constraints. To put it in dry economic terms, when these assumptions do not hold, society will suffer inefficient outcomes.

Initial rights matter very much to the current holders of those rights. The person starting the market with more rights, leading to more tradable allocation than needed, can sell them, and can enjoy the confidence of sufficient allocation in case it is needed. The person starting with fewer rights than needed must buy allocation, and that person is likely to fear not getting enough at a reasonable price. Discussions about the initial distribution of water rights for a smart market are likely to get tied into a discussion about what would constitute reasonable prices.

This chapter proposes mechanisms to manage the initial conversion of users' water rights to tradable allocations.

This conversion process needs to be done only once, before market implementation. Then, at every auction, the market manager must adjust allocations to match the available water.

Economists have used laboratory experiments to test for "endowment effects," where a seller demands a higher price than the "rational" selling price, which Sect. 7.2.5 explores. We have heard anecdotally that some water users could be so accustomed to inflexible water allocations, that when they first are able to bid for water they could have a tendency to avoid selling. This could be from risk aversion, distrust that the system will allow them to receive water later, or dislike of selling to outsiders.

If implementers can convert the license regime to a new regime with tradable allocations, where the old quantities map intuitively to new quantities, market implementation should not disrupt users. People should feel a smooth transition into the market with no sense of loss. If the existing rights regime is unreasonable, it should be overhauled anyway.

We don't have all the answers here. While we cannot hope to explain how implementers would convert every license regime to our smart market allocations, we will examine some typical rights regimes, and we will suggest ways to convert those licenses into initial allocations for the market.

The smart market requires an initial rights definition that is highly specific. Up to this point, we have emphasized that the tradable thing is a lease for allocation for a specific period of time at a specific location for a specific use. The legal allocation is only a proxy for actual water. The holder must be allowed to earn income from the water specified by the water allocation, must be allowed to transfer the allocation to other people and be paid for doing so, and must have reasonable government protection from anyone attempting to take the water without compensation to the user. To willingly participate in the market the user will want these essential assurances.

8.1.3 Is This Privatization?

To many observers, especially business people needing actual water for their operations, water markets seem to require "privatization" of the water, while other observers fear such a thing. The business person wants control of the resource and security of supply; the environmentalist fears that users will steal the commons. So "privatization" is a loaded word, sometimes meaning "security" to a business person and sometimes "theft of the commons" to an environmentalist. We need to unpack a few ideas here.

We will define "privatization" as a transfer from the government of a secure right to use water for a period of time. In this sense, privatization applies to ownership by a business that operates for profit (we could include non-profit organizations in this definition, but we want to focus here on the issue of privatization and businesses). *Ownership* implies exclusivity.

For water markets to work, the regulator must allow users to privatize something. The rules have to specify a tradable thing that the holder retains with sufficient certainty, and that buyers can receive with sufficient certainty, i.e., the thing is tradable. Otherwise a seller has nothing to sell, because a prospective buyer could not be sure of taking possession of the thing.

In this sense of the word, privatization allows the government to separate a part of the commons exclusively to some users. If government can separate some resources cleanly, prevent externalities, and account for everyone's rights, then the issue of private use and trading is no longer an impingement on the commons. Because water is a complex resource, this separation is difficult to accomplish. Under a crude water allocation mechanism, such as Australia's, government officials may think their users no longer impinge on the commons, but the allocation mechanism still has externalities, so the privatization process is incomplete.

For all practical purposes, users have already privatized water resources. This is not cynicism, but an observation. Users, especially farmers, have strong claims to water, even when, in many countries, that water does not exist. And what is ironic is that although users require privatization of something for trade, and government has allowed users to privatize water rights, trade happens rarely.

Perhaps the *lack* of tradability has incentivized users to hold on to their water rights tightly. The existing form of water rights is not proving very useful to their owners. Many water users already have firm legal rights to more water than actually exists, and when these users speak of wanting "privatization," they mean access to water in a marketplace. These owners realize that they need a form of privatization that they can use to manage risk to water access.

Some people say they oppose privatization when they really mean they do not want holders of water rights to "profit" from their water rights, by selling a resource that should be common. This makes no sense, as the water users already profit from their water rights; that is the point of having the rights. By preventing users from selling those rights, the critics would have artificial shortage for others and for the environment. Prohibiting water sales encourages waste. In any case, holders of water rights in most cases already have the legal right to also sell those rights.

After some more thought, some people might instead say that holders of water rights should not profit *unfairly* from their water rights. The users will say that taking their existing rights without compensation would be unfair, and to the extent that this results from an auction of initial rights, implementers must address this matter.

Yes, the market requires privatization of use of the water. The minimum "privatization" requirement for a smart market is only that the right be *sufficiently* firm. As Tietenberg (2006) indicates, the water user needs assurance that government will not arbitrarily take the right. Tietenberg further points out that the tradable permit scheme must not violate the public trust doctrine: that the government holds certain resources in trust for the people, and cannot give them away. Referring to the U.S. sulfur dioxide trading scheme, Tietenberg writes:

> In practice this means that administrators are expected to recognize the security needed to protect control investments by not arbitrarily confiscating rights. They do not, however, give up their ability to change control requirements as the need arises. In particular they will not be inhibited by the need to pay compensation for withdrawing a portion of the authorization to emit as they would if allowances were accorded full property right status. It is a somewhat uneasy compromise, but it seems to have worked. So a market can be successful even if government does not guarantee the quantity of water associated with a long-term license or with an allocation.

The market does not require "ownership of the water" in the sense that the holder of allocation has absolute control over the water. The market requires a much less stringent definition of ownership than you may require over your cell phone. "Ownership of the water" in a smart market, to the extent that it is relevant, belongs to the water regulator that empowers the market manager. We could say the same for all government water resources systems: to the extent that it makes sense to say that the water has an owner, the water resource manager is the owner by holding water in trust for society as a whole.

8.2 The New License System Must Manage New Entrants

Since the market-clearing model requires that allocations match the actual expected water, the new license system must have rules to manage new entrants.

We have assumed that water is generally short in the catchment, i.e., users would take more water if it were cheaper. In the unlikely event that the catchment is under-allocated, the regulator can grant the new rights. But in such a case users will have little need of a spot market. This situation is rare and will get more unusual in the future, unless the human population declines significantly. The regulator under a poorly managed license regime may be tempted to grant new water rights even though the catchment is over-allocated. The regulator of a well-managed license regime, including that required by a smart market, may grant new water rights only if the water appears to be available.

In the precision-managed regime that we develop in this book, a gain in rights to one user usually results in a loss in rights to another: any water granted to a new entrant must come from existing users. The existing water users will therefore rightfully protest any new grant of a long-term license. If the regulator allowed such new licenses, users in a fully allocated catchment may object to implementation of the smart market in the first place. Given a context of existing users holding long-term or perpetual licenses, the regulator should probably never allow new entrants without means of compensating existing holders of tradable allocation. The difficulty in making such an arrangement for new entrants is one reason for market developers to consider removing all long-term licenses in favor of short-term ones (see Sect. 8.4).

The smart market will tend to result in fully allocated catchments for most periods of the year. Interestingly, this observation suggests that the regulator has leeway to allocate new licenses for periods that are under-allocated, thus incentivizing storage. But this observation makes a stronger point about long-term licenses. If the regulator grants a new long-term license, the terms should specify the period of the year. For example, the regulator could grant a 10-year license for X megaliters of week 24 of the hydrological year. This new license translates cleanly into allocation.

We discuss trading of long-term licenses in Sect. 8.5. Reforming the license system and enabling the spot market enables trading of long-term licenses.

8.3 Degree of Security

The security of an existing water right can take many different forms. At one extreme, the water right could be in the form of a guarantee from society that the user retains the right in full and forever, like property. At the other extreme, the water right could be merely administrative assurance that the user retains the right only for a short period, and subject to a range of conditions. These different types of rights provide the holder with different attributes and degrees of security. Other attributes include government's ease in changing any of these conditions without awarding compensation, and other parties' ability to reduce the holder's water rights without paying compensation. Here, we discuss the degree of security generally.

Government should protect users' rights just as government protects rights generally. When a user has lawfully obtained an allocation, the government should protect the user from injury to that allocation by third parties, and the government should not arbitrarily take that allocation. However, the regulator justifiably needs the ability to modify the water allocation. Real water is variable and imperfectly predictable. In a smart market design, the thing traded is a concept called allocation, not water per se. The water associated with an allocation is uncertain, so the right to the water must be uncertain. Government should not be responsible for the uncertainty of nature, and the regulator must have authority to adjust the users' allocations to the available water. Users must realize that they bear some risk. Thus, the allocation is currency for real water. Users will want the allocation to give a good indication of the actual water quantity, and the market manager should strenuously endeavor to give them this good indication. We think that the smart market, if implemented with care, will provide holders of allocation with highly reliable supplies of water, in the sense that the quantity of allocation is likely to reflect the available quantity of water.

Some jurisdictions may wish to provide a stronger right, in which the allocation is a promise from government, whether water is available or not. When the water is insufficient for the agreed allocation, then government would compensate the user for the mismatch. This stronger right would work with a smart market design, but the design does not need it. We do not think this approach is practical in the long run, because it is likely to result in a permanent stream of payments from society to water users, and it is likely to incentivize unwanted gaming.

Allocation through almost all existing permit processes amounts to a gift from society to selected businesses, and that gift tends to be inefficient for society as a whole, because it is almost never reflects the most economic allocation. An inefficient allocation made for a long period is worse than an inefficient allocation made for a short period. So, have we any solutions? We do, but our solutions may not be politically palatable.

8.4 A Case Against Long-Term and Perpetual Licenses Within Smart Markets

Partly because the permitting process has a high transaction cost, and partly because users think they need long-term security, existing license regimes tend to have term lengths that are long or even permanent.

Users may believe that the term of the license must be as long as the life of associated capital assets. If farmers depreciate capital investments such as irrigators or new fruit trees based on a 30 year period, then they may feel the need for 30 year licenses. Furthermore, long-term licenses decrease administrative costs for re-assessment of water rights and with changing titles. However, long-term licenses raise the need for a mechanism that allows new users to obtain their rights in a way that compensates existing users.

To see how the licensing should work, consider the question: "What license term would business people buy if they had to?" Farmers recognize fertilizer and tractors as normal operating costs. They also recognize that water infrastructure, such as irrigators and on-property pipes, are assets with fixed useful lives. They recognize that many systems for creating value from water have fixed useful lives, e.g., fruit trees, or thermal electricity stations requiring cooling water.

Water should be a normal operating cost. Indeed, society needs users to view water as a short-run variable cost, so those users are correctly incentivized by that cost. Just as a truck owner has no need for a permanent right to diesel fuel, a farmer should have no need for a permanent right to water. The truck owner has confidence in the spot market for diesel, even though its price varies. Once a smart market for water has been operating for some time, users will gain confidence in it as well. Users will find that they can get water when needed, even though its price varies.

We recommend that licenses have limited term lengths, while recognizing that users will pressure regulatory bodies to create long-term or perpetual term lengths. Government should give users just enough security to enable the users' operations, and no more. Giving permanent standing rights to the commons is more than businesses need, because any water user (whether an industry, a farmer, or a city) is likely to change its use of water over time. Reallocation requires one user to surrender rights in some fashion, and reallocation of license is more difficult when the licenses are long-term or perpetual.

The regulator's least-political mechanism for allocating rights is the auction. Governments already use auctions to manage the commons, as for timber logging rights, for mining rights, for radio spectrum, and fishery rights. Formal auctions would give business their needed security while compensating the rest of society for supplying that security. An auction for limited-term licenses brings fairness to allocation, in contrast to the political approach. With existing perpetual licenses, periodic auctions allow new entrants to bid for water rights alongside existing water users.

The government may wish to end rights at some event, such as after 25 years, or the sale of the property, or the death of the user, or some other criterion. The government could also diminish the long-term or perpetual rights by some formula, e.g., 5% per year, while creating limited-term licenses to take their place.

8.5 Trading Long-Term Licenses

Efforts to start a smart market could falter because of an inability to agree on conversion of long-term to short-term licenses. Fortunately, such a conversion is not necessary for a smart market. Assuming that licenses remain long-term under a smart market, how might the market work? With no changes in long-term licenses, users' needs for water will become distorted over time when compared with their fixed long-term license quantities. As economic conditions change, people will want new long-term licenses. Climate change and changes to environmental constraints will lead to changes in land use, so some landowners will wish to sell both water rights and land. Some buyers and sellers will tire of always buying or selling the same amount. The smart market's spatial prices will encourage new entrants to seek land with cheap water. The long-term rights should change hands as society's needs for water change, so a smart market will have limited benefits if the conversion of water rights happens only once. In short, users need a system to trade long-term licenses.

Many researchers have written about permanent water right transfers, but those markets suffer a high transaction cost similar to that that our market reduces for the spot market. Implementers should develop a mechanism to allow trading of long-term licenses within a smart market.

Implementation of the smart market enables active trade. The smart market for water is a spot market, for short-term allocation. Having a spot market enables transfer of long-term rights too, and the rules need not be complicated. Commodity traders understand that an active spot market also allows a range of other useful financial services (see Sect. 10.4). These mechanisms for trading long-term license need not be centralized, though a central broker brings advantages. The mechanisms need not be run by the government, though government operation brings advantages. So, implementers have a range of options, depending on the type of new license regime that the implementers choose. How can these trades work?

Vesting The regulator could allow new entrants to become vested in a long-term license by a combination of a long history of purchases in the spot market with an application process to the regulator, who could convene a community board to review the applications. The rate of new vesting would depend on the degree that water in the catchment was under-allocated, perhaps indicated by the market manager often having positive revenue. Conversely, in over-allocated catchments, decisions to vest new users would be more restrictive. A user's history of purchases in the spot market should assure other participants that the new license would not disrupt the other users. This method has the disadvantage of leading users interested in vesting to raise their bids in order to demonstrate a record of purchases needed to support an application for vesting, thereby increasing water prices in the market.

Simultaneous Auction The market manager could allow new users an opportunity to obtain long-term licenses through an auction using a special version of the market-clearing model as a "reference model." The reference model would use long-run expected inflows by period, rather than near-term forecasted inflows by period. Bidding and clearing would function virtually the same as in the spot market, but the traded commodity would be a long-term license rather than a short-term allocation. The purchased long-term license would confer a weekly schedule of expected water over the year. The schedule could differ from user to user, and would be determined by their bids and by the reference model.

The inflows used in the reference model have direct implications for scaling allocations in the subsequent spot markets. If the reference model uses long-run average inflows, buyers of the long-term license would expect the market manager to scale down their initial allocations in the spot market about half the time. The regulator could use the reference model with inflows lower than average, implying that these flows would be relatively more certain than the average flows, and then users' allocations would be scaled down less often. Whatever inflows the regulator chooses for the reference model, buyers must understand that the market manager will scale down the resulting licenses, with a frequency that depends on the inflows used in the reference model. Users may have some preferences for the level of inflows in the reference model, depending on their tolerance for having their resulting allocations scaled down. Sophisticated versions of such a reference model could employ stochastic programming models. The market manager could grant different reliabilities of water at different prices.

To make this auction effective, a sufficient number of holders would have to offer to sell part or all of their license quantity. Holders of existing licenses who do not wish to participate would be represented in the reference model as constant abstractors.

To simplify this auction, the traded long-term licenses should all have the same expiration dates, if any. Buyers would acquire licenses with those same expiration dates. Similarly, while allowing perpetual licenses for water under a smart market, the market regulator could require that users rebid for a certain percentage of those rights at a specified frequency. For example, 10% of all rights expire every 10 years, and users would need to bid in an auction along with any new users, and the adjusted licenses would then continue for 10 years before the next adjustment auction. This method of withdrawing rights also provides a convenient way to manage the issue of over-allocation developing over time.

Standing Sale of Allocation The regulator could encourage a system of permanent transfers via the spot market. Suppose Ann wishes to buy a permanent license from Bob in a pair-wise trade. Ann and Bob could have different hydrological effects, and may not even be in the same catchment. Bob agrees to stop taking water, and signs over to Ann ownership of his allocation, at his location and for his existing terms. Ann then posts a permanent standing sell bid for Bob's allocation, agreeing not to use that water. Ann bears some risk regarding the market value of Bob's allocation, but with a few years of price history and allocation records, and knowledge of the expiry date of Bob's license, Ann could easily decide the value of buying Bob's license. Ann and Bob can agree privately to terms. Ann now has a stream of income to do with as she wishes. The tradable water allocation has turned into a variable annuity—uncertain payments but certain term length. Ann could use it to buy water at her own location and in different time periods than Bob's allocation. Ann and Bob should inform the market manager of the deal, at least Bob's allocation on sale, but the manager need have no role in it otherwise.

Ann would have a particular interest in buying Bob's allocation if Bob were her neighbor. She would have better information about the local availability of water than would other parties. If her pattern of water use matched his, then the stream of income from his allocation is likely to nearly match her need to buy allocation.

Ann could decide to manage the sale of Bob's allocation passively, such as by asking the market manager to always sell at the current market rate. If Ann posted a standing bid just above zero, e.g., "I will sell all allocation for $0.01/ML," the market manager will pay her any going rate above that, because the market clears at the marginal price. Ann might think she could manage the sale actively by attempting to raise the bid above the usual going rate, but adding a sell bid to the existing set of bids cannot raise the clearing price.

Ann and Bob can make these contracts within the smart market, though they add to the transaction cost for Ann. After a few years of this arrangement, one would imagine that Ann would ask to somehow amalgamate the two licenses into one. She might even agree that it would be better in the long-term to simply have a short-term license rather than one or more long-term licenses (as long as other water users in the smart market agree).

Private or public third parties could offer to buy long-term licenses from users who no longer need part or all of their allocation. But implementers should not appoint one agent with a monopoly on purchase of long-term licenses, because the agent could have significant market power. More sensibly, the market manager or another public agent could offer to buy or even be required to buy long-term licenses, perhaps as a buyer of last resort, at a price predetermined by some formula. In this case, implementers must specify rules about how the agent should handle the money, how the agent would raise funds for the purchase, how the market manager would sell the excess water and disburse the associated income.

8.6 Converting Disparate Existing Water Quantity Rights to Smart Market Allocations

8.6.1 Establishing Initial Quantities

If each user's water is metered (which is a prerequisite to the market anyway), then implementers can make the conversion formula for quantity depend on actual use. Meter records also help the implementers establish allocations based on time of year, e.g., by week. Users will be more willing to give up rights for a period in which they never used those rights.

To get an adequate historical record, users should start metering several years in advance of the market implementation. Recall that we envision that a smart market would be voluntary and would start with only the largest water users. This focus on large users should make it easier to establish a historical record for water use.

If users do not have meters before market implementation, users may benefit if they increase or exaggerate their water use. Section 5.5 underlines the importance of water metering in setting initial rights. The regulator will need to be especially careful about auditing meters during this period. Nevertheless, the more users who have a history of use, the easier it will be for implementers to develop a formula to convert water use and existing water rights into new smart market water rights. Implementers could find alternative metrics, if necessary, such as calculating water abstraction based on electricity use. As long as the regulator has authority to adjust the users' allocations to the available water, as a prerequisite to the market, implementers can afford to be somewhat lenient in their volume conversion formula.

Fortunately, the implementers and the regulator can use the hydrological model to help calculate the initial rights. For example, a water regulator could use a hydrological model with assumed crop types, climate data, and soil data, to estimate how much water users abstracted during a growing season. More simply, the implementers could use allocations implied by current licenses, and then scale them. This will bring some numerical precision to the problem. As long as the hydrology model has the force of law, then a path exists to derive new rights meeting some fairness criteria. This would be similar to the process of "adjudication" used in the past to have water rights better meet water availability (Porse et al. 2015).

Because water rights will need to change from an existing system to a smart market system, users could call for compensation for the taking of their pre-existing rights. This issue needs consideration, but is not likely to be severe. A right to a large quantity of water that is often reduced considerably in dry years may be worth less than a right to a smaller quantity of water that is much more certain, especially when the latter right has the flexibility of market actions. Particular cases will vary, but the loss associated with the switch needs to be considered alongside the gains for users when deciding on any compensation due to existing water rights holders.

8.6.2 Conversion Without Initial Rights

Where all users have no initial rights (say, on Mars), the regulator can require all users to pay in full for allocation at every auction. This may be the easiest possible case. This method would be highly efficient and fair to all participants. Society would be better off with a simpler rights system, and government could choose to use the water charges to offset income taxes. No conversion of initial rights is necessary. Implementers need not worry about the term length of the allocations.

Where users have initial rights of some kind, the regulator could decide to cancel those rights and require all users to pay in full for their allocation. The conversion of initial rights requires a difficult legal process of zeroing out everyone's rights. This may be the hardest possible course of action for the implementers. Beginning the market this way would make losers of existing rights holders. According to Coase, the economic benefit to society would be about the same, but that is no consolation when users thwart implementation.

In this case of "no initial rights," buyers of future allocation obtain initial rights for the associated periods. The regulator must recognize their rights to allocations, just as if they had obtained them in any other way. Section 9.3.3 considers the implications on tradable allocations and market operation of this approach to assigning initial rights.

8.6.3 Conversion of Riparian Rights

In a riparian rights regime, ownership of land confers the right to use water on or adjacent to the land. The user's physical access to water conveys unlimited rights to that water. "I dug this well, so I can take as much water from it as I wish." All upstream and downstream users have equal rights. Users are usually required to use water "reasonably," and people argue about what "reasonably" means, especially in dry years, with resolution through the courts.

At the time of this writing, many California groundwater districts work this way. Unfortunately, a riparian rights regime fails to address the true physics of water, because use of groundwater under one person's property will lead to less water available for neighbors. Consequently, riparian rights also impose externalities on other users and the environment. Riparian water rights cannot efficiently manage a common pool resource.

Users with essentially unlimited use of water have no reason to buy rights. Just as seriously, in the extreme case the users may have no standing to sell their rights, because no one believes that these users have unlimited water to sell. Buyers and sellers are unsure just how much anyone has to trade.

If each user has some firm upper limit, whether quantity per day or year, then the conversion formula could depend in part on this established upper limit. But upper limits are unlikely with riparian rights. The best approach is to persuade users to determine initial market allocations based on the users' records of historical use.

Furthermore, the regulator must separate allocation from the land for the smart market to work. The ownership of the land must not confer the right to water. Rather, the regulator must confer the right without regard to the land. Section 5.7 covered this.

8.6.4 Conversion of Appropriative Rights

Appropriative rights are sometimes called prior appropriation rights, first-come-first-served, or squatter's rights. Users acquire rights to water by use, not simply by land ownership. The user that started taking the water first has the senior right, up to a certain "reasonable" amount. When the senior user agrees to use no more than a maximum amount, then a second user can make a claim on water, and so on, until the regulator believes that the water is fully allocated. In dry years, senior users get water first.

Appropriative rights can be lost through abandonment. This differs from riparian rights, because senior users have stronger rights to water than others, and the rights mechanism implicitly has the ability to avoid overexploitation. Appropriative rights are often limited, or severely limited, in their tradable value. The rights regime might restrict users to historical uses of water, or impose specific restrictions on use to conform to "reasonable" use of the water. A loss in quantity of water rights that would likely come with the change to smart market rights could be counterbalanced by an improvement in the freedom to use water by the user and to trade for more or less allocation.

This allocation system can permit trades of allocated water. If the senior user is allocated water that is rarely used, they are likely to receive a windfall gain by trade. On the other hand, if the senior user exchanges their right for a smaller tradable share, then they may view the change as a loss of security, and they may expect compensation. These difficulties diminish if the regulator specifies water rights into groups with equal seniority. Then users can trade rights of the same seniority within a smart market.

There are ways that the market manager can recognize seniority explicitly in the initial allocations (Raffensperger 2011). Those methods rely on iterative use of the market model to estimate the maximum water available to each appropriative user. Briefly, the market manager can modify the water allocation model to maximize the allocation of each user (up to that user's maximum), starting with the most senior user and solving again with each next most senior user. We recommend implementers review that document.

A catchment can become damaged when implementers transfer appropriative rights into tradable rights. Because the appropriative rights are capped, most users rarely use their full allocations in any given year. In the absence of smart markets, water regulators can attempt to control water use by estimating the catchment's sustainable yield, which is based on past use rather than the maximum allowable pumping. If these maximums become tradable, then users can sell their unused allocations to new users, over-using the resource. Allowing trading of appropriative rights during a drought without a process of adjudication can lead to severe environmental impacts, as for example has happened in Australia. This risk exists even when all users have the same seniority. The problem is that the regulator based the old allocations on maximum water use, while a smart market allocation must work on actual water use.

Alternatively, implementers could convert the appropriative rights based on historical use. Because appropriative use must be "reasonable," users should accept the historical use criteria.

Interestingly, appropriative rights could probably work with the smart market. Implementers would have to specify a priority or queue system based on the probability distribution of inflows. Users could plausibly buy and sell their priority or place in the queue when the inflows were uncertain. New users could enter the market in an economically efficient manner.

8.6.5 Conversion of Reservoir Rights

Reservoir rights are usually clear cut, based on actual quantities in a reservoir and expected inflows. The implementer should be able to convert these directly to market allocations. Implementers should convert these rights in a way that incentivizes users to think about how much water to use now versus in the future. Allocation for a quantity of storage should attenuate at some rate, due to evaporation and losses to groundwater. Such losses would correctly encourage users to take water sooner rather than later. But the model must have some mechanism for valuing water at the end of the market planning horizon, so that the model saves an adequate amount of water for the following periods.

8.6.6 What to Do with Hold-Outs

Some users may not want to convert their existing rights to market allocations. They may wish to keep their rights under the old rights regime.

If the user's maximum take is sufficiently small, the market manager can treat the user's abstraction as a small fixed constant in the market-clearing model. The example of Marlborough in Sect. 5.5 showed that many users take so little water that the market manager can ignore them, even in aggregate.

If the user's variability in water abstraction is large enough to affect other users' abstractions, then the implementers will need to persuade or require the user to convert. The implementers could establish conversion rules, e.g., "All users must convert," or "All users with average take above *X* ml/year must convert."

Conversion of an earlier right to a new smart market right does not imply a requirement to trade. This change will affect users who will not trade, at least not initially, but many users could see the change as beneficial. The user would no longer be under the riparian system's uncertainty of security, or under the appropriative system's restrictions on historical or reasonable use of water.

The implementers could also allow some tolerance in the market-clearing model at the beginning, and encourage users to join. Conditions for joining the market would include all the necessary requirements, such as metering. As users realize the value of participating in the market, they are likely to agree to the conditions.

References

Easter, K. W., & Huang, Q. (Eds.). (2014). *Water markets for the 21st century: What have we learned?* (Global issues in Water Policy 11). Dordrecht: Springer Science+Business Media.

Porse, E., et al. (2015). Pumping for the masses: Evolution of groundwater management in metropolitan Los Angeles. *Geojournal*, 17 pp. doi:10.1007/s10708-015-9664-0, first online 7 August 2015.

Raffensperger, J. F. (2011). Matching users' rights to available ground water. *Ecological Economics 70*(6), 1041–1050.

Tietenberg, T. H. (2006). *The evolution of emissions trading: Theoretical foundations and design considerations.* http://www.colby.edu/personal/thtieten/edintro.pdf. Accessed 19 March 2017.

Chapter 9
Examples: Hypothetical and Real

Abstract This chapter gives examples, based on hypothetical hydrology data and real hydrology data. The first section describes how marginal cost pricing can help better allocate water. The second section contrasts assignment of water allocation with assignment of environmental impact quota. The third section covers different ways to scale allocations during a drought. The final two sections describe cases for groundwater and one reservoir, and surface water and one reservoir.

Keywords Marginal cost • Water allocation • Constraint rights • Examples

9.1 How Marginal Cost Pricing Can Help Better Allocate Water

The Arcadian Water Stewardship Agency (AWSA) allocates water in the Arcadian catchment. Society has agreed that the Willow River should have a minimum flow of 10, as measured at the town of Pinetown, and a minimum flow of 11 further downstream at Oak City.

However, Arcadians are concerned that the region's three water users, Ann, Bob, and Cal, are taking too much water. So AWSA investigated, and found that the river flow is too low at certain times of year. They have decided to institute a smart market for tradable water allocations. Figure 9.1 is a diagram of the Arcadian catchment.

The amount that the river flow decreases depends on the distance from each assessment point to each farmer's well. AWSA has used its water allocation model to estimate the decrease in river flow when each farmer takes 1 unit of water per day:

	Pinetown	Oak City
Ann	0.10	0.05
Bob	0.04	0.16
Cal	0.22	0.09

J.F. Raffensperger, M.W. Milke, *Smart Markets for Water Resources*, Global Issues in Water Policy 12, DOI 10.1007/978-3-319-55008-4_9

Fig. 9.1 Schematic of the Arcadian catchment

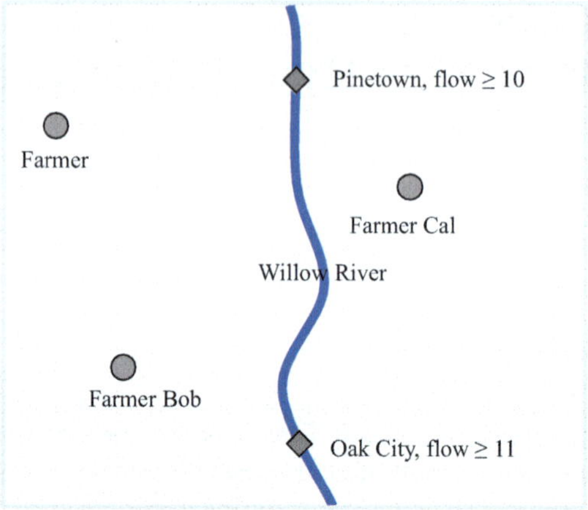

The purpose of this chapter is to demonstrate the flexibility of the smart market mechanism by examining a set of examples while developing some concepts related to market design and implementation.

So how much water should AWSA allocate to each farmer? Let's do some analysis, with A, B, and C being the quantity that each farmer takes.

$$\text{Flow at Pinetown} = 12 - 0.10\text{ A} - 0.04\text{ B} - 0.22\text{ C}.$$
$$\text{Flow at Oak City} = 13 - 0.05\text{ A} - 0.16\text{ B} - 0.09\text{ C}.$$

To be within the law regarding the environmental requirements:

$$\text{Flow at Pinetown} = 12 - 0.10\text{ A} - 0.04\text{ B} - 0.22\text{ C} \geq 10.$$
$$\text{Flow at Oak City} = 13 - 0.05\text{ A} - 0.16\text{ B} - 0.09\text{ C} \geq 11.$$

Simplifying, the legal requirement is:

$$\text{Flow at Pinetown} : 0.10\text{ A} + 0.04\text{ B} + 0.22\text{ C} \leq 2.$$
$$\text{Flow at Oak City} : 0.05\text{ A} + 0.16\text{ B} + 0.09\text{ C} \leq 2.$$

The regulator still doesn't know who should get the water, but knows only the result of a particular decision on water allocation. The regulator has no way to "blame" any farmer for taking too much water. All of them are part of the problem, and the regulator should consider all of them to protect the river. For example, if each farmer pumps 8 units, for a total flow of 24 units, the abstractions violate both environmental flow requirements. Should each farmer get an equal amount? What would be best for Ann? For the flow at Oak City? For the economy? Let's consider four scenarios for water allocation.

9.1.1 Scenario 1: Maximize Total Pumping

The regulator can write the problem like this:

$$\begin{aligned}&\text{Maximize A} + \text{B} + \text{C},\\&\text{Flow at Pinetown}: 0.10\text{ A} + 0.04\text{ B} + 0.22\text{ C} \leq 2.\\&\text{Flow at Oak City}: 0.05\text{ A} + 0.16\text{ B} + 0.09\text{ C} \leq 2.\end{aligned}$$

The result is that A = 17.1, B = 7.1, and C = 0. The total water taken is 24.2 units. The computer solution ensures that Pinetown and Oak City each get their required flows. However, Cal doesn't get any water. Is this fair? Probably not. Taking the maximum amount of water is not society's real objective.

9.1.2 Scenario 2: Maximize Total Pumping, Giving Each Farmer Equal Amounts

The regulator can write problem like this:

$$\begin{aligned}&\text{Maximize A} + \text{B} + \text{C},\\&\text{Flow at Pinetown}: 0.10\text{ A} + 0.04\text{ B} + 0.22\text{ C} \leq 2.\\&\text{Flow at Oak City}: 0.05\text{ A} + 0.16\text{ B} + 0.09\text{ C} \leq 2.\\&\text{A} = \text{B} = \text{C}.\end{aligned}$$

Solution: A = B = C = 5.55.

Again, the computer solution ensures sufficient flow at Pinetown and Oak City. The total amount of water taken is only 3*5.55 = 16.67 units, less than with the previous method. This allocation is probably a bad idea. The flows at Pinetown and Oak City are the same, so the environment is no better off. Also, maximizing pumping ignores the value of the water taken.

9.1.3 Scenario 3: Maximize Economic Value of Water, Letting Users Bid for Water, with Price as Bid

Suppose each user has an initial allocation of 5 units of water. Suppose Ann bids \$3 per unit, Bob bids \$4 per unit, and Cal bids \$11 per unit. Assume the bids are private, so only the bidders and the market manager can see them. The regulator, now a market manager, can write the problem like this:

$$\text{Maximize } 3\,A + 4\,B + 11\,C,$$
$$\text{Drawdown at Pinetown}: 0.10\,A + 0.04\,B + 0.22\,C \leq 2.$$
$$\text{Drawdown at Oak City}: 0.05\,A + 0.16\,B + 0.09\,C \leq 2.$$

The best solution is this: A = 0, B = 8.2, C = 7.6. For these allocations, the drawdown at Pinetown is 2.0 and the drawdown at Oak City is 2.0. Using the initial allocation of 5 for each user, the market manager can calculate the trades.

- Ann bought 0–5 = −5 units of water, a sale, and pays −5*$3 = −$15, taking care with the signs.
- Bob bought 8.23–5 = 3.23 units of water, and paid 3.23*$4 = $12.92.
- Cal bought 7.59–5 = 2.59 units of water and paid 2.59*$11 = $28.49.

Each user paid based on the price the user bid. Total water use is now 15.82, less than for Scenario 2. How can this be better? For the water values of $3 (Ann), $4 (Bob), and $11 (Cal), and Scenario 1 allocations, the economic value of the water used is (17.1*$3) + (7.1*$4) + (0*$11) = $79.70. For Scenario 2 allocations, the economic value of the water used is: (5.55*$3) + (5.55*$4) + (5.55*$11) = $99.90. For Scenario 3, the economic value of the water used is: (0*$3) + (8.23*$4) + (7.59*$11) = $116.41. So, yes, the economic value is higher in Scenario 3. Why is that? In Scenario 1, society assumes that water has equal economic value to everyone, while in Scenario 2 society assumes that each unit of water use has the same environmental impact. Scenario 3 allows society to consider users' value of water as well as the variable environmental impacts of water abstraction. The model will maximize the value of water when it has the correct objective.

Assuming that users bid their true value for water, Scenario 3 gives the greatest economic value from the water resource, even though users take less water than in the other scenarios. All three water users have benefited. Even Ann, who uses no water, has benefitted because she was able to sell her water allocation.

The market manager's revenue is −$15 + 12.91 + 28.54 = $26.45. Most likely the users will not be happy about making net payments to government.

9.1.4 Scenario 4: Maximize Economic Value of Water, Letting Users Bid for Water, with Marginal Cost Pricing

In the previous scenario, users paid as bid. With a price-as-bid auction, users are not likely to bid their true values. Sect. 6.4.4 and Sect. 6.6 discuss this behavior, but the issue of bidding behavior may be clearer with this example. Cal and Bob may wonder whether they had overbid. Ann might be wondering whether she had underbid, and she would be pretty sure she had underbid if she found out what the others had bid. All three users will be tempted to shade their bids away from their true value for water. Buyers will prefer to bid below, and sellers above, their honest reserve prices. The problem is worse if the market managers keep users' bids secret, which they will want to do. Users have no easy way to determine the market price or going rate for water. They know only whether their bids were accepted or not.

Thus, users do not reveal their true demand values for water. Further, they will be tempted to try to find out what other people are bidding or have bid previously, to avoid bidding incorrectly, and they will be tempted to coax other users to change their bids. The result is that the model will misallocate water, as users try to game the market, and some people will cause mischief in seeking information. To eliminate this poor incentive, the market manager can charge users based on the dual prices from the linear program, more in keeping with a second-price auction.

The solution to the linear program in the previous scenario gives dual prices of \$44.30 for the Pinetown constraint, and \$13.92 for the Oak City constraint. To calculate the trades with market prices, the market manager applies the price of each constraint to each user's coefficients, and then calculates the sum. The manager calculates the total value of Ann's impact as the price at Pinetown times her impact there, plus the price at Oak City times her impact there: 0.10*\$44.30 + 0.05*\$13.92 = \$5.13, better than the \$3 that she bid. The true value of her sale is therefore −5*\$5.13 = −\$25.65. She could indeed have bid higher than \$3 per unit.

Similarly, the market manager calculates the value of Bob's impact as 0.04*\$44.30 + 0.16*\$13.92 = \$4, exactly what he bid. The value of Cal's impact is 0.22*\$44.30 + 0.09*\$13.92 = \$11.00, exactly what he bid. A model with two constraints tends to have a solution with two positive variables.

Thus, the market manager could publish the "market rates" as \$44.30 for Pinetown and \$13.92 for Oak City. These prices will make sense to users as the *market* prices, the "going" rates. By offering Ann the higher price of \$5.13 instead of the \$3 that she bid, she can bid her true value, and she will not be injured. She has no reason to try to guess other people's bids, or to spend much effort getting them to change their bids.

The market manager's revenue would then be \$15.82 instead of \$26.45. Section 9.3 examines how to adjust initial rights so the market manager breaks even, but we invite the reader to think about the problem at this point.

9.2 Assignment of Water Allocation Versus Assignment of Environmental Impact Quota

To now, we have described the allocation as corresponding to water taken by the user. If the regulator wishes, the regulator can define users' allocation of water based on the *impacts* of the user on the environmental assessment points. The market-clearing model in this case would be based on bids for shares of the impact on a given assessment point. In this way, it would be like fishing companies bidding for fishing quota. Just as a fishing company would bid for quota for tuna separately from haddock, with this method water users would bid for environmental impact quota on river flow at point A as well as quota for seawater intrusion impact at point B.

This idea is not new. Montgomery (1972), in the context of air pollution, defined “ambient” permits (perhaps an unfortunate name), where the regulator specified separate rights for each assessment point. A user wishing to discharge at his or her own location had to obtain sufficient separate rights for each assessment point that his or her discharge would impact. Montgomery’s method allowed for multiple assessment points. Full details of this approach (originally suggested by E. Grant Read) have been previously published (Raffensperger 2011).

Consider our simple example in Sect. 9.1, with Ann, Bob, and Cal, and the following market-clearing model.

$$\begin{aligned}&\text{Maximize } 3\text{A} + 4\text{B} + 11\text{C}.\\&\text{Flow at Pinetown}: 0.10\text{A} + 0.04\text{B} + 0.22\text{C} \le 2.\\&\text{Flow at Oak City}: 0.05\text{A} + 0.16\text{B} + 0.09\text{C} \le 2.\end{aligned}$$

The solution is A = 0, B = 8.2, C = 7.6. The two constraints are associated with the flows at Pinetown and Oak City. For the allocations A = 0, B = 8.2, C = 7.6, the total drawdown at both Pinetown and Oak City is 2. The market manager can calculate the drawdown from each user.

$$\begin{aligned}&\text{Flow at Pinetown}: 0.10*0 = 0, 0.04*8.2 = 0.328, 0.22*7.6 = 1.672.\\&\text{Flow at Oak City}: 0.05*0 = 0, 0.16*8.2 = 1.312, 0.09*7.6 = 0.684.\end{aligned}$$

Ann has no impact on Pinetown or Oak City, because she is taking no water. Bob is taking 8.2 units of water, which draws down Pinetown by 0.328, and Oak City by 1.312. Cal is taking 7.6 units of water, which draws down Pinetown by 1.672, and Oak City by 0.684.

To define rights in terms of impacts instead of quantity of water abstracted, the market manager can specify users’ allocations to the environmental assessment points as follows:

	Ann	Bob	Cal
Pinetown	0	0.328	1.672
Oak City	0	1.312	0.684

Each user now has a right to two things instead of one. The insight here is that constraints define rights—as the availability of a thing and the ability of people to consume that thing. This type of market would make it easier for users to understand changes in their initial allocation over time. If a river needed extra water to protect an endangered species, for example, the market manager would reduce rights to the river, and affected users would receive an economic signal to reduce their use of river water. Just as a new study of tuna breeding might show that tuna reproduction is slower than expected, leading to a decrease in tuna quota, the environmental impact quota at one assessment point could change independent of each other assessment point.

Depending on how the regulator implements the environmental impact quota, understanding and managing that quota could overwhelm users. A groundwater model with ten assessment points would mean that users would need ten different environmental impact quotas. The auction might require bids for water in 78 future periods, implying 780 quotas that the user has to track, and 780 bids that the user has to enter in each auction. Only in rare situations would users be willing to track the additional allocations and bids needed for a smart market based on environmental impact quota. One way to use an allocation system based on impact quotas without overwhelming users would be to have users bid for water and then have the bidding platform convert those bids into bids for impact quotas before the smart market optimization. The implied impact quotas could appear on a separate tab of the bidding web page to aid in transparency.

The constraint-based allocation method has many advantages. Should one constraint need changing, users face the minimum adjustments necessary. When the market manager runs the market-clearing model, the allocations will rebalance. The attention of users would be on the real impacts of the underlying water allocation problem, rather than on volume of water when each unit of water used has a different impact depending on where and when it is abstracted.

9.3 Different Ways to Scale Allocations During a Drought

9.3.1 A Groundwater Example with 18 Wells and Three Types of Constraints

We present here a simple groundwater example, modified from Greenwald (1998), McDonald and Harbaugh (1988), and Harbaugh and McDonald (1996). Figure 9.2 is a diagram of the catchment with 18 wells. At the left is the sea, a stream runs through the middle of the area, the other boundaries do not allow flow, and the lines are the contour lines of equal water pressure prior to the auction. Arrows and triangles indicate environmental constraints.

Some pairs of wells are so close that they are in the same MODFLOW grid, and are considered identical hydrologically. These are wells 2 and 17, wells 1 and 16, and wells 3 and 18. Wells 2, 3, 17, and 18 are in layer 2. Wells 1 and 16 are in layer 3. Wells 4 through 15 are in layer 1.

This small catchment receives water only from rainfall, and all water users have 1 unit of water allocation under normal conditions. The hydrologist modeled the aquifer with three layers. Layer 1 is unconfined and the other two are confined. Aquitards limit water flow between the layers, with no flow below the third layer. At the left side the water level (or head) is at 0 m, representing an ocean. The other

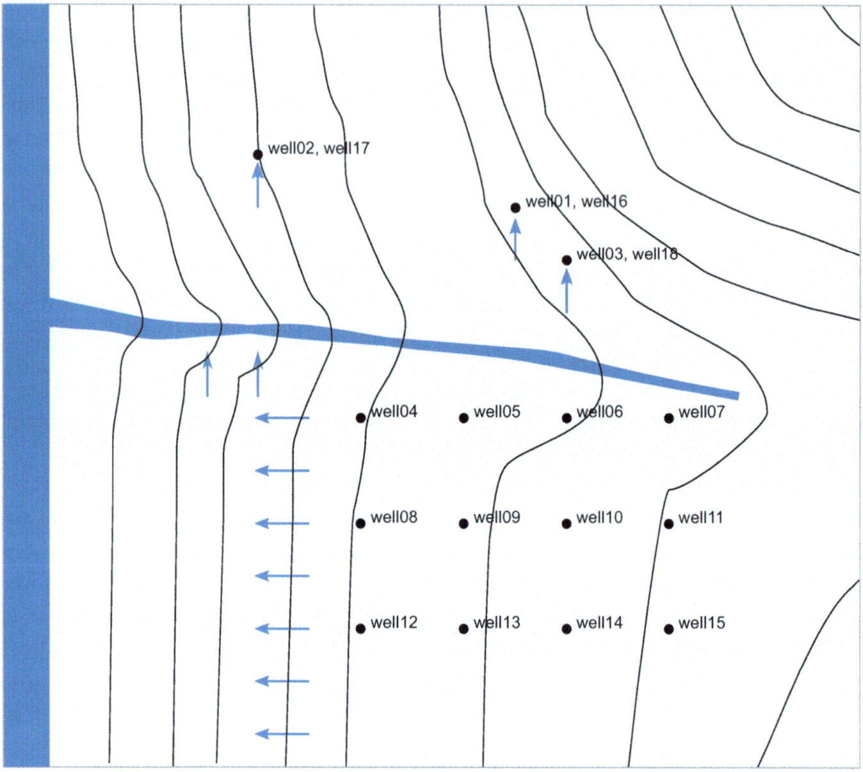

Fig. 9.2 A catchment with 18 wells and a stream, and a coast at left

three sides allow no flow of water and represent groundwater divides or impermeable formations (e.g., rocks). The top two layers connect to the surface water boundary. A small stream runs down to the left from the middle of the catchment. Only the top layer of the aquifer connects to the stream. The whole catchment receives recharge to layer 1 at $9.14 * 10^{-7}$ m/sec ($3 * 10^{-6}$ ft./sec), which is a value that results in a drought condition. This recharge is not enough to meet the combined demands of water pumped from the wells and the environmental constraints. Harbaugh and McDonald (1996) list other inputs.

Water flows will be in units of 142 L (5 cubic feet) per second. (The original example is in cubic feet.) Thus, if a user gets 0.5 units of water allocation, then the user has allocation for 71 L per second. The contour lines in Fig. 9.2 show the steady-state groundwater levels (heads), rising from 0 at left to 120 in the top right.

The regulator uses three types of environmental constraints in the example. First, to maintain stream flow, a head constraint ensures a minimum discharge to the stream from Layer 1. The flows will satisfy the minimum discharge requirements if the aquifer head remains at least 0.3 m (1 ft) above the stream elevation at two key assessment points, shown as the two short vertical arrows in Fig. 9.2.

Second, to prevent saltwater intrusion from the coast (at left in Fig. 9.2), the regulator specifies a minimum discharge of water in Layer 1 towards the constant head boundary. Therefore, the model requires a minimum head difference of 1.14 m (3.75 ft) in Layer 1 at seven control locations, as the horizontal arrows show in Fig. 9.2.

Third, total drawdown is limited to a maximum of 9.14 ms (30 ft) for layer 2 and layer 3 wells. These constraints limit the long-term exploitation of the resource. Total drawdown is based on steady-state heads prior to the drought. Triangles in Fig. 9.2 indicate these locations. (We omitted the fourth constraint used in Greenwald's original example.)

Suppose users in the region have agreed to use a smart market. To develop the market-clearing model, the market manager needs to know the marginal effect of extraction at each well on each of the constraint locations. These marginal effects depend on the character and behavior of water flow. When a farmer pumps water from the ground, the level of water—the head—falls nearby. Over hours, days, and weeks, this effect radiates out away from the well, wave-like, with a smaller effect further away and later in time. These effects of pumping on head will be the coefficients in the LP constraints.

Consider the constraint associated with maintaining flow in the stream. The stream flow depends on the aquifer around it. If irrigators take groundwater near the stream, the stream will tend to fall as the stream recharges the aquifer. Similarly, if the aquifer rises, the stream tends to rise as it is recharged by the aquifer. The aquifer head, at each location and time period, is therefore the key state variable, and thus the market manager needs to know the effect of pumping at one location on the aquifer head at other locations. The local hydrologist can tell us the maximum allowed drawdown in head to ensure that the sustainability constraints will be satisfied. Figure 9.3 is part of one row in our linear program: total drawdown effect at the assessment point at well 16 must be less than or equal to 5.43.

The constraint says that total drawdown must be less than 5.43 units of head. The coefficients are the drawdown at the assessment point at well 16 in month 3. Note that pumping in all of months 1, 2, and 3 affects the assessment point in period 3. Well 18 appears to have one of the largest impacts on this assessment point—1.817—since it is adjacent to assessment point 16, located at well 16, as shown in the top center of Fig. 9.2.

If the variable $q4,1 = 1$, the meaning is that well 4 takes 1 unit of water now. As a result, the response matrix in Table 9.1 tells us that the head at assessment point 1 would immediately fall by 0.0602. Next month, it would fall by 0.1545. In the month after that, it would fall by 0.1765.

For simplicity, assume that all users have the same demand and price curve as shown in Table 9.2 and Fig. 9.4. Each user is willing to sell from 0.17 down to 0.05 units of water only if the price is $4 or more. Each user is willing to buy water to reach 0.4 units if the price is $3 or less, and willing to buy water to reach their pre-drought allocation of 1.0 units if the price is $2 or less, and willing to buy excess water, if available, to 5.0 units if the price is $1 or less.

Effects from pumping in	Drawdown row for period 3, assessment point at well 16, variables q[well],[period].
period 1	0.177 q4,1 + 0.308 q5,1 + 0.344 q6,1 + 0.315 q7,1 . . .+ 0.665 q18,1
+ period 2	+ 0.155 q4,2 + 0.308 q5,2 + 0.329 q6,2 + 0.242 q7,2 . . .+ 0.972 q18,2
+ period 3	+ 0.060 q4,3 + 0.164 q5,3 + 0.171 q6,3 + 0.083 q7,3 . . .+ 1.817 q18,3
	≤ 5.43, allowable drawdown in period 3 at the assessment point at well 16.

Fig. 9.3 Part of the drawdown constraint for the assessment point at well 16

Table 9.1 Part of the linear program matrix for the groundwater problem

	q4,1	q4,2	q4,3	q5,1	q5,2	q5,3
Assessment point 1, period 1	0.0602	0	0	0.1644	0	0
Assessment point1, period 2	0.1545	0.0602	0	0.3077	0.1644	0
Assessment point 1, period 3	0.1765	0.1545	0.0602	0.3077	0.3077	0.1644
Assessment point 2, period 1	0.0288	0	0	0.0145	0	0
Assessment point 2, period 2	0.0825	0.0288	0	0.0602	0.0145	0
Assessment point 2, period 3	0.1040	0.0825	0.0288	0.0960	0.0602	0.0145

Table 9.2 Values for user demand and price curve

Price	$1	$2	$3	$4	$5
Quantity	5	1	0.4	0.15	0.05

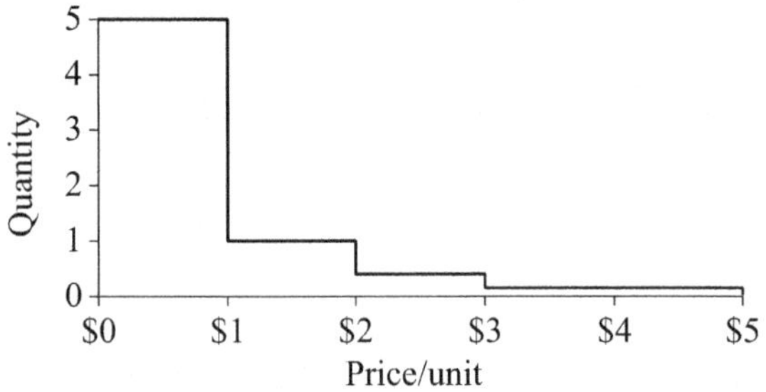

Fig. 9.4 User demand and price curve for water

We assume further that these values are true reservation prices. If the price of water were more than $5 per unit, a user without water would buy no water. Conversely, this reservation price means that if a user had allocation for water and the price were more than $5, the user would sell all the water. In the real world, users are likely to have different value functions for water, and we would not expect a user to have such large steps in the blocks. In a real bidding situation, the user will have a reasonable idea of the likely price and quantity (especially if the market manager runs trial auctions before a final auction) and will be able to bid around those values with much more precision.

We will use this example to explore various methods to manage over-allocation within a smart market. We define over-allocation as when the regulator finds that flows will not satisfy an environmental requirement when all users take their entire allocation during some time period. We pointed out that catchments are over-allocated almost everywhere. The first reason to convert rights to allocations is so that the market manager can adjust those allocations to match the available water, to prevent over-allocation.

The smart market requires that allocations match the actual expected water. Of course, expected water does not always appear. The market manager cannot predict the inflows exactly. Any misallocation in a smart market is likely to be small compared with the misallocation and over-allocation that is common at present throughout the world. Nevertheless, the market manager must be able to manage it. The smart market design cannot prevent over-allocation, but it has tools to manage it. How should the market manager adjust these allocations when the expected water does not appear, or when some other factor results in allocations misaligned with actual water?

We will consider several methods of proportional adjustments using the parameter α, which we define as the fraction of the water rights that the market manager allocates to users. We have already used α in Sect. 7.2.4. We will sometimes show α with various subscripts:

α_w = the fraction of user w's water rights that the market manager allocates to user w, or.

$\alpha_{w,t}$ = the fraction of user w's water rights that the market manager allocates to each w in period t.

Proportional adjustments are a common method used in water basins for adjusting water allocations (e.g., in Southern California as described by Porse et al. 2015). Section 9.3 covers much of the same material (with different examples) as previously published work (Raffensperger 2011), which we encourage implementers to review.

9.3.2 Method 1: Proportional Reduction Based on Tightest Constraint

In this first method, we explore what would happen if all users have their allocation decreased by the same percentage. The market manager has multiple ways to choose the percentage. In this method, we calculate the proportional reduction by running the LP for the catchment with the objective of maximizing the proportion that the manager can give to all users, subject to the environmental constraints. All users have the same α, which may seem fair to many people. This selection of α is independent of the bidding and trades and could be readily calculated by the market manager prior to auction.

After solving this problem for our example, the proportion is 17.4%. Figure 9.5 shows the price map and water trades for this case. Because we give all users an

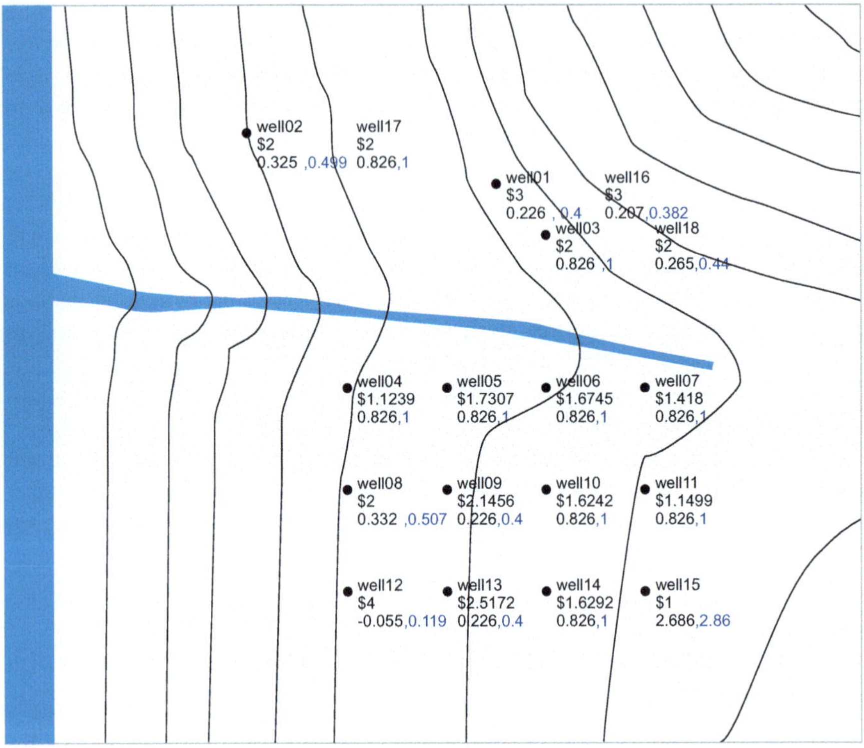

Fig. 9.5 Price map for the scenario where allocations are reduced proportionally to meet the tightest constraint ($\alpha = 17.4\%$). Each well shows the price, the trade, and the final allocation (italics)

initial allocation of 1.0, the proportional reduction results in users having a new allocation of 0.174, implying total water allocation for 18 wells of 3.14, far less than the 18 units used without a drought; still, most water users purchase water to bring the total consumption up to 15.

Because all users have had their allocations reduced, they are probably willing to buy water, if the price were low enough. Table 9.2 shows that users with 0.174 units of water will buy up to 0.226 units of water as long as the price is not greater than $3 per unit, up to 0.826 as long as the price is not greater than $2, and up to 4.826 if the price is not greater than $1. The users still have sell bids, if the price is sufficiently high. Figure 9.5 shows that all users except well 12 buy back the water they need.

The user named well04 buys 0.826 units of allocation from the common pool at a cost of $1.124 per unit. Because of the drought, the market manager reduced the user's allocation from 1 to 0.174, and this user has purchased 0.826 of allocation to bring the total allocation post-market to 1.0 unit. In contrast, user well12 sells 0.055 units of allocated water back to the pool at a value of $4 per unit. After the market, this user's allocation has reduced from 0.174 to 0.119.

Everyone is buying or selling water allocation. No series of one-to-one transactions could plausibly replicate this result. The trading going on is far too complex for that. The buying and selling are from and to a common pool. This rich and complex map of water values and trades is a unique and important feature of a common pool market for exchanging water allocation.

Trades occur even though all users have the same value for water. Wells 8, 9, 12, and 13 get only a little water. The saltwater intrusion constraints are binding. Water is more expensive near environmentally sensitive locations, and users close to locations are likely to pay more. Of course, this is exactly what the regulator wanted—users with the biggest impact should back off first. The value of water for well 12 is $4, while the value of water for well 15 is only $1. However, the stream flow constraints are loose even during the drought. The aquifer drawdown constraints near wells 16, 17, and 18 are tight, and so is the seawater intrusion constraint near well 12. The $4 cost for water at well 12 reflects its marginal value. If well 12 were to take another unit of water, other users would have to give up sufficient water to ensure that salt water did not intrude, and this loss would cost the economy $4. Well 12 has a large financial incentive to conserve water, and to use it efficiently.

This variability in price may strike some as being inequitable, but it signals that users should consider only the most productive uses of water where environmental impacts are likely. This shifts water use to where it causes less harm, further away from the coast. In this example, the high price at well 12 has led this user to sell water. The willingness of well 12 to sell has allowed the neighbors to use more water than they otherwise would in a drought, but this required the neighbors to pay well 12. All users have traded, and all users have benefited from trading, while respecting the environmental requirements.

Water availability depends on where and when it is taken. The catchment's quantity of water available is not fixed. Rather, the quantity that users can sustainably take depends on how close the withdrawals are to environmentally sensitive locations. Withdrawals close to those sensitive locations tend to disproportionately reduce the amount of water that is available further away and so have a higher price, while the price decreases for users further from these sensitive locations.

With this approach, the market manager will take in more funds from purchases than he or she will pay for sales of water from the common pool. With this method, the market manager has revenue of $18.51. The proportional adjustment of the allocation is the key factor in deciding whether users or the market manager get the money.

The allocation has been reduced so much because the proportional reduction has been set by the well most impacted by the drought, well 12, because it is hemmed in between the sensitive coast and the impermeable barrier. Of course, if well 12's neighbors were to give up some water, one might reasonably expect that well 12 could take a bit more, and this is true, but still only to a certain limit. All other well users are penalized in this scenario because of an environmental constraint in one corner of the catchment. The other users have their allocation cut and must purchase water allocation from the market, even though water use did not need to be so limited in their area. So the proportional reduction sounds fair at first, but users are likely to complain that it is not fair.

9.3.3 Method 2: Pure User-Pays

In this scenario, assume that all users' water allocations are zero. Users must purchase all the water, so $\alpha = 0$. This corresponds to a pure user-pays market, where users pay for all water at the time of use. While it is unnecessary for proper economic allocation with our market design, and may be untenable politically, the example is instructive.

Figure 9.6 shows the solution, with spot prices and water allocated. Well 2, for example, was allocated 0.499 units of water, at a spot price of \$2 per unit, and therefore paid \$0.998. Following standard optimization theory, a user's spot price is the marginal cost to all users if that user were to take another unit of water. The user at well 2 would have been willing to pay \$5 for the first 0.05 units of water, \$4 for

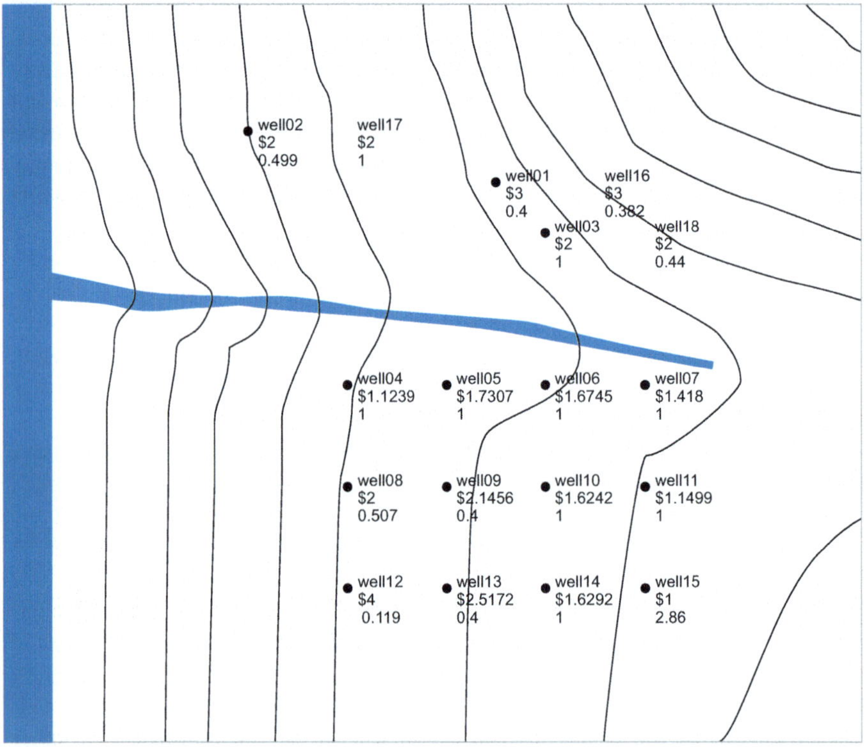

Fig. 9.6 Price map for the pure user-pays scenario, $\alpha = 0\%$. The \$ value shown is the well's spot price. The second value is the units of water allocated. Each well shows the price and the final allocation

the next 0.1 units, \$3 for the next 0.25 units, and \$2 for the remaining 0.099 units up to the allocation of 0.499 units. In total, the user would have been willing to pay \$1.598 (\$5*0.05 + \$4*0.1 + \$3*0.25 + \$2*0.099), but had to pay only \$0.998. The consumer surplus at well 2 is therefore \$0.60.

Most users received 1 unit of water allocation; however, well 12 was allocated only 0.119 units of water, was charged \$4 per unit of water, and had to pay \$4*0.119 = \$0.476. The consumer surplus here is low, but positive, at \$0.05. The price was high at well 12 because it is near the coast and near the impermeable boundary, and thus faces the saltwater intrusion constraint directly.

Wells 8 and 13 paid more for the same reason, and because of the flow constraints on the stream. Wells 1 and 16 face a higher price due to active aquifer drawdown constraints. Thus, even though all users bid identically, the price varies over the aquifer, because the hydrology varies, and because different users are nearer more environmentally sensitive regions.

Imagine, now, a new user who wishes to enter the market. A reasonable course of action would be to move in near wells 11 and 15, where water is cheap. The new user will have to bid higher than \$1.15 or so to take water away from the existing users, but a bit of competitive bidding is expected as the market adjusts to the new user. The user would naturally avoid the environmentally sensitive area near well 12.

Summing the total amount that all users pay, as their allocated quantities times their spot prices, the result is that the market manager nets \$24.80. The amount paid to the market manager would vary with the user demand curves and the other assumptions, with the manager's revenue rising as the resource becomes relatively scarcer. This is more than the \$18.51 that the market manager earned in Method 1, and is the maximum amount that the market manager could earn by varying α.

Notice though that the water allocations after trades are the same as for the first method, and the total water use in the catchment is still 15 units as with Method 1. Users would object to a user-pays smart market; they will much prefer a revenue-neutral trading system.

9.3.4 Method 3: The Market Manager Buys Excess Allocation

In this example, we analyze the case where users have rights to more water than is available, and where the market manager has no authority to decrease water allocations during a drought but still wants to run a smart market. Before the auction, all users have an allocation of 1 unit and so the total of the rights is 18. The total water allocated after the auction is 15 as with Methods 1 and 2. Nevertheless, the market manager runs the auction with $\alpha = 100\%$, meaning no across the board reductions in allocations prior to running the market optimization. This means that the market manager has had to buy 3 units of water from users within the market under this method.

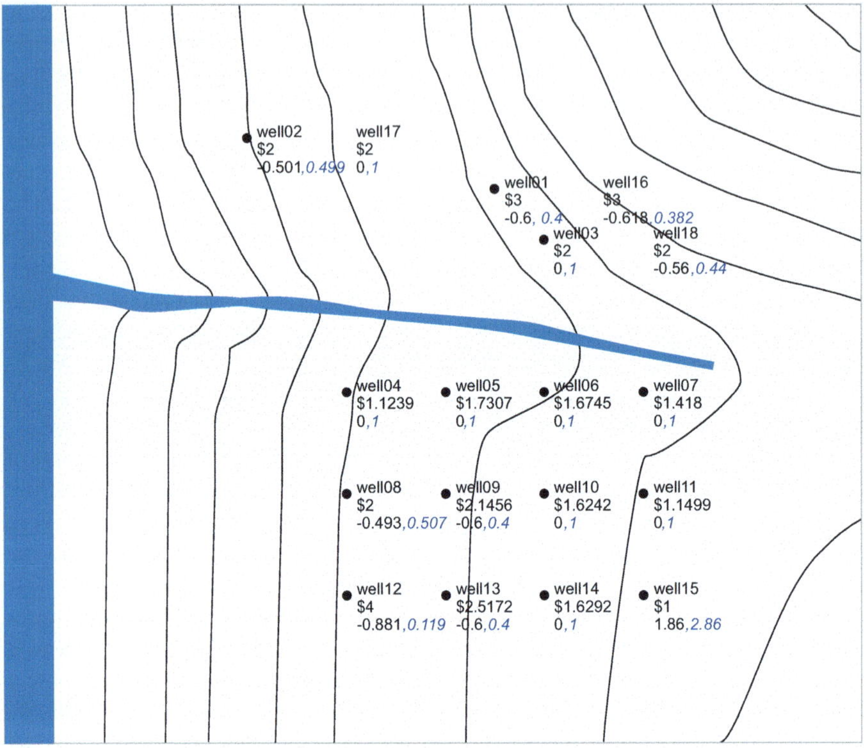

Fig. 9.7 Price map for the over-allocation scenario (α = 100%). Each well shows the price, the trade, and the final allocation (italics)

Figure 9.7 shows the spot prices and the water allocations. Because the hydrology has not changed and nor has the users' value of water, the set of water allocations that optimizes consumer surplus, given these environmental constraints, is the same as in the previous case. Figure 9.7 shows purchases from the water manager (positive values) or sales to the water manager (negative values). For example, well 2 sold 0.501 units of water at $2 per unit, resulting in a final allocation of 0.499 units. Well 15 purchased 1.86 units of water at $1 per unit, resulting in a final allocation of 2.86 units.

With the user demand given in Table 9.2, users will purchase additional water only when the price is $1 or less, and will sell it only when the price is $2 or more. Those users who did not trade, such as wells 4, 5, 6, and 7, used their existing allocation of 1 unit of water without money changing hands. Their local prices are within their indifference limits for this quantity of water. The optimization procedure is able to provide an estimate of the marginal value of water at these locations even without their participation in the auction. This time, the market manager had to pay $11.23 because the market manager had to buy water from the users to satisfy the environmental constraints. This money may be viewed as a subsidy from government to users for refraining from using water to protect the environment, which is a payment that people outside the water market might find objectionable. Users profit from using the water, but they are also profiting from *not* using water. Those people who think "users should not profit from selling water" will be outraged.

In some parts of the world, including Australia (McGauran 2006; NSW DNR 2006) and the state of Georgia in the U.S. (Cummings et al. 2004), government has recognized that catchments are over-allocated, and have bought water permits to bring them closer to the physical availability as adjusted for environmental needs. In theory, if users know that government *must* buy back a certain quantity, and they get a chance to sell to government, users could bid up the price as high as they wish. In the worst case, water users could collude and decide to all double their bids, resulting in double the payouts to water users for not using water. Section 10.2 explores more fully this case of collusion and over-allocation. Even without collusion, the payouts to not use water can be significant and ongoing.

That would not be politically tenable. Buying water allocation from users as a one-off agreement (whether to protect the environment or to match permits with physical water) may be politically feasible or even expedient, as it was in Australia and Georgia, but regularly buying water allocation from users is probably not politically or economically feasible. Fortunately, users have difficulty bidding up the price of water, simply because of the number of bidders. We take up this issue again in Chap. 10.

We invite the reader to review the price maps in each of the four methods of allocation. Note that the price and final allocation at a given well are the same between the four methods. The model produces identical optimal solutions for all feasible initial allocations. The total value of the water consumed is the same. This demonstrates the theorem by Coase (1960) that, without a transaction cost (which our system assumes), any pre-trade allocation is equally efficient.

9.3.5 Method 4: Users Trade

The full-allocation $\alpha = 100\%$ method ended with a loss for the market manager, while the proportional reduction method ended with positive revenue for the market manager. In between those two approaches, the market manager could adjust α after everyone has bid, taking into account users' willingness to sell, to find α where the market manager's revenue is zero. At this value of α, users will buy and sell with each other, and the market manager will serve only as a broker. We term this a "users trade" condition for the market.

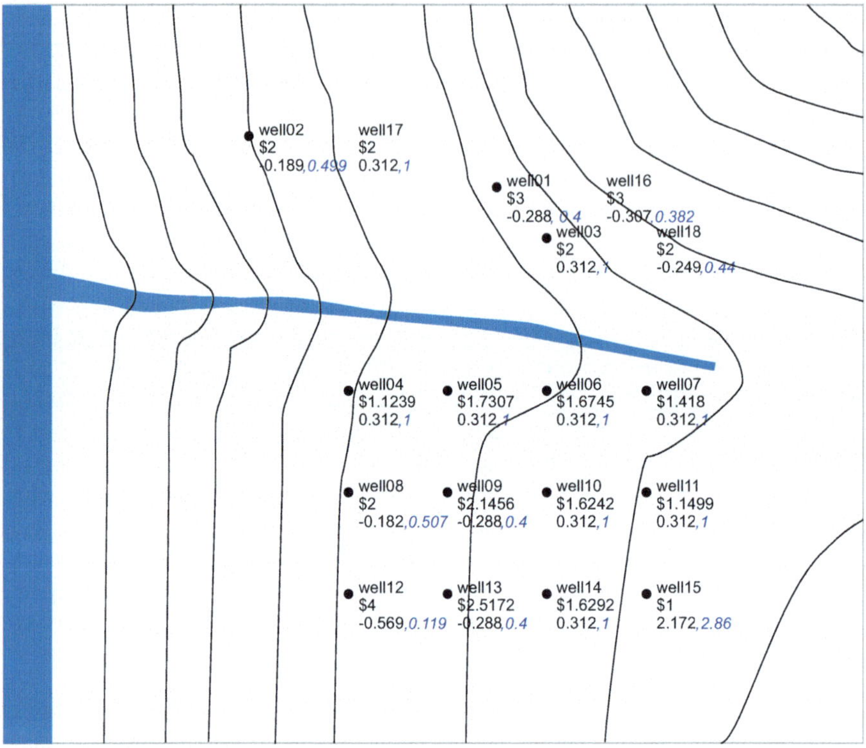

Fig. 9.8 Price map for the users-trade scenario ($\alpha = 68.8\%$). Each well shows the price, the trade, and the final allocation

For our hypothetical example, a users-trade condition requires $\alpha = 68.8\%$. Figure 9.8 shows the resulting price map. With the previous proportion $\alpha = 17.4\%$ in Fig. 9.5, well 8 had an initial allocation of 0.174, and had to buy 0.332 units of water. In this scenario, well 8 has an initial allocation of 0.688, then sells 0.182 units of water, and finishes with 0.506 units of water. Users who did buy, such as well 4, had to buy much less than they would have had to buy with the normal proportional reduction scenario.

As with the previous method, in this method the amount bought or sold depends heavily on the most-constrained part of the catchment. Users will not seem to think so badly of it, because they are not nearly as far out of pocket as in the proportional reduction case, and because users as a group do not pay out to the market manager. But the problem remains that, if one part of the catchment has reached an environmental constraint, prices will go up and users nearer that point will need to pay more to keep their allocation at pre-drought levels.

This method has the disadvantage of relying on users' bids. Users might feel that their drought-adjusted initial allocation depends on what they are willing to pay for it, and that this is a back-door user-pays mechanism, even though the market manager's revenue is zero. They would ask with some concern, "Why do you need my bid to determine my initial allocation?" This method, or a variation on it, could work if the market manager can earn users' trust.

9.4 Groundwater and One Reservoir

This case is modified from an example by Ahlfeld et al. (2005), with one reservoir upstream from a valley with ten farmers, and we have used it to help show water users and water policy experts how a smart water market functions. Here is the story to go with it: The farmers take groundwater, which interacts with a river, as shown in the schematic in Fig. 9.9. The river splits at the bottom of the catchment. Four meters record river flow. As often happens, some farmers want more water, and some farmers say reservoir releases are timed badly. The valley catchment is over- or under-allocated, depending on reservoir releases. The water agency has observed that flows are often too low, especially near farmers 9 and 10 as measured by the two meters near those farmers. The power company says it must plan for winter, so it cannot release whenever farmers wish. Perfect hydrological data cannot resolve the contention—the basic problem is not one of uncertainty, but rather lack of a good mechanism to negotiate among many players.

We treated the reservoir as "out of the market" (see Sect. 6.3.3)—the reservoir operator ensured feasibility of reservoir operations, not the market manager. Using GWM (Ahlfeld et al. 2005) and MODFLOW, we then developed the response matrix, which specifies the relationship between the pumping wells and the aquifer heads at each assessment point. We specified bounds for aquifer heads at the assessment points, implying minimum flows in the river, as if they had been set by an environmental authority. We therefore had the complete linear program, but without the users' values for water.

In Teasley et al. (2010), we used this case to determine whether the reservoir operator would have significant market power. When the reservoir operator restricted releases, prices went up, but the reservoir operator received less money because the outflow was much less. Because the farmers could still take from groundwater, prices for the farmers rose unevenly and slightly; in fact, some prices were unchanged. Realistically, the reservoir operator would probably have to release water to meet its power generation commitments, and may be responsible for the downstream environmental flows.

We have used this case to demonstrate to an audience how the market works, in a simplified format. We invited the audience to submit bids via cell phone, accepting them via Google Forms, and then we solved the model with a spreadsheet. The planning horizon has eight periods; we assume conveniently that bids, reservoir releases, and abstractions are identical across time periods.

Figure 9.11 shows the bid form for Farmer 10. Figure 9.12 shows part of the optimization spreadsheet with a solution. We also showed the audience some of the numerical constraints in Excel, to convince them that the environmental requirements were satisfied, as in Fig. 9.10.

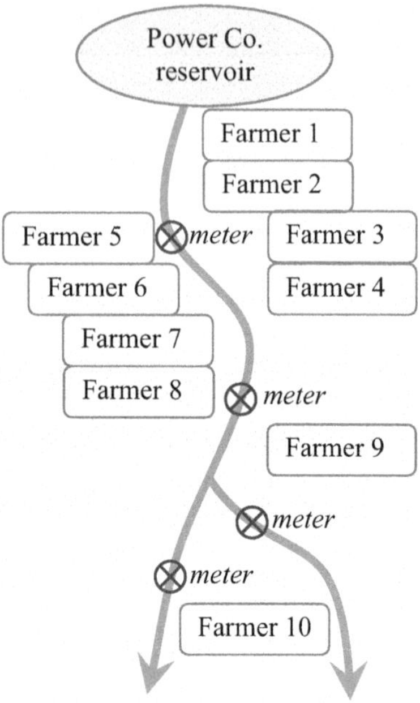

Fig. 9.9 Schematic of a reservoir, river, and groundwater catchment. Farmers take water from groundwater downstream of the reservoir

We pointed out that the environment was protected, that no users would lose (if they had bid rationally), that this solution could not have been found by a community committee, and that users would lower their risk with trading. The reservoir's price gives strong signals about future storage. The power generator can better decide how much water to save for winter, and has a signal about whether the reservoir is big enough. One-to-one trading wouldn't work, because the water wouldn't balance. We ended the demonstration by asking each audience group whether they had made money in trading. If they had done it correctly, they all said "Yes" or that they had broken even.

Space does not permit a full examination of this case study. We introduce it here to show how implementers can create hypothetical case studies to help in the process of explaining how a smart market for water would work. Because a smart market approach is so different from other approaches to water allocation, and because of long-standing use of (and heated disagreements arising from) other approaches, there is a great need for development of case studies such as this to inform water communities that are considering the introduction of smart markets.

Users		Res	1	2	3	4	5	6	7	8	9	10	Res	1	2	3	4	5	...	7	8	9	10			
Period		1	1	1	1	1	1	1	1	1	1	1	2	2	2	2	2	2	...	8	8	8	8		Allowed	Flow over
Location		R	a	a	a	a	b	b	b	b	c	d	R	a	a	a	a	b	...	b	b	c	d		drawdown	env limit
Meter 1	1	-0.97	0.43	0.43	0.43	0.43	0.20	0.20	0.20	0.20	0.01	0.00	0	0	0	0	0	0	...	0	0	0	0	<=	1,500	40,088
Meter 2	1	-0.97	0.55	0.55	0.55	0.55	0.53	0.53	0.53	0.53	0.27	0.02	0	0	0	0	0	0	...	0	0	0	0	<=	2,000	20,020
Meter 3	1	0.00	0.00	0.00	0.00	0.00	0.01	0.01	0.01	0.01	0.39	0.29	0	0	0	0	0	0	...	0	0	0	0	<=	2,100	345
Meter 4	1	-0.97	0.55	0.55	0.55	0.55	0.60	0.60	0.60	0.60	0.76	0.87	0	0	0	0	0	0	...	0	0	0	0	<=	3,000	15,383
Meter 1	2	-0.01	0.10	0.10	0.10	0.10	0.05	0.05	0.05	0.05	0.01	0.00	-0.97	0.43	0.43	0.43	0.43	0.20	...	0	0	0	0	<=	1,500	34,398
Meter 2	2	-0.01	0.14	0.14	0.14	0.14	0.09	0.09	0.09	0.09	0.03	0.00	-0.97	0.55	0.55	0.55	0.55	0.53	...	0	0	0	0	<=	2,000	10,893
Meter 3	2	0.00	0.01	0.01	0.01	0.01	0.00	0.00	0.00	0.00	0.03	0.01	0.00	0.00	0.00	0.00	0.00	0.01	...	0	0	0	0	<=	2,500	242
Meter 4	2	-0.01	0.14	0.14	0.14	0.14	0.11	0.11	0.11	0.11	0.07	0.02	-0.97	0.55	0.55	0.55	0.55	0.60	...	0	0	0	0	<=	3,000	4,709
Meter 1	3	0.00	0.03	0.03	0.03	0.03	0.01	0.01	0.01	0.01	0.00	0.00	-0.01	0.10	0.10	0.10	0.10	0.05	...	0	0	0	0	<=	1,500	32,684
Meter 2	3	0.00	0.04	0.04	0.04	0.04	0.02	0.02	0.02	0.02	0.01	0.00	-0.01	0.14	0.14	0.14	0.14	0.09	...	0	0	0	0	<=	2,000	8,297
Meter 3	3	0.00	0.00	0.00	0.00	0.00	0.00	0.00	0.00	0.00	0.01	0.00	0.00	0.01	0.01	0.01	0.01	0.01	...	0	0	0	0	<=	2,800	329
Meter 4	3	0.00	0.05	0.05	0.05	0.05	0.03	0.03	0.03	0.03	0.02	0.00	-0.01	0.14	0.14	0.14	0.14	0.11	...	0	0	0	0	<=	3,000	1,628
Meter 1	4	0.00	0.01	0.01	0.01	0.01	0.00	0.00	0.00	0.00	0.00	0.00	0.00	0.03	0.03	0.03	0.03	0.01	...	0	0	0	0	<=	1,500	32,077
Meter 2	4	0.00	0.02	0.02	0.02	0.02	0.01	0.01	0.01	0.01	0.00	0.00	0.00	0.04	0.04	0.04	0.04	0.02	...	0	0	0	0	<=	2,000	7,425
Meter 3	4	0.00	0.00	0.00	0.00	0.00	0.00	0.00	0.00	0.00	0.00	0.00	0.00	0.00	0.00	0.00	0.00	0.00	...	0	0	0	0	<=	2,890	336
Meter 4	4	0.00	0.02	0.02	0.02	0.02	0.01	0.01	0.01	0.01	0.01	0.00	0.00	0.05	0.05	0.05	0.05	0.03	...	0	0	0	0	<=	3,000	600
Meter 1	5	0.00	0.00	0.00	0.00	0.00	0.00	0.00	0.00	0.00	0.00	0.00	0.00	0.01	0.01	0.01	0.01	0.00	...	0	0	0	0	<=	1,500	31,859
Meter 2	5	0.00	0.01	0.01	0.01	0.01	0.00	0.00	0.00	0.00	0.00	0.00	0.00	0.02	0.02	0.02	0.02	0.01	...	0	0	0	0	<=	2,000	7,114
Meter 3	5	0.00	0.00	0.00	0.00	0.00	0.00	0.00	0.00	0.00	0.00	0.00	0.00	0.00	0.00	0.00	0.00	0.00	...	0	0	0	0	<=	2,930	345
Meter 4	5	0.00	0.01	0.01	0.01	0.01	0.00	0.00	0.00	0.00	0.00	0.00	0.00	0.02	0.02	0.02	0.02	0.01	...	0	0	0	0	<=	2,800	31
Meter 1	6	0.00	0.00	0.00	0.00	0.00	0.00	0.00	0.00	0.00	0.00	0.00	0.00	0.00	0.00	0.00	0.00	0.00	...	0	0	0	0	<=	1,500	31,778
Meter 2	6	0.00	0.00	0.00	0.00	0.00	0.00	0.00	0.00	0.00	0.00	0.00	0.00	0.01	0.01	0.01	0.01	0.00	...	0	0	0	0	<=	2,000	6,992
Meter 3	6	0.00	0.00	0.00	0.00	0.00	0.00	0.00	0.00	0.00	0.00	0.00	0.00	0.00	0.00	0.00	0.00	0.00	...	0	0	0	0	<=	2,940	344
Meter 4	6	0.00	0.00	0.00	0.00	0.00	0.00	0.00	0.00	0.00	0.00	0.00	0.00	0.01	0.01	0.01	0.01	0.00	...	0	0	0	0	<=	2,940	49
Meter 1	7	0.00	0.00	0.00	0.00	0.00	0.00	0.00	0.00	0.00	0.00	0.00	0.00	0.00	0.00	0.00	0.00	0.00	...	0	0	0	0	<=	1,500	31,746
Meter 2	7	0.00	0.00	0.00	0.00	0.00	0.00	0.00	0.00	0.00	0.00	0.00	0.00	0.00	0.00	0.00	0.00	0.00	...	0	0	0	0	<=	2,000	6,948
Meter 3	7	0.00	0.00	0.00	0.00	0.00	0.00	0.00	0.00	0.00	0.00	0.00	0.00	0.00	0.00	0.00	0.00	0.00	...	0	0	0	0	<=	2,700	100
Meter 4	7	0.00	0.00	0.00	0.00	0.00	0.00	0.00	0.00	0.00	0.00	0.00	0.00	0.00	0.00	0.00	0.00	0.00	...	0	0	0	0	<=	2,940	0
Meter 1	8	0.00	0.00	0.00	0.00	0.00	0.00	0.00	0.00	0.00	0.00	0.00	0.00	0.00	0.00	0.00	0.00	0.00	...	0.20	0.20	0.01	0.00	<=	1,500	31,745
Meter 2	8	0.00	0.00	0.00	0.00	0.00	0.00	0.00	0.00	0.00	0.00	0.00	0.00	0.00	0.00	0.00	0.00	0.00	...	0.53	0.53	0.27	0.02	<=	2,000	6,946
Meter 3	8	0.00	0.00	0.00	0.00	0.00	0.00	0.00	0.00	0.00	0.00	0.00	0.00	0.00	0.00	0.00	0.00	0.00	...	0.01	0.01	0.39	0.29	<=	2,600	0
Meter 4	8	0.00	0.00	0.00	0.00	0.00	0.00	0.00	0.00	0.00	0.00	0.00	0.00	0.00	0.00	0.00	0.00	0.00	...	0.60	0.60	0.76	0.87	<=	3,000	54

Fig. 9.10 Part of the linear program constraints, in a spreadsheet, for the reservoir and groundwater example. Cells in gray, "Flow over env limit," are calculated based on the allocations

You are Farmer #10.

Your current water allocation is 2,500 cubic meters. You can use this for free.
If you use none of your water allocation, rainfall alone will let you to break even this year.

If you use your whole water allocation of 2,500 cubic meters,
then you would make a profit of $44,000, about $17.60 per cubic meter.
Or you can sell some or all of your allocation. Selling for less than $17.6/cubic meter is dumb.

Your farm can use up to 2,500 more cubic meters, which would let you make an additional profit of $22,000, about $8.80 per cubic meter (excluding the cost to buy this water).
Paying more than $8.8/cubic meter is dumb, as is buying more than 2,500 cubic meters.
Beyond the additional 2500 cubic meters, more water would be a waste.

Your bid form

What is your user name? I am Farmer #10.
Sell price, $ per cubic meter of water. $ ____________________

I would be willing to sell this quantity (cubic meters of water) at the above price per cubic meter.
____________________ cubic meters.

Buy price, $ per cubic meter of water. $ ____________________

I would be willing to buy this quantity (cubic meters of water) at the above price per cubic meter.
____________________ cubic meters.

Fig. 9.11 Demonstration bid form for a simple demonstration market. Water values are random

User name	Sell bid, $/cumec	Sell bid, cumecs	Buy bid, $/cumec	Buy bid, cumecs	Initial right, cumecs	Sell qty actual	Buy qty actual	Final allocatn	Price	User pays
Reservoir	$1.00	5,000	$0.50	5,000	-36,000	5,000	0	-41,000	-$7.73	$38,650
Farmer 1	$7.00	1,500	$1.10	6,000	6,000	0	0	6,000	-$5.57	$0
Farmer 2	$2.17	6,000	$0.43	6,000	6,000	0	0	6,000	-$5.57	$0
Farmer 3	$3.00	6,000	$2.10	6,000	6,000	0	3,444	9,444	-$5.57	-$19,183
Farmer 4	$5.00	1,000	$0.25	6,000	6,000	0	0	6,000	-$5.57	$0
Farmer 5	$7.00	6,000	$1.50	6,000	6,000	0	0	6,000	-$4.65	$0
Farmer 6	$4.17	6,000	$2.50	6,000	6,000	0	6,000	12,000	-$4.65	-$27,900
Farmer 7	$2.50	6,000	$1.75	6,000	6,000	0	0	6,000	-$4.65	$0
Farmer 8	$0.33	6,000	$0.17	6,000	6,000	6,000	0	0	-$4.65	$27,900
Farmer 9	$17.00	2,500	$4.50	2,500	2,000	0	2,470	4,470	-$4.56	-$11,263
Farmer 10	$1.00	5,000	$0.50	5,000	5,000	5,000	0	0	-$4.71	$23,550
					Total:	16,000	11,914		Gov't net:	$31,754

Fig. 9.12 Spreadsheet of financial results from one auction. Gray cells are formulas. Columns "Sell qty actual" and "Buy qty actual" are the decision variables. Note that total sell quantities do not equal total buy quantities

9.5 Surface Water and One Reservoir, East Anglia

This case is based on actual hydrology, from East Anglia, UK. The work was part of a larger project, reported in Lumbroso et al. (2014). In this major project, we developed a smart market design for surface water, including reservoirs and non-consumptive water use, among other things. We tested the market design with real water users in a presentation format. Another publication from this project is Erfani et al. (2014).

The larger project gave water users in East Anglia the opportunity to put in their own values via a website. This project had all the work of developing a market simulation, including guessing users' values for water as defaults for users who did not attend the workshops, while also creating a working market website for people to try.

We began developing the market model with a GIS map of the region and an understanding of the overall catchment. We will not dwell on difficulties with obtaining data. Figure 9.13 is a map of the catchment. The outlet is at the top far right. Left of the outlet in the map, the Grafham Water reservoir is an important feature. The reservoir is owned by a water agency, Anglian Water Services Ltd. (Anglian Water), which pumps water from the Great Ouse River into the reservoir, and then from the reservoir, through processing facilities, to various potable urban supplies. Urban areas use this water, releasing some through sewage treatment works (STW) back into the river upstream from the reservoir. These STW flows proved to be an interesting feature of this case.

We chose an auction schedule of a 52-week year, with auctions every week, starting 1 January for the model year. Most of the users in the catchment are farmers, who generally thought badly of the large water agency, fearing that it might take their water. The government prohibits Anglian Water from releasing from the reservoir to the river, to prevent release of exotic shrimp. Thus, the reservoir only takes in water from the river. The catchment includes non-consumptive users, including hydroelectricity generators and gravel processors.

We treated licensed volumes as initial rights, with some adjustment. The regulator recorded licensed volumes as annual volumes, so we converted them into monthly initial rights, relying heavily on actual volumes used.

Figure 9.14 is a schematic of the catchment. The operations researcher will recognize this as a tree network. We realized that we could simplify the market-clearing model mathematically, with no loss in the fidelity of the results, by joining reaches that had no confluence, as shown by the heavy lines in Fig. 9.14. This mathematical simplification implies a simplification of the graph by eliminating nodes with degree 2. We calculated and considered the natural inflows, but also the significant inflows from the STWs.

The UK water regulator imposed a range of relatively complicated environmental requirements, based on whether the natural flow was below a certain percentile of natural flows, and the week of the year. We coded these directly into the market-clearing software, and we added constraints for the environmental requirements at the gauges shown in Fig. 9.14.

We developed computer code that created the river network from the GIS file, cleaned data, added user locations to the network, added nodes for assessment points, and simplified the graph by aggregating nodes and inflows. We used the market-clearing formulation described in Sect. 7.5. The resulting linear program had about 25,000 rows and 78,000 variables. We easily solved this with an open-source scripting language and solver. The slowest part of the process of running the smart market was assembling the linear program from the database.

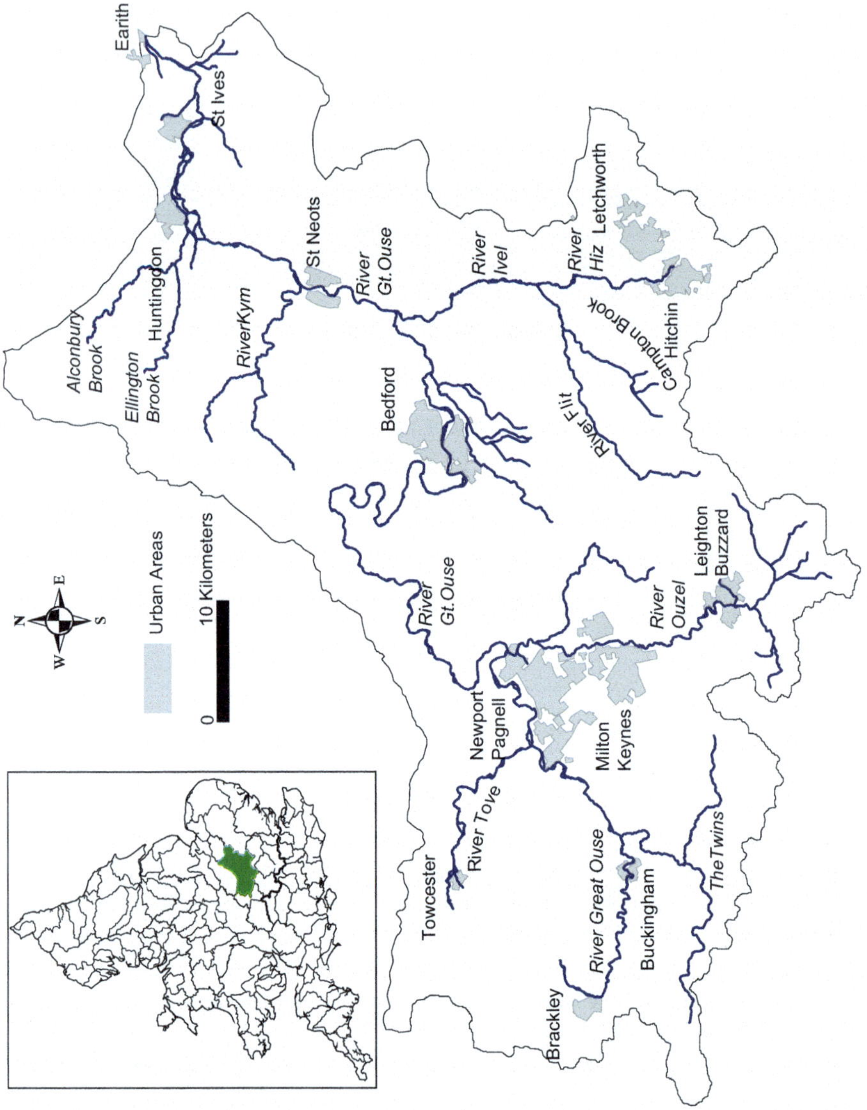

Fig. 9.13 Map of the Upper Ouse catchment, East Anglia, UK (Modified from Environment Agency 2001 and Environment Agency 2005)

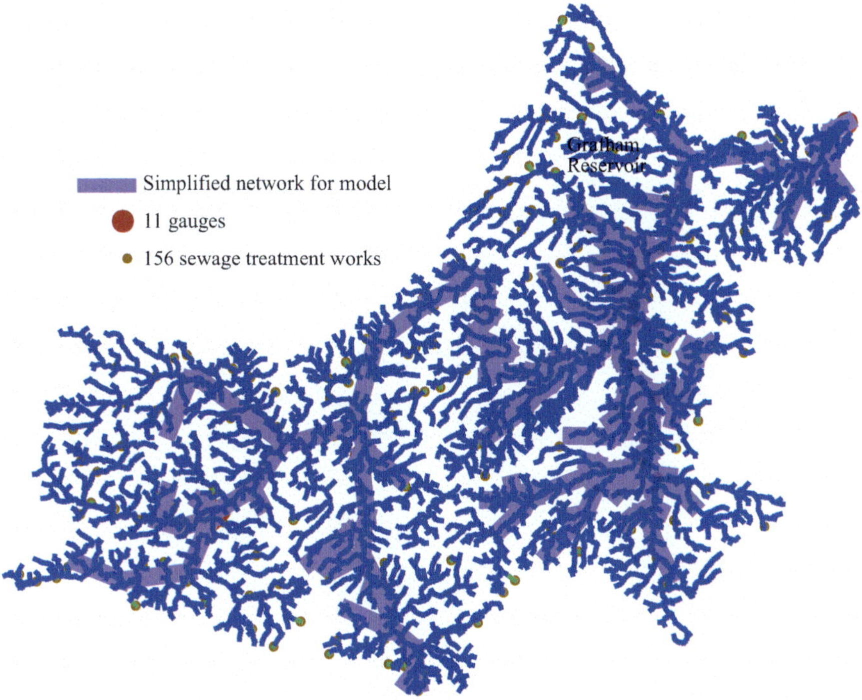

Fig. 9.14 Schematic of the Upper Ouse catchment

Figure 9.15 shows the price map for the market for 1 week during the simulation. The model calculates a price at every user's point and every gauge. At most points, the price is £25, driven by the assessment point at Offord. Other assessment points (the 11 larger dots) have more than required, so their prices are all £0. The prices of £1 and £5 at top left indicate local shortage.

Although Fig. 9.15 shows the prices for 1 week, the market outcomes showed almost perfectly flat prices over the course of the year, due to the sufficiently large size of Grafham Water reservoir. The case study highlights that a reservoir will greatly influence the design of a smart market for water and the resulting prices. Sections 7.3, 7.5, and 7.7 discuss the influences, but it is worth reconsidering them in the context of this specific example.

With a smart market, the timing of inflows and outflows from the reservoir could be managed either by the smart market manager or the owners, Anglian Water.

If the reservoir were operated by the market manager, the results of the smart market optimization model would determine the reservoir intakes. All users from the reservoir would be "stacked" into a single user, represented by Anglian Water in the bidding. In this case, the optimization must be multi-period, because the model must account for the storage from each period to the next.

If the reservoir were operated by Anglian Water, then Anglian Water must bid for intakes of water to its reservoir, offer to sell reservoir releases back to the

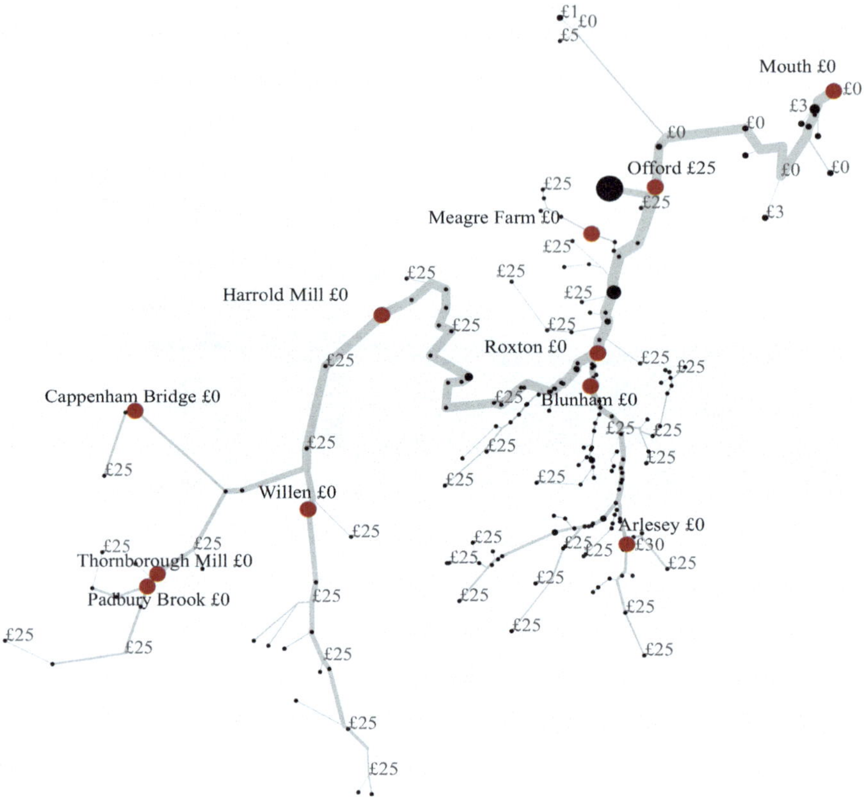

Fig. 9.15 Price map from week 25 of the Upper Ouse demonstration market. Segment widths indicate sizes of flows. Larger dots represent assessment points, and the largest dot represents the Grafham Water reservoir

market, and manage the reservoir outflows to various water supplies outside of the market. All users from the reservoir would then buy and sell directly from Anglian Water, not the public market. This implies that Anglian Water would likely have long-term fixed contracts with specific water suppliers on the cost of supply while not knowing what price it might need to pay for the water intake to the reservoir in a smart market, raising risk to Anglian Water. Anglian Water might choose to develop its own optimization model to help set variable prices for delivery of water to public supplies. In this case, the smart market optimization for the river catchment as a whole could be a single-period run-of-river model without storage.

Because the reservoir operation flattened prices over the year, one would expect that in a smart market the reservoir operator would buy when the price was low to such an extent as to remove all opportunity for time-based arbitrage (i.e., taking more water when it is cheap, and releasing more water when it is expensive). The flat prices would also give users certainty about the price of water.

One notable finding was that the STW inflows improved reliability to water river users, most importantly when natural inflows were low—we simulated both a "normal" year and a drought year. In effect, Anglian Water (through its public supply water users) was recycling river water for farmers. So it appeared that Anglian Water already provided substantial service to the users in the catchment, despite those users' apparent distrust of the big water agency.

Could Anglian Water exercise market power in this catchment? It already has a regulated monopoly for supply of water to users, but would its use of the reservoir in a smart market allow it to extract monopoly earnings from those users currently taking river water? It would appear to have little ability to do so, because its contracts with public water supplies prevent it from varying supply based on cost. The required outflows from the reservoir limit the ability to drive down the price by curtailing outflows or drive up the price by hoarding water. In addition, the required outflows of the reservoir to public supplies will lead to STW inflows to the river system, again limiting the control that Anglian Water can have over water availability during a drought.

An interesting lesson learned from this market development process was that the STW flows were large enough to play an important role in the hydrology, and therefore the market. This case study differs from other market scenarios with reservoirs, such as in Section 9.4, because the reservoir here supplies water users higher in the catchment rather than below it. Anglian Water pays to pump water uphill, to its use for public water supply. That means that STWs return flow to other users in the same market. Market implementers in this case, and perhaps in others, will have to make a decision about the associated rights for the return flows. Here are some options that seemed relevant to this particular case:

First, the market could treat STW flows as though they were free natural inflows. In this case, the river users would eventually take rights to the natural inflows, through the market manager's allocation.

Second, the market manager could pay Anglian Water for the STW flows as for return flows from irrigation. The amount paid would vary based on the market prices for water in the catchment, and the market manager would in turn collect income from water users who would be allocated water from a larger common pool of allocable water. If STWs have little or no control over the quantity they release to the river, the market manager could not expect Anglian Water to commit to any return flows. Modeled STW flows could then explicitly link to Anglian Water's pumping out of the reservoir. The market manager would have to pay Anglian Water for the return flows at the nodal price where each STW discharged to the river.

Third, if the STWs could control the outflows, they may actually serve as storage points in the system. The smart market would then operate where the STW has an assumed release rate in the optimization. Anglian Water might control these release rates on behalf of the STWs or by the STWs directly. The relevant agency could then offer to sell a greater release, or could buy the right to release less. We note that this option would rarely work, because STW facilities are usually in urban areas without room for reservoirs; we mention this to illustrate some of the thinking that goes into market design.

If Anglian Water or the STWs could not get enough water to supply the outflows, they would be considered in default of their contract to release, and would have to buy allocation on the spot market to make up for it. On the other hand, Anglian Water's ability to store the STW discharge would allow agencies to release in periods when the price of water was higher, helping everyone in the market.

References

Ahlfeld, D. P., Barlow,P. M., & Mulligan, A. E. (2005). *GWM – A ground-water management process for the U.S. Geological Survey modular ground–water model (MODFLOW–2000).* U.S. Geological Survey Open–File Report 2005–1072. http://pubs.usgs.gov/of/2005/1072. Accessed 15 July 2006.

Cummings, R. G., Holt, C. A., & Laury, S. K. (2004). Using laboratory experiments for policymaking: An example from the Georgia irrigation reduction auction. *Journal of Policy Analysis and Management, 23*(2), 341–363.

Environment Agency. (2001). *Managing water abstraction map of CAMS areas.* Environment Agency, Bristol, UK.

Environment Agency. (2005). *The upper Ouse and Bedford Ouse catchment abstraction management strategy.* Environment Agency, Bristol, UK, https://www.gov.uk/government/uploads/system/uploads/attachment_data/file/289835/LIT7708_df73f8.pdf. Accessed 22 Mar 2017.

Erfani, T., Binions, O., & Harou, J. J. (2014). Simulating water markets with transaction costs. *Water Resources Research, 50*(6), 4726–4745.

Greenwald, R. (1998). *MODMAN: An optimization module for MODFLOW Version 4.0.* Freehold: GeoTrans.

Harbaugh, A. W., & McDonald, M. G. (1996). *Programmer's documentation for MODFLOW-96, an update to the U.S. Geological Survey modular finite-difference groundwater flow model.* (Open-File report 96–486). U.S. Geological Survey; Branch of Information Services. https://pubs.er.usgs.gov/publication/ofr96486. Accessed 19 Mar 2017.

Lumbroso, D. M., Twigger-Ross, C., Raffensperger, J., Harou, J. J., Silcock, M., & Thompson, A. J. K. (2014). Stakeholders' responses to the use of innovative water trading Systems in East Anglia, England. *Water Resources Management, 28*(9), 2677–2694.

McDonald, M. G., & Harbaugh, A. W. (1988). *A modular three-dimensional finite-difference ground-water flow model.* TWRI 6-A1, U.S. Geological Survey Reston. http://pubs.usgs.gov/twri/twri6a1

McGauran, Peter. 2006. *New Murray water plan a breakthrough for farmers.* http://www.maff.gov.au/releases/06/06047pm.html, 28 April. Accessed 2 Oct 2006.

Montgomery, W. D. (1972). Markets in licenses and efficient pollution control programs. *Journal of Economic Theory, 5*(3), 395–418.

NSW DNR, Dept. of Natural Resources, New South Wales, Australia. http://www.dlwc.nsw.gov.au/water/recovery.shtml. Accessed 2 Oct 2006.

Porse, E., et al. (2015). Pumping for the masses: Evolution of groundwater management in metropolitan Los Angeles. *Geojournal.* doi 10.1007/s10708-015-9664-0. 17 pp, first online 7 August 2015.

Raffensperger, John F., "Matching users' rights to available ground water," Ecological Economics, v70, n6, pp.1041–1050, 2011.

Teasley, R., Raffensperger, J., & Milke, M. 2010. *A smart market for ground and surface water with hydropower generation.* ASCE World Environmental and Water Resources Congress, 16–20 May, Providence, Rhode Island.

Chapter 10
Market Behavior

Abstract Many people fear that a water market will lead to monopolization, and, for this reason, believe that government should not implement a market. In this chapter, we aim to convince the reader that a water market is unlikely to result in monopolization. We acknowledge that researchers have not fully examined the issue of market power, but we think that problems are likely to be isolated to unusual situations.

This chapter discusses the likely market behavior within a smart market for water, including the chance of price manipulation, market power when the catchment is over-allocated, and market power when the catchment has a single rich bidder. For market participants who feel that they face too much risk, and many of them will, Sect. 10.4 explains how the spot market can lower that risk by enabling a range of financial products. The final two sections discuss effects of limits to water conveyance capacity, and lack of competition through lack of bidding.

Keywords Market power • Monopoly • Competition • Capacity

10.1 Market Power and Price Manipulation

10.1.1 Price Elasticity

Asserting market dominance requires inelastic demand, among other things (Markham 1951). With inelastic demand, demand falls only slightly with a price increase. If this were the case, a user with a big allocation may be able to command a high sell bid during drought.

> We remind the reader that the smart market we are describing is intended for wholesale commercial water, especially water for farming and industry, in a developed country. In this context, people are unlikely to be able to exert market power, due to the elasticity of demand for water, the large number of users, spatial dependency of abstraction, the temporal delays, and considerable regulatory attention.

J.F. Raffensperger, M.W. Milke, *Smart Markets for Water Resources*, Global Issues in Water Policy 12, DOI 10.1007/978-3-319-55008-4_10

Studies show variable price elasticities for water. A typical example shows a price elasticity of agricultural water demand in California of −0.79 (Schoengold et al. 2006), indicating that a 10% increase in price would result in a 7.9% decrease in demand, due to improved water management, crop changes, and better irrigation technology. Such high elasticity is associated with highly competitive markets, which suggests that a water market with a diverse set of users is not likely to suffer from any user having much market power.

Demand for water is elastic because users can usually reduce their abstractions considerably in the short run. The tendency is to water in a way that ensures the plants will always have more than enough water, rather than to give them the least they likely need. More significantly, commercial users' demand for water is elastic because users can reduce water use in the long-term when faced with higher prices for water months or years away. Agricultural users can change the crop grown or the acreage tilled, while industrial users can invest in new water-saving technologies, or shift water-intensive processes to other facilities.

10.1.2 Number of Users

Research on protecting the commons has shown that market power declines quickly with the number of market participants (McAfee 2010; Montero 1998). Disegni-Eshel (2005) showed how to allocate initial rights to a group of competitive and non-competitive firms (for free) in order to balance efficiency with market power. Again, more participants resulted in less market power, and the numbers required are small, seven or eight. Because of the large number of participants in the typical water catchment, gaming is unlikely. In Sect. 5.5, we point out that one small catchment in New Zealand has hundreds of users.

Perhaps users could collude? Georgia, U.S., conducted an interesting experiment in a water buy-back. Cummings et al. (2004) developed a web-based auction system for the government to pay farmers to reduce irrigation during drought. This work did not use an optimization, and the hydrology was simplified. The researchers tested their system in pilot experiments, and then the government used their system for the real auction. Their auction had one budget-constrained buyer, the government, facing many sellers. The authors were concerned that sellers might collude to force up the price.

Based on results from their pilot auctions, the subjects played the game better when there were a few preliminary auctions before the final auction. In addition, policymakers wanted assurance that prices would be "reasonable," and allowing revisions before the final auction "gives farmers a chance to think about the situation and to respond to policymakers' decisions and the bidding behavior of others." The authors found that sellers attempted to collude, but the collusion had little effect on prices. We note that the Cummings et al. system is not a water-trading system, but an irregular drought management system. In any case, even when sellers tried to collude, the researchers saw little effect.

More users results in a more competitive market in part because users attempt to out-think each other. Suppose that in March Ann wants to have 100 units of water in July, but guesses that July will be drier than forecasted. She thinks the market manager will scale her July allocation with $\alpha = 85\%$, so she might plan to buy $100/0.85 = 118$ units of July allocation in March. On further thought, she guesses that her neighbor Bob is not so forward thinking, and she could make some extra money by buying extra now and selling it back to Bob. Let's say that she gets a total of 130 units of July allocation in March, expecting to have it scaled down a bit, but with still enough left over to have some to sell on the spot market in July. This is risky speculation. She might win if her forecasts outperform the market manager's. If Bob hears word of Ann's strategy, he could bet that the market manager is correct, and sell July allocation in March at the high price Ann has bid for it, hoping to buy it back at a reduced rate in July when Ann realizes she cannot use the extra water.

More often, though, with many market participants, no user is large enough to have significant sway over the market.

10.1.3 Spatial and Temporal Dependency of Abstraction

Groundwater users in particular would find market manipulation to be difficult because of the dispersed nature of groundwater. A well owner has the most control of water nearby, and less control of water further away. Anyone who somehow obtained control of the dispersed groundwater would also need control of sufficient land to employ the water productively—after all, paying even a small amount for water makes no sense unless the water user has a way to turn that marginal extra water into extra income. Because the allocations are time-based, the person holding the allocation must either use it or sell it.

Consider a somewhat extreme case, that of the reservoir operator in the example of Sect. 9.4. The reservoir operator is a hydroelectricity company. When the reservoir operator releases water, the water flows into a valley where farmers irrigate crops. Would the reservoir operator have significant market power because of its control of the reservoir releases? Most likely not. The reservoir operator has minimum and maximum reservoir capacity constraints, which would limit market power at extremes. The reservoir operator will be concerned about the value of water in the power market, which might operate on a half-hour basis, while the water market operates on a weekly basis. So the reservoir operator must choose releases based on half-hourly differences in water value. The farmers might be able to abstract groundwater, which would provide an alternative supply if the reservoir operator restricts supply. Finally, the market manager will require that allocations satisfy the environmental flow requirements in the streams, which will almost certainly require releases from the reservoir as an obligation on the operator. Should the operator wish to reduce those releases to earn more in the electricity market, the operator may have to buy allocations from the farmers, highlighting how the smart market is more likely to lead to arrangements of mutual benefit rather than one party exerting market power over others.

Market power could appear where only a few water districts trade allocations in a water resources on behalf of their water users. If such market power became obvious, the regulator could require the water districts to disaggregate their users' bids rather than submit overarching bids.

10.1.4 Regulatory Attention

Interestingly, the market manager will need monopoly power. The government's commerce regulator will likely have to permit the market manager to operate the smart market as a monopoly over the water resource.

Dalziel (1987) used optimal control theory to study how a government manager should manage stored water over time, comparing a welfare-maximizing solution with the monopolist profit-maximizing solution. The study showed that the price of stored water cannot rise faster than the prevailing interest rate. It also showed that a monopolist would have two resources, water and storage, and the monopolist cannot exploit monopoly power in both resources simultaneously. The market manager that we have described works under the close watch of the water regulator, and should be revenue neutral. But the concern exists that a private entity serving as the market manager without strong governmental oversight could exert monopoly power over water users.

Should any user attempt to gain market power, the market manager would likely see it. In the rare cases when a user could exert monopoly power—few users, short response time in the water system, few users with storage, inelastic demand — implementers could attempt to give the market manager authority to impose restrictions on water use.

10.2 Market Power and Over-Allocation

In this section, we continue with the example from Sect. 9.3 to explore the situation where some users attempt to benefit from a situation where the catchment is over-allocated, and the market manager has no authority to scale users' allocations. The market manager must buy back allocation to bring it in balance with the actual water.

The situation is exactly as in Sect. 9.3.4, but the users are much more sophisticated. To maximize their total profit, the users could collectively increase their user demand and price curves, as shown in Table 10.1. Each user still has an allocation of 1, but they will sell only if the price is $2000 or more. We recognize that this is not a full game-theoretic treatment; we do not model each user's behavior as they might act in such a market. Rather, we simply vary the bids and observe the results.

Table 10.1 Alternate values for user demand and price curves for over-allocation scenario where users try to gain from government's protection of the environment

Price	$1	$2000	$3000	$4000	$5000
Quantity	5	1	0.4	0.15	0.05

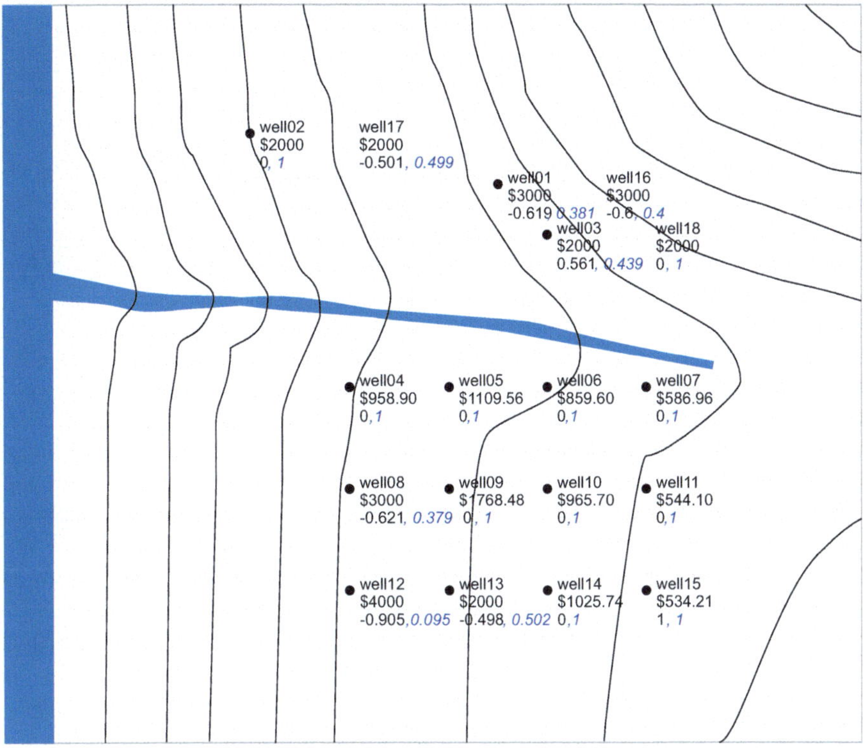

Fig. 10.1 Price map for the over-allocation scenario ($\alpha = 100\%$) using the user demand and price curve in Table 10.1. Each well shows the price, the trade, and the final allocation

The price map in Fig. 10.1 shows that the only trades were sales, and all at the market manager's expense. The market manager had to pay $12,256, almost a third of which went to well 12.

Because the prices paid for water are public, even without collusion the users could change their bids to sell very little, and force the price of water to high values. In fact, if a common-pool smart market operates with over-allocation, users could choose to sell so little that the model will become infeasible: the users insist on taking more water than is available, and the flows would not satisfy the environmental constraints. Clearly, this is an untenable method of operation. But it is also artificial, as it assumes that the market manager cannot scale the allocation, and that users are sophisticated enough to collude in this extreme way.

10.3 Market Power and One Rich Bidder

What happens when one user has deep pockets and bids up the price "unreasonably" high? Will the environment be damaged? Will other users be bought out or denied water? Again, we recognize that we do not model each user's behavior as they might act in such a market. Rather, we will simply vary the bids and observe the results.

With the smart market, the environmental requirements must still be satisfied, because they are hard-coded constraints in the market-clearing model. The market mechanism protects the environment when one bidder has deep pockets.

To see whether other users would lose somehow, suppose that the operator of well 12 were willing to pay for water at a much greater relative rate. To exaggerate the effect we intentionally place the high bidder at the point where water is the scarcest. Assume all users' bids are the same smaller numbers as earlier, except well 12's, which is changed to the demand and price curve in Table 10.2. Assume further that the market manager will scale allocations to get revenue neutrality (the case of "users trade" explored in Sect. 9.3.5). The result in this case is $\alpha = 18.3\%$.

Figure 10.2 shows the price map for this scenario. The high demand at well 12 means that the proportional reduction factor increases only slightly in this scenario compared with Method 1 in Sect. 9.3.2. Figure 10.2 shows that, in this scenario, the high bidder does take water from others: the only other wells to get water are wells 2 and 17. The very high bid from well 12 reverberates throughout the catchment, with decreasing effect further away from well 12. Indeed, if well 12 were to raise the bid by another factor of 10, no other users would get water.

On the other hand, the monopolization of water in this scenario has put money in the selling users' pockets. Well 4, for example, was willing to sell all its water for anything over \$5 per unit, and after the proportional reduction to $\alpha = 17.4\%$, this would have netted well 4 about \$0.87. But since well 12 bid up the price so high, well 4 instead goes away with \$99.5638*0.183 = \$18.22.

Thus, under the users-trade auction, a wealthy user could bid up the price of water and "take it away" from others nearby. With the user-trades method of operation, the other users can choose to be compensated handsomely, and at the wealthy user's expense. And this is only for the one period. Should the wealthy buyer want to play the same game in later periods, the buyer will have to front up the money.

But why would anyone bid up the price so high for so little water? The shrewd business person would simply move away from the binding environmental constraint of well 12 to a location with less tight environmental constraints and cheaper water.

Table 10.2 Alternative values for demand and price curve for well 12, a hypothetical monopolist

Price	$1000	$1500	$2000	$2500	$3000
Quantity	5	1	0.4	0.15	0.05

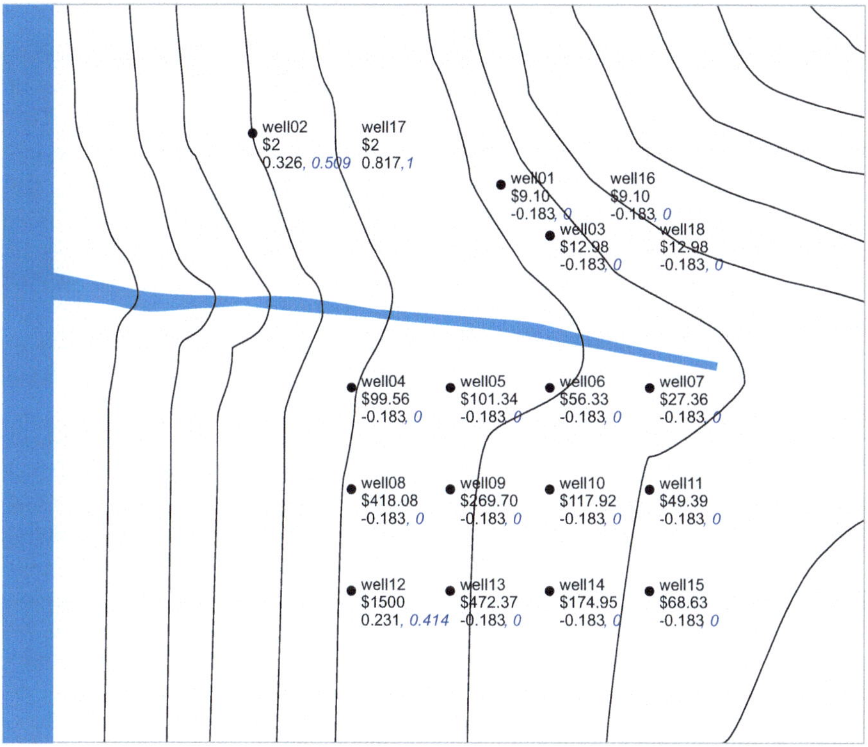

Fig. 10.2 Price map with users trade (i.e., rights scaled so the auction is revenue neutral, $\alpha = 18.3\%$) and where well 12 bids up the price of water. Each well shows the price, the trade, and the final allocation

10.4 Financial Products

Implementation of the smart market creates the spot market, which in turn enables a range of financial products that market participants will want. We have already described the ability to transfer the long-term rights as one example of a behavior that will naturally flow from the spot market. Participants will naturally develop a range of other contracts. These contracts will often depend on how the market manager adjusts users' allocations in each auction, for example by changing the value of α as described in Sect. 9.3. With the spot market, people can make these arrangements themselves, in pair-wise fashion. Government need not be directly involved in these arrangements. We will describe in only broad terms how they might work.

One common arrangement that might arise within smart markets would be insurance. When Bob considers his allocations for the year, he will be concerned that the conditions in important weeks may be drier than expected, in which case the market manager would scale his allocations downward. Depending on the degree of scaling, his operations, and his costs, he could take it on the chin or he could attempt to obtain more allocation at the time. If he has trouble sleeping at night out of concern about this, he could find someone who is willing to sell insurance. This insurance could be ordinary crop insurance that might protect against the impact of droughts, or it could be specifically about water availability. The terms of the insurance contract could stipulate hydrological and financial trigger conditions, such as the value of α imposed by the market manager and the spot price of allocation. The terms could further stipulate the type of relief: whether the insurer would make a payment to the farmer, or be required to supply water through purchase in the spot market at the going rate, or be required to bring in water in tanker trucks. A simple payout of money would probably be the simplest for a financial services company to arrange.

A market participant may wish to take up an option, similar to insurance, of buying the right to have additional allocation at a fixed price. Bob could pay a premium to someone for the option of buying extra allocation at a strike price agreed in advance. When the auction day arrives, Bob notifies the option seller whether he wants to exercise the option. If so, the option seller is responsible for buying allocation on Bob's behalf in the spot market. If the spot price is higher than the strike price, the option seller loses money. If the spot price is lower than the strike price, the option seller makes money. In the long run, Bob's use of options would probably cost him some money. Otherwise the option seller would not do business with him. However, Bob has a sense of certainty about the cost of the water.

10.5 Effects of Limits to Water Conveyance Capacity

We have noted several times that our smart market design does not account for the fixed costs of water-conveying infrastructure (e.g., canals between users). Users would have to pay for maintaining infrastructure through some other payment mechanism. Fortunately, such mechanisms are possible through the market-clearing model, the market-clearing process, and operation of the spot market.

As with an electricity market-clearing model (see Sect. 4.2), the water market-clearing model produces price information that indicates the value to participants for an increase in the capacity of the constructed conveyance. As a result of the prices, market clearing can also produce revenue when capacity constrains the market solution.

Owners of conveyance are likely to want to exercise control over that conveyance. They may be tempted to restrict capacity to generate revenue for themselves, creating "congestion," as the electricity market literature calls it. The regulator may view such restrictions as attempts to exert market power.

Electricity markets and their regulators have learned to manage these behaviors in several ways. First, the smart markets for electricity naturally led to smart markets for power line capacity. Similarly, we expect that markets for water conveyance capacity would arise from spot markets for water. Briefly, the market manager can arrange a market for conveyance capacity in which traders buy the right to have flow along a particular segment. This right brings the buyers any associated revenue from the spot market. If someone's behavior causes congestion on the conveyance segment by bidding for extra water from a canal that is already at its maximum flow capacity, they will have to pay the conveyance owner. The one causing the congestion then may wish to buy some capacity right in order to hedge their costs. The infrastructure owner may be motivated to sell the capacity rights in case the conveyance is likely to be uncongested. In case the infrastructure owner is not willing to sell capacity rights, and other stakeholders accuse the owner of exerting market dominance despite reasonable offers from users, the water regulator may need to work with government to develop a proper market for capacity rights.

Electricity markets have a wide variety of other financial products, such as contracts that depend on differences in prices at different locations. Thus, if a conveyance capacity results in the inflow point price differing from the outflow point price, the contract can result in a payment. These contracts allow market participants to hedge their risks.

10.6 Lack of Competition Through Lack of Bidding

Another type of market behavior should also concern those implementing a smart market for water: a lack of bidding. Users facing a new smart market for water might find it difficult to accept the uncertainty of unknown prices in a market that has not begun, so they may not be willing to join the market. These potential entrants might want an estimate of prices before agreeing to join a smart market. An economic analyst might be able to bracket likely prices for users who already trade, but the issue of whether individuals will trade in a market at a "rational price" is not an easy one to resolve.

We can only speculate on the motivations behind a reluctance to bid or enter the market. Perhaps water users might perceive a loss of power when entering into a smart market. Without a smart market, users are price makers—they can determine the price for any water transactions, and can control when they buy and sell and, often, to whom. Within a smart market, because a given user is likely to be a small player among many, the market is likely to be competitive, and the water user becomes a price taker, not a price maker. The water user will adjust behavior and decisions to the market price. In order for a trader to willingly participate, the trader would need to believe that the benefits of a smart market outweigh what they might perceive as a loss of power.

The lack of trust is likely to impede initial participation in a smart market. Users may distrust the market manager and water regulator in their ability to guarantee delivery and enforce the rules on others. A smart market for water will not work through being forced on individual water users, but rather needs to build trust with willing traders, and to then expand to include those initially reluctant to join. We would expect reluctance to enter the market to lessen over time.

References

Cummings, R. G., Holt, C. A., & Laury, S. K. (2004). Using laboratory experiments for policymaking: An example from the Georgia irrigation reduction auction. *Journal of Policy Analysis and Management, 23*(2), 341–363.

Dalziel, P. C. (1987). Optimal water storage and pricing: The effect of monopoly. *New Zealand Economic Papers, 21*(1), 3–16.

Disegni-Eshel, D. M. (2005). Optimal allocation of tradable pollution rights and market structures. *Journal of Regulatory Economics, 28*, 205–223.

Markham, J. W. (1951). The nature and significance of price leadership. *The American Economic Review, 41*(5), 891–905.

McAfee, R. P. (2010). Four issues in auctions and market design. *Revista de Análisis Económico – Economic Analysis Review, 13*(1), 7–24. http://www.raeear.org/index.php/rae/article/view/117/226. Accessed 18 Mar 2017.

Montero, J. P. (1998). A simple auction mechanism for the optimal allocation of the commons. *The American Economic Review, 98*, 496.

Schoengold, K., Sunding, D. L., & Moreno, G. (2006). Price elasticity reconsidered: Panel estimation of an agricultural water demand function. *Water Resources Research, 42*(9), W09411.

Chapter 11
Water Markets in a Stochastic World

Abstract To this point, the smart market models have been deterministic, using one estimate of future inflows into the catchment. This chapter explores sources of uncertainty and discusses which of those the market should directly address and which it should not, why the deterministic approach may not calculate the best allocation, and ways to address uncertain inflows explicitly within the market-clearing mechanism using stochastic optimization. We will see that a stochastic optimization will probably be too hard to solve, and could complicate bidding and market clearing. We discuss some alternatives to the stochastic market model.

Keywords Uncertainty • Stochastic optimization • Risk

11.1 Which Sources of Uncertainty Should the Smart Market Incorporate Directly?

Market participants face uncertainty in many forms. Market participants may want the market manager to help them address these various uncertainties explicitly within the market-clearing mechanism. Stakeholders should not expect the regulator nor the market manager to manage all of them.

Users face uncertainties from:

- variable price for water, perceived as a variable supply,
- loss of supply due to government's exercise of existing regulation,
- loss of supply due to new regulation,
- loss from another user's new abstraction,
- change in water use requirements due to climate change,
- change in water use requirements due to other market forces, such as prices for crops and fuel,
- unknown future inflows of water.

> We do not have all the answers here. The problem of clearing a multi-period stochastic market model remains open for research.

J.F. Raffensperger, M.W. Milke, *Smart Markets for Water Resources*, Global Issues in Water Policy 12, DOI 10.1007/978-3-319-55008-4_11

The above list is by no means exhaustive. A farmer may worry about soil erosion and herbicide resistance; a hydroelectricity generator may worry about electricity prices and equipment downtime; a city water manager may worry about major leaks and excess lawn watering.

In managing water for users and for the environment, the water regulator also faces uncertainties such as:

- uncertainty of users' demands, because the market manager has a detailed operational plan for the catchment,
- lack of visibility on actual environmental flows, such as river flows and aquifer heads,
- uncertainty in estimates of preferred environmental flows (requirements), in both the science of what the environment actually needs, and the policy as to society's preferences for those flows,
- uncertainty about political and policy changes that affect regulation of the water,
- uncertainty in future inflows of water, combined with little ability to match those uncertain inflows to demand.

In many cases, regulators don't know the past and current flows, much less future flows, because users don't meter their water use. The regulator cannot be sure whether the water will satisfy the environmental requirements. Even with perfect metering, the regulator has only hindsight.

We should add to the list the uncertainties of all stakeholders regarding the implementation of the smart market. Will it work? Will the price of water be high or low? Will the environmental flow requirements be satisfied? Will people game the system? We address matters such as these in Chap. 13.

Thompson (2005) describes a range of uncertainties faced by water users. These uncertainties include:

- loss of water supplies due to the various statutes,
- government curtailing water rights based on their powers of reasonable use,
- climate change,
- growth in demand,
- forfeiture of long-term rights due to a temporary trade, and
- another user's exercise of previously unused rights.

He points out that "uncertainty raises the chance that people will make the wrong decision." Further, he writes, "By providing the legal infrastructure for water markets and actively encouraging such markets, the government can help reduce the harm from uncertainty in water rights and deliveries." An important point to his paper is that the law can help allocate uncertainty to minimize its impact on society. He then gives seven guidelines for allocating risk, and sensitively demonstrates how those guidelines could be used. A key point is that "water markets allow water users who face uncertainty to reallocate the uncertainty to individuals or entities that can better bear the risk of the uncertainty." Implementers should read Thompson's paper.

Which of these uncertainties should the market directly incorporate? A subtle point of Thompson's is that a water market can best address uncertainties associated most directly with the use of water. In line with his key points, the market should seek to incorporate uncertainties that directly affect water use and lead to externalities. For most other uncertainties, the market should only inform users of other uncertainties without necessarily incorporating them.

To see this, consider those tasks that the smart market does well. The smart market provides a complete schedule of planned water flows, and users will have an incentive to adhere to that schedule. The regulator can be sure that users will take their water allocations. The regulator can be sure that the planned flows will satisfy environmental requirements. Thus, some of the regulator's key uncertainties are resolved.

Within a smart market, users will have greater confidence in their ability to change allocations conveniently, which improves their certainty of supply, and they will have greater confidence in the regulator's behavior. Users are no longer directly responsible for nearby public environmental features, because the regulator can manage them through the market model.

The smart market could directly manage features such as crop prices, climate change, or generator equipment breakdowns, only if the smart market model could incorporate those features. An electricity market incorporates generator downtime directly, because electricity generation is a direct component of the market-clearing model for electricity. The electricity market does not directly address climate change, because climate change is most likely to affect the market through a change in demand for electricity. Those participants making bids for electricity need to consider many factors affecting their operations, including the effect of climate change on demand. Similarly, the smart market for water would not directly address climate change. Climate change will affect the market through changes in inflows and demand for water, so additional model machinery is unwarranted.

Implementers should not design the smart market to eliminate uncertainty associated with regulatory changes, political changes, changes in legal interpretation, changes in demand, or crop prices. Participants will respond to matters such as forecasts in crop prices by changing their bids for water. To see why, consider future crop prices. Crop prices depend on the forecasted rain, but they also depend on many other factors, such as changes in demand, rules about genetically modified crops, subsidies for biodiesel fuel, transport costs, and spread of plant diseases. Farmers can consider these factors when bidding for water, so implementers can leave those uncertainties to farmers' bidding decisions.

Working through the lists above:

- Metering will dramatically reduce the regulator's uncertainty of users' current use.
- Market incentives will dramatically reduce the regulator's uncertainty of future users' demands.

- Metering of the environment (a market prerequisite), with near real-time simulations, and market incentives to disclose future abstractions, should greatly reduce regulatory uncertainty in meeting environmental flows.
- Uncertainty about the flows the environment really needs will remain. Ecologists must continue their work in that regard.
- Users' uncertainty about political and policy changes that affect regulation of the water should reduce, as policymakers observe better water management.
- Regulator's uncertainty about future inflows of water would not reduce, but the regulator would have full ability to manage the demand for those inflows.
- Uncertainty over prices will exist because prices for water will vary, but users will have better visibility of those prices than before.
- Users should no longer face uncertainty with regard to when or whether government might curtail water rights based on "reasonable use" policies, because the market determines whether use is reasonable, not the government.
- Users should no longer face uncertainty because of lost supply due to new regulation outside market rules, because the regulator can manage the hydrology through the market mechanism.
- Users should not face uncertainty because of lost supply due to another user's new abstraction, because the smart market protects against externalities.
- Uncertainty from climate change will still exist. Climate change will affect inflows, which will affect water use requirements, which will change users' bids for water.
- Other market forces, such as prices for crops and fuel, which affect water use requirements, will still exist and will change users' bids for water.
- The market will manage uncertain inflows through pricing.

In short, only the rather near-term factors of hydrology—in particular the unknown future inflows—are endogenous to the market. Other factors are exogenous.

The market model must incorporate uncertainty in inflows in some fashion, because the inflows directly impact the hydrology. Will rain come as forecasted next week? Next month? Next year? Will the snow pack be below normal, normal, or above normal?

To now, we have presented only deterministic models with point estimates for inflows. While the market manager has a variety of tools to study inflow variability and to convey that uncertainty to users, the deterministic models do not explicitly consider the uncertainty. We touch on the problem of forecasting throughout the book, especially in Sect. 5.6.3. The modern water manager should have excellent forecasting tools available, and those tools will continue to improve, but a forecast contains uncertainty, and the market manager cannot predict the hydrological flows exactly. The regulator should require the market manager to use best practices for forecasting, but the market manager cannot be responsible for forecast error. When a forecast error occurs, prices will adjust, perhaps drastically, and when forecasted water fails to arrive, over-use of water will cause externalities.

Ideally, a smart market would consider the variability in the future inflows, and the price of water should reflect the variability of water delivery. When users plan in March for July allocations, they must know that the market manager can adjust those allocations. With the combination of short-term planning, flexibility from scaling, and trade in the spot market, a deterministic model will serve users far better than existing water allocation mechanisms. Nevertheless, the users may wish for better, and considerable theory and experience suggest that a deterministic model can sometimes miss the best allocations.

11.2 Optimal Allocations Differ Depending on the Degree of Uncertainty

Deterministic optimization sometimes gives poor results when the real problem has uncertain parameters. For water, the problem of uncertainty appears most emphatically with storage. Taking more water today reduces water for tomorrow.

Consider a simplistic reservoir problem. The reservoir manager must decide how much water to allocate to farmers this year, and how much to save for next year. The local reservoir has a capacity of 10 Mm^3 and starts the year empty. This year has a known reservoir inflow of 12; farmers would get \$1 profit/Mm^3. Next year has a 50% chance of being wet with inflow of 12; farmers would get \$1 profit/Mm^3. Next year has a 50% chance of being dry with inflow of 0; farmers would get \$2 profit/Mm^3. The farmers never want to take more than 20 Mm^3. Let's consider only periods 1 and 2 for simplicity. How much water should users take this year and how much should remain in the reservoir for next year?

Assuming the *average* values for next year results in a profit of $0.5 * \$2 + 0.5 * 1 = \$1.5$ profit/Mm^3, with an expected inflow of $0.5 * 0 + 0.5 * 12 = 6$ Mm^3. Let q1 be the water used this year, and q2 be the water used next year, while Inventory1 and Inventory2 are the water left in the reservoir after the first and second year. This can be modelled as:

$$\begin{aligned}
&\text{Maximize } q1 + 1.5 * q2,\\
&12 - q1 = \text{Inventory1},\\
&6 + \text{Inventory1} - q2 = \text{Inventory2},\\
&q1 \le 20, \text{Inventory1} \le 10,\\
&q2 \le 20, \text{Inventory2} \le 10.
\end{aligned}$$

The solution is q1 = 2, q2 = 16, Inventory1 = 10, Inventory2 = 0. So the reservoir operator tells the farmers they can take 2 Mm^3 this period, and saves 10 Mm^3 for the next period. Note, however, that if next period is wet, farmers will have 22 Mm^3 available, but will want only 20 Mm^3, so 2 Mm^3 will be wasted.

Using a stochastic optimization, the result is different.

$$\begin{aligned}&\text{Maximize } \mathrm{q1} + 0.5*2*\mathrm{q2dry} + 0.5*\mathrm{q2wet},\\&12 - \mathrm{q1} = \mathrm{Inventory1},\\&0 + \mathrm{Inventory1} - \mathrm{q2dry} = \mathrm{Inventory2dry}, \text{if the weather is dry},\\&12 + \mathrm{Inventory1} - \mathrm{q2wet} = \mathrm{Inventory2wet}, \text{if the weather is wet},\\&\mathrm{q1} \le 20, \mathrm{Inventory1} \le 10,\\&\mathrm{q2dry} \le 20, \mathrm{Inventory2dry} \le 10, \mathrm{q2wet} \le 20, \mathrm{Inventory2wet} \le 10.\end{aligned}$$

The solution is q1 = 4, q2dry = 8, q2wet = 20, Inventory1 = 8, Inventory2dry = 0, Inventory2wet = 0.

Using the stochastic optimization, the reservoir operator tells the farmers to take 4 Mm^3 this period, and leave 8 Mm^3 for next period. If the weather is wet, farmers will have exactly the right amount, 8 + 12, so no water would be wasted. Today's best choice should not be based on an average future.

With uncertain inflows, the market-clearing model will misallocate in different degrees depending on a wide range of factors. Implementers should study their catchments to determine the potential importance of uncertainty to the market. Starkey et al. (2012) developed a normalized set of experiments to quantify the value of stochastic optimization. A small reservoir (relative to demand) is of little value to the market in any case; stochastic optimization provides little improvement. An enormous reservoir removes any need for stochastic optimization. When the market relies on a medium-sized reservoir, as with our example above, stochastic optimization has great value. Starkey et al. also characterize the value with respect to the relative size of possible release. We invite the reader to redo our reservoir example above assuming a much larger reservoir with a large initial volume.

11.3 Difficulties in Modeling Stochasticity

The operations research community well knows that stochastic optimization is more difficult than deterministic optimization. The difficulty arises in the stochastic model's larger number of variables and constraints. The number of variables and constraints grows quickly with the number of scenarios ("wet" or "dry," "wet" or "normal" or "dry," "very wet" or "wet" or "normal" or "dry" or "very dry" . . .) and the number of time periods (4–6 seasons, or 12–18 months, or 52–78 weeks . . .). The variables and constraints lead to a tree-like growth in the combinations of variables over time. Three scenarios over 52 time periods leads to $3^{52} \approx 6.4 * 10^{24}$ scenario branches, with many bidding and hydrology variables per branch. Considering that a typical "big" stochastic optimization might have $2 * 10^{12}$ variables, we do not think that solution of a full stochastic optimization will be feasible for some years.

Implementers may wish to simplify the stochastic optimization, perhaps with just two or three stages. For example, the first 26 weeks may be treated deterministically, followed by a second stage of 26 weeks of "wet," "normal," or "dry," followed by a third stage of 26 weeks of "wet," "normal," or "dry." Maybe these simplifications will provide the risk reduction that the market needs, but that degree of risk reduction depends on local factors, especially the hydrology. Implementers will need to address further complications, such as whether the specification of environmental flow requirements must account for the uncertainty of flows at assessment points.

Many researchers have developed useful stochastic reservoir optimization models, such as Jacobs et al. (1995), and Tilmant and Kelman (2007). But these models are unlikely to have the fine-grained time resolution (e.g., weekly) for a sufficiently large number of periods (say, 52 or 78 weeks), nor the hydrological generality needed to work for the general case. Hollinshead and Lund (2006) developed a stochastic optimization that California's CALFED agency could use to minimize the cost of buying water on behalf of the environment for the Sacramento–San Joaquin River Delta. The model works in annual time steps, which should be daily time steps for a smart market application. Another use of stochastic optimization is to study climate change. Those stochastic models typically use years for period lengths; the climate change uncertainty is decades ahead. A year will usually be much too long for market operations, because users will be trading in weeks, months, or seasons. The market is allocating water for the near future, when the market manager can predict weather with more confidence, not for the distant future with its long-term effects of climate change.

In some specific circumstances, in particular for a single reservoir, the stochastic model is easy, even for many scenarios and time periods. Dye et al. (2012) and Mahakalanda et al. (2012) showed easy stochastic models for simple single reservoir systems specifically in a potential market context. So in simple enough cases, the market manager could plausibly use a stochastic model. Pritchard et al. (2010) developed a two-stage market model to manage uncertainty in an electricity market. This type of stochastic model (see also Pinto et al. 2009; Pinto 2013) is tractable because it has only two stages, and a few stages may be enough to capture most of the value of modeling uncertainty in smart markets for water. Zhu et al. (2015) develop a two-stage model to study water transfers under uncertainty; perhaps a number of deterministic stages followed by a single stochastic stage could be enough for the smart market. Gorelick and Zheng (2015) give a modern motivation and review of stochastic optimization of hydrology models.

11.4 Difficulties of Bidding for Stochasticity

If implementers successfully develop a stochastic market-clearing model, they still face the problem of designing a workable bidding mechanism. Implementers will have to develop some way to allow traders' bids to track the set of scenarios as the scenarios expand in tree-like fashion. A stochastic model imposes a greater burden for information on users, who would somehow have to provide information about their future values for water under different trajectories of future scenarios. Consider a market with different bids based on whether each of the next four quarters of a year is normal, wet, or dry. The farmer will have different water values for the sequence normal, wet, dry, wet, compared with dry, normal, wet, wet, even if the total water over that year is the same. With only three scenarios and four future periods, the market manager would have to ask users for 243 bids. Implementers could not plausibly ask traders for all water values over all scenario sequences. To be workable, the bidding framework would have to be simplified.

Finding a tractable method of bidding to a market engine with a stochastic model would allow users to bid explicitly for higher-level security. A stochastic model would allow the market manager to develop and price a range of contracts that would be more awkward with a deterministic model.

To test a stochastic market-clearing model, we see the need for first-rate stochastic optimization tools, an adequate hydrology model, and a laboratory for experimental economics. The researchers would choose a realistic catchment that they can model with stochastic optimization. The experiments could then test different simplified bidding mechanisms to determine which gets *enough* information to reap the benefits of the stochastic modeling. The problem remains open to research.

11.5 Alternatives to the Stochastic Market Model

The market manager has various alternatives to a full stochastic market. First, the manager should point users to financial products, as we discuss in Sect. 10.4, which are exogenous to the market. Perhaps others will find ways to develop risk products for water with a deterministic market model that allow for some of the benefits of a stochastic market.

Second, the market manager should provide information about the future scenarios. Producing information about scenarios should be straightforward. First, the market manager should post a searchable database of historical prices, so users can review price behavior in similar past situations. Second, the market manager should post ranges of future prices. The market manager can do this by using the hydrology model, historical water flows, and historical bids, modified for each scenario. Based on forecasted conditions, the market manager should be able to provide a probability distribution on price by location, for the short and long range. For each location l and future period t, the market manager can give a set of prices p_{ilt} that will occur with probability i at location l in period t.

Third, the market manager can choose the inflow parameters to give users greater confidence that the planned water will be available. The market manager's selection of inflow parameters in the actual market clearing could raise users' certainty about the expected inflows. Forecasting the inflows will be an important part of market operations, and the market manager should have good information on the probability distribution of inflows. If the market manager uses the expected inflows in the market model, then the manager will have to scale users' allocations downward about half the time. Instead, the market manager could clear the market with a lower inflow, perhaps a half standard deviation below the mean, so that users will be roughly 70% sure of getting the resulting planned allocation. As the future period approaches, the market manager can update the expected inflow with greater certainty. This approach will tend to raise future prices, encouraging low-value users to wait to buy water until later, while assuring high-value users that they will be roughly 70% sure of getting the water they need. Users may wish to speculate that the inflow will be higher or lower than the market manager has stated, and such speculation will tend to improve market activity.

References

Dye, S., Read, E. G., Read, R. A., & Starkey, S. R. (2012, June 14–15). *Easy implementations of generalised stochastic CDDP models for market simulation studies*. 4th IEEE and Cigré International Workshop on Hydro Scheduling in Competitive Markets, Bergen.

Gorelick, S. M., & Zheng, C. (2015). Global change and the groundwater management challenge. *Water Resources Research, 51*, 3031–3051. doi:10.1002/2014WR016825.

Hollinshead, S. P., & Lund, J. R. (2006). Optimization of environmental water account purchases with uncertainty. *Water Resources Research, 42*(8), W08403.

Jacobs, J., Freeman, G., Grygier, J., Morton, D., Schultz, G., Staschus, K., & Stedinger, J. (1995). SOCRATES: A system for scheduling hydroelectric generation under uncertainty. *Annals of Operations Research, 59*(1), 99–133.

Mahakalanda, I., Dye, S., Read, E. G., & Raffensperger, J. F. (2012, December 10–11). *Intra-period market clearing for a multi-use catchment via CDDP*. Paper presented at the ORSNZ Conference, Wellington.

Pinto Rodriguez, A. A. (2013). *Market mechanisms to allow trading of impervious cover*. PhD thesis, University of Canterbury, Christchurch.

Pinto Rodriguez, A. A., Raffensperger, J. F., Cochrane, T. A., & Dye, S. (2009, December 3–4). A proposed smart market for impervious cover runoff under rainfall uncertainty. In *Proceedings of the ORNSZ conference*, Christchurch, (pp. 190–199).

Pritchard, G., Zakeri, G., & Philpott, A. (2010). A single-settlement, energy-only electric power market for unpredictable and intermittent participants. *Operations Research, 58*(4), 1210–1219.

Starkey, S. R., Dye, S., Read, E. G., & Read, R. A. (2012, June 14–15). *Stochastic versus deterministic market design: Some experimental results*. 4th IEEE and Cigré International Workshop on Hydro Scheduling in Competitive Markets, Bergen.

Thompson Jr., B. H. (2005). Uncertainty and markets in water resources. *McGeorge Law Review, 36*, 117–138.

Tilmant, A., & Kelman, R. (2007). A stochastic approach to analyze trade-offs and risks associated with largescale water resources systems. *Water Resources Research, 43*(6), W06425.

Zhu, T., Marques, G. F., & Lund, J. R. (2015). Hydroeconomic optimization of integrated water management and transfers under stochastic surface water supply. *Water Resources Research, 51*, 3568–3587. doi:10.1002/2014WR016519.

Chapter 12
Smart Markets for Water Quality, Sediment, and Impervious Cover

Abstract This chapter discusses smart markets for water quality, sediment, and impervious cover. The market designs for these resources are similar to the design for a water quantity market. Implementers must work through prerequisites, including setting initial allocations, which may need to be subject to scaling of some kind for revenue neutrality. But the designs differ in important ways from the water market design. Briefly, users cannot precisely know and control their run-off or groundwater recharge quality; they can control only their use of the land. Thus, implementers must design the bid structure to allow users to buy and sell the right to change their land use, not to buy and sell the water resource impact.

Keywords Water quality • Sediment • Impervious cover

12.1 Water Quality

12.1.1 The Water Quality Setting

As with water quantity, an activity that alters water quality in one place will also affect quality in many other places. Many researchers feel that problems of water quality, like the problems of land development and allocation of water quantity, should be amenable to a market solution (Morgan et al. 2000; Collentine 2005; Hung and Shaw 2005). Some researchers studying water quality are developing combined hydrology–contaminant transport–optimization models.

With colleagues, E. Grant Read, Thomas Cochrane, Antonio Pinto, and Ranga Prabodanie, we have investigated designs for smart markets for water quality, for sediment, and for impervious cover. All of these issues will be of increasing importance with anticipated population growth, industrial development, and, most significantly, climate change.

The literature on each of these topics is extensive. The existing markets suffer from a large transaction cost and blunt regulatory processes and they could benefit from a smart market approach.

Because we and our colleagues have written about these in the academic literature, we will include only summaries here.

J.F. Raffensperger, M.W. Milke, *Smart Markets for Water Resources*, Global Issues in Water Policy 12, DOI 10.1007/978-3-319-55008-4_12

Currently, the U.S. Environmental Protection Agency regulates point source discharge, but does not regulate non-point source (i.e., agricultural) discharge. Politics, agricultural lobbyists, and public ignorance all play a role in the deterioration of water quality. The tragedy part of the tragedy of the commons is not the result of bad or unethical people; quoting Alfred North Whitehead, Hardin (1968) writes, "The essence of dramatic tragedy is not unhappiness. It resides in the solemnity of the remorseless working of things." So it is with water quality deterioration.

When a farmer fertilizes the land, nutrient levels in the water are increased. These nutrients may help some people, but they usually hurt other people (such as with contaminated drinking water) and the environment (such as with ocean dead zones). The increased nutrient levels are externalities of the farmer's commercial operations. These externalities are priced at either zero (users can do what they like) or, less often, infinity (users are prohibited from a given action). Users have limited incentive to reduce their externalities, and society has limited ability to improve water quality in economically efficient ways.

Each farmer rationally seeks to maximize his or her own gain. The farmer's marginal utility for fertilizer is positive until the marginal profit of production of extra crop equals the cost of the extra fertilizer needed to obtain the extra crop. The existing incentive to a farmer is to discharge more than the environment can tolerate, because the farmer must apply fertilizer in an unnaturally high amount to obtain unnaturally high agricultural productivity. The farmer bears only a fraction of the marginal cost to society.

12.1.2 A Market for Quantity with Side Constraints for Quality

As we describe in Sect. 6.9, trading water quality within a spot market for water quantity makes sense only if the quality attribute has similar physics to the quantity attribute. The user's behavior that affects the water quantity must directly correlate with the water quality. This would not be appropriate for nitrate and phosphorus in groundwater when farmers apply those nutrients seasonally as fertilizer, because the abstraction of water is independent of the application of the fertilizer. It would be appropriate for a non-consumptive use that adds salt to discharge water, or one that increases the temperature of the water, affecting downstream users and the environment.

To examine the market design issues, let us start with our generic gross pool formulation from Sect.7.2.4, adding constraints for a generic quality attribute.

$$\text{Maximize} \sum_{\text{users } w} \sum_{\text{bids } b} BidPrice_{b,w} bid_{b,w},$$

$$q_w = \sum_{bids\ b} bid_{b,w} \text{ for each user } w,$$

$$0 \leq bid_{b,w} \leq BidQty_{b,w}, \text{ the quantity in bid block } b \text{ for user } w,$$

$$Aq \leq d, \text{ arbitrary quantity constraints in generic matrix form.}$$

$$Bq \leq f, \text{ arbitrary quality constraints in generic matrix form.}$$

The coefficients in $Bq \leq f$ must have direct physical meaning. The elements of matrix B are the per-unit quality impacts of user w on each environmental assessment point. The right-hand-side vector f is the list of assessment point limits.

Let us extend our example from Sect. 9.1 to include a constraint on quality. The hypothetical story here is that abstraction draws down the aquifer, affecting the local river, but the same abstraction also causes a return flow of unwanted selenium.

Maximize $3\text{ A} + 4\text{ B} + 11\text{ C}$,
Drawdown at Pinetown: $0.10\text{ A} + 0.04\text{ B} + 0.22\text{ C} \leq 2$.
Drawdown at Oak City: $0.05\text{ A} + 0.16\text{ B} + 0.09\text{ C} \leq 2$.
$\text{A}, \text{B}, \text{C} \geq 0$.

The solution is A = 0, B = 8.2, C = 7.6. Assume units of cubic meters of water per day.

Now suppose that the return flow from Ann's water use adds 25 mg of selenium per cubic meter of water used (mg/m^3), Bob's adds 100 mg/m^3, and Cal's adds 30 mg/m^3. Further, suppose the regulator has set a limit of 400 mg/day for return flows. Adding a new constraint as $25\text{A} + 100\text{B} + 30\text{C} \leq 40$, or equivalently:

Limit on selenium: $25\text{ A} + 100\text{ B} + 30\text{ C} \leq 400$.

The new solution is A = 0, B = 1.3, and C = 8.8. Prices have changed, too, and the new constraint produces a price. We leave this simple numerical example to the reader to explore.

The water allocations imply a selenium allocation. Since Bob has a right to 1.3 m^3 of water per day, he must also have a right to impact the river with 1.3*100 mg of selenium. The market manager could unbundle these rights as separate environmental impact quota, or could treat them as a single commodity with two attributes. In either case, the market manager is making a rather sophisticated agreement with Bob—they are making a contract, not simply trading water.

The reader may then want to think more about the physics. If the user's return flows affect the river differently at different times, we should embellish the market model with time subscripts on the variables and a response matrix for both water quantity and selenium; these matrices are likely to differ. See Lefkoff and Gorelick (1990) for a detailed example of a hypothetical water rental market with constraints for salinity.

If the regulator is concerned with a range of quality factors, e.g., selenium and salt and temperature, and if users' abstractions directly affect those factors, the market manager can build those factors into the market model. Setting quality constraints for each assessment point and time period is identical mathematically to setting quantity constraints. Though the optimization will not be much more complicated in this joint quality–quantity market, the hydrologist will have a more challenging task in developing the expanded model. Even more challenging would be the role of policymakers and politicians who would have to codify, explain, and provide leadership on a more complex market.

12.1.3 A Market for Nutrient Run-Off

Agricultural phosphorus and nitrate return flows, whether groundwater recharge or surface run-off, can damage waterways and cause ocean dead zones. The government could develop a smart market for water quality, operating similarly to the market for water quantity, to reduce nutrient run-off. As with the market for quantity, a local regulator would manage the quality market. The main participants are likely to be farmers and wastewater treatment plants.

At first look, it appears that the market design requires (1) specifying the allocations of each user to discharge nutrient, (2) measuring the impact of the local discharge on the nutrient level at a set of assessment points via a nutrient transport model, and (3) solution of a linear program. The design needs those, but the physics of the problem makes these steps a bit harder than they might seem at first.

Scaling initial allocations for nutrient run-off will most likely be at least as contentious as for water allocation. The U.S. has especially perverse rules that implementers will need to address.

Ideally, land users would buy and sell nutrient allocation specifying the allowable nutrient to discharge for a season, in some commensurate units such as kilograms per time period. Unfortunately, most users, especially agricultural users, cannot control their run-off directly in the same way that they can control the water they take. Agricultural run-off has no convenient discharge tap that the regulator can meter. The mass of nutrients leaving a property as return flow depends on soil type, soil slope, local surface and subsurface hydrology, crops, plowing, rate of fertilizer, grass buffers near waterways, and uncertain rainfalls. Implementers must design a bidding system for agreements that the regulator can enforce, for acts that the user can control. Implementers must then provide means for the auction manager to convert those controllable acts to commensurate units of kilograms run-off per time period. Thus, instead of bidding to discharge kilograms of nutrient, a bid is for a contract, an agreement, as we describe in Sect. 6.4.2.

Hydrologists understand how to estimate these conversions. For each possible land use, the regulator can estimate the periodic nutrient leaching from soil to the aquifer and nearby waterways. Many environmental authorities already have tools to estimate the leaching as a function of loading. The conversion from user-controllable actions to kilograms per period would need to be transparent to the user. Users might use approved calculations in a spreadsheet or website to prepare bids. The user selects land use changes, the area planted with each crop, the plowing method, etc., and the web page converts these selections to the kilogram per period.

In effect, the regulator is telling farmers, "If you grow corn on your soil type, then your nutrient leaching rate per kilogram nutrient in period 1 will be reckoned as L_t for each future period." After the regulator has determined leaching rate (in kilograms per hectare per year) from each possible land use, users can bid to buy and sell allocations based on the number of allocations required and potential profit of each land use option. But these conversions are essentially notional; users are bidding for the land use changes. Thus, implementers have considerable leeway to invite users to bid for almost any kind of action that might affect nutrient loads, and to offer a wide range of actions to reduce nutrient loads in waterways.

The above logic concerns the emissions side of the market, but implementers will also need to think carefully about the impact side of the market. The amount of nutrient at each location, given each user's land use and bid, would (under this system) be determined by a nutrient transport model. The nutrient transport model measures the impacts of nutrient loading by land users on water quality at a set of assessment points. Each impact is represented by an impact coefficient, which is the increase in nutrient at a particular assessment point at a particular time, if a given user discharges one allocation of nutrient in the current period. The regulator should require the market manager to constrain the cumulative nutrient at a set of control locations for those time periods in the future where the impacts are sufficiently large as to be of importance.

Those time periods can be far in the future; long nutrient transport times will complicate the market. If land use today impacts the assessment points in many decades, no buyer is likely to be on the other side of the transaction. The problem here resembles the problem of valuing water in a reservoir at the end of the auction schedule. Current users may worry little about discharging so much now as to inhibit the next generation's discharge, but the implementers and the regulator should worry about it. To see this, consider a catchment with 10 farms and a single lake, where the transport time from the farms to the lake varies from 5 to 50 years, depending on the distance from the lake. The market model would constrain the future inflows to sustainable levels. But in a few years, farmers with short transport times could no longer discharge, because previous discharges still flowing in the catchment would use up all available capacity. The market would benefit from encouraging users to save capacity for the future.

We can imagine that the regulator appoints a trust agent to trade on behalf of the future years. The agent would buy sufficient future rights to raise the price enough to restrict current discharge to a sustainable level. Participation of a trust agent might incentivize other people to buy and hold future discharge rights too, thus thickening the market.

An alternative approach would be to use a steady-state model, one in which all impacts are treated as if they continued forever. We have not explored this option. It appears to trade in permanent rights, not in seasonal or annual rights, because a buyer acquires the rights to the steady state. However, this design would appear to force current use to be sustainable without the need for a trust agent.

From this logic, implementers can specify the auction schedule, including how often each auction will run, the number of future periods traded in each auction, and the number of future periods constrained in each auction. The auction should run as often as users can change their actions. Most likely, farmers will be managing their nutrient loading on a seasonal basis. Interestingly, point sources such as factories and wastewater treatment plants could change discharge on a daily basis, resulting in a potentially highly flexible market design. Private and commercial users may want to trade only a few years into the future, while the auction manager will be obligated to include constraints to protect the environment for many years, depending on the residence times of the nutrient loading.

As with the model for water quantity, the nutrient market model could be either net pool or gross pool. The basic net pool market-clearing model is as follows:

Max $\sum_{\text{users } i} \sum_{\text{blocks } b} \sum_{\text{periods } t} \left(buy_{i,b,t} \cdot PriceBuy_{i,b,t} - sell_{i,b,t} \cdot PriceSell_{i,b,t} \right)$,
Bids: $buy_{i,b,t} \leq MaxBuy_{i,b,t}, sell_{i,b,t} \leq MaxSell_{i,b,t}$, for each bid block b, time period t, and user i.
Rights: $discharge_{i,t} = InitialPermit_{i,t} + \sum_{\text{blocks } b} buy_{i,b,t} - \sum_{\text{blocks } b} sell_{i,b,t}$, for all users i; dual price p_i.
Sustainability: $\sum_{\text{users } i} \sum_{u=1}^{t} H_{i,k,u} \cdot discharge_{i,u} \leq S_{kt}$, for all assessment points k, $t = 1, \ldots, T$, where period 1 is the current period, and T may be many years.

For further information on this market design, see Prabodanie et al. (2009, 2010, 2011). In that market design, the model represents trades with continuous variables. Agricultural run-off and wastewater treatment discharges can be continuous, but some land use changes (such as to wetlands) have larger fixed costs, implying that the trades involving them must be represented with binary variables to reflect the all-or-nothing nature of the bid. More recently, Kostel et al. (2014) extended this work to incentivize landowners to build wetlands, taking into account the all-or-nothing nature of either building a wetland or not building it. The market-clearing model selected the optimal locations of the wetlands as well. These papers give good background on nutrient markets, and describe at length reasons that nearly all nutrient markets have had few trades—the high transaction cost of making a trade.

Implementers could extend the market for nutrient run-off, probably easily, to consider the uncertainty of return flows as a function of rainfall patterns, following Pinto's (2013) market design for impervious cover.

12.2 Sediment

Construction and land disturbance can cause streams and rivers to deliver too much sediment, with effects as diverse as stress on local ecosystems and flooding impact on property. Dams and other man-made waterway obstructions can prevent outflow of sediment, contributing to coastal land loss. As with the other market designs, government could decide to start a market for sediment. The purpose of the market would be to manage sediment run-off in waterways. The main participants are likely to be farmers, local municipalities, and construction firms.

The market design requires (1) specifying initial allocations of each user to discharge sediment, (2) measuring the impact of the local load on the sediment at assessment points via a transport model, and (3) solution of a linear program.

As with the nutrient market, implementers must design a bid structure for contracts that users can control and that the regulator can enforce. Ideally, participants would trade discharge allowance, specified as the nominal right to discharge 1 kg of sediment from a particular location in a particular period. Users cannot control sediment discharge directly; they control their land use and their sediment control methods. Thus, traders might select their land use type, sediment control methods, and quality assurance methods from a list on the bid page. Their selection then determines the run-off coefficients associated with their bid. The web page should show the run-off coefficients.

Implementers would have to choose the auction schedule based on requirements of the local catchment and the needs of the users. Most likely the auction schedule would be seasonal, with auctions often enough for construction firms and farmers to change their actions based on prices.

For the proposed land uses, implementers would obtain impact coefficients $F_{i,k}$, the sediment that reaches assessment point k from landowner or firm i, from a hydrology discharge model. Users might use approved calculations in a spreadsheet or website to prepare bids. The basic gross pool market-clearing model for a single time period is as follows. Implementers could easily extend it to multiple periods.

$$\text{Maximize} \sum\nolimits_{i \text{ firms}} \sum\nolimits_{b \text{ blocks}} Price_{i,b}\, discharge_{i,b},$$

$$discharge_{i,b} \leq Demand_{i,b}, \text{ for each firm } i \text{ and step } b.$$

$$totalDischarge_i = \sum\nolimits_{b \text{ steps}} discharge_{i,b} \text{ for each firm } i.$$

$$\sum\nolimits_{i \text{ firms}} F_{i,k} \cdot totalDischarge_i \leq sedimentLimit_k \text{ for each assessment point } k.$$

For further information on this market design, see Pinto et al. (2013). When we wrote that paper, we thought of sediment only as a bad thing. We did not think about downstream demands for sediment, especially the value of sediment in preventing coastal loss. Some assessment points could have lower bounds for sediment rather than upper bounds. Nor did we think about the various participants that might demand sediment. The market design would extend easily to those cases, with enormous benefit to coasts.

12.3 Impervious Cover

Impervious cover tends to increase the risk of flooding, as it prevents infiltration of rain and surface water into the ground, resulting in short intense storms overwhelming local waterways. Flooding is a common and carefully managed risk, with established insurance mechanisms to deal with risk, and people generally believe that government is the insurer of last resort. Compared to water quantity, the public seems to have less appreciation of the risk associated with impervious cover, and how to manage and distribute that risk.

In this complicated context, government could choose to develop a market for impervious cover to mitigate flood risk. The purpose of the market is not simply to reduce impervious cover, nor to reduce flooding per se, but to reduce the costs from flooding. The market would incentivize people to reduce impervious cover and to move out of flood plains. It would signal engineers where to put flood mitigation infrastructure. The main market participants are likely to be farmers, large urban landowners, and city managers. Urban regulators could decide to push participation to relatively small parcels of land, so that the market covers a large fraction of the urban area. These changes would reduce the common risk, help to price landowners' insurance more precisely, and lower the cost to society of infrastructure and catastrophe. The market design requires (1) specifying initial allocations of each user of impervious cover, (2) measuring the impact of the impervious cover on assessment points via a transport model, and (3) solution of a linear program.

Initial allocations should be easy, because concrete and lawn are facts on the ground. The difficulty will be that the current flood absorbance capacity is almost certain to be—to use an analogy from water allocation markets—over-allocated. People will believe they already have permission for their land use, and yet when taken together all land users will have been given too much permission to create impermeable cover. So the interesting part comes in scaling down of initial allocations into tradable allocations. Once land users find that they do not have enough tradable allocation for their current land use, every landowner must either reduce impervious cover or buy rights from the market. For a catchment at serious risk of flooding, the regulator may scale everyone's rights a bit each year, thus reducing the anticipated cost from floods over time until it reaches a more socially acceptable value.

A smart market for impervious cover could be either deterministic or stochastic. A deterministic smart market would be relatively simple to develop (Raffensperger and Cochrane 2010). However, because rainfall and flooding are so uncertain, the market should be cleared by a stochastic model. Fortunately, a stochastic model for impervious cover would be easier to solve compared to a stochastic water allocation market (Pinto et al. 2009; Pinto 2013). We know of only one other proposed smart market with a stochastic market-clearing model, the paper by Pritchard et al. (2010).

Land use changes slowly over decades, but hydrological effects vary over minutes; the market in this case must constrain land use for what *might* happen.

As with nutrient and sediment run-off, users cannot completely control the water run-off from their land use (short of placing a large tank on their property to hold the storm water from the largest possible storm). So implementers must design a bidding mechanism that allows people to trade contracts for land use, based on activities they can control. The activities include conversions in area from one type of land cover to another. The types of land cover can include concrete, crops, and forest: anything that the regulator believes should be managed for flooding. The market should incentivize concrete removal, tree planting, and construction of storm water storage systems (including wetlands). Most often, users will wish to change the area of a given type of land cover. Total area does not change much, so reducing area of one type of land cover means increasing another, which the bidding software will have to enforce. The simple way to think about this market is that people are trading permits to change area of land cover type. Figure 12.1 shows a schematic of how Pinto's market design treats the random rainfall events given the different types of impervious cover on property in the catchment.

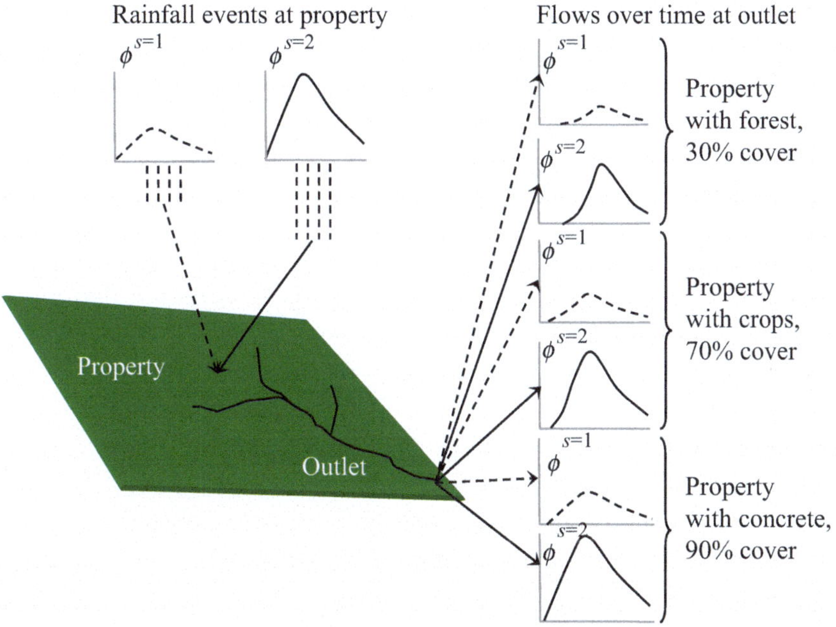

Fig. 12.1 Different rainfall events at the property induce different flows at the outlet, depending on rainfall event s (with probability ϕ^s), and the differing impervious cover type at the property (Modified from Pinto 2013)

Given the types of land cover in a catchment, a qualified hydrologist can determine run-off hydrographs for different rainfall events for each type of cover, the coefficients $H_{i,j,k,s}$ in the model below. The market-clearing model has constraints associated with total run-off in each rainfall scenario. The hydrologist then has a chore to develop appropriate constraints for a wide range of rainfall scenarios. Because no one can prepare for a cataclysmic flood, constraints for extreme scenarios should have variables, $C_{s,k}(f_{s,k})$ in the model below, that allow the constraints to be violated, but also violations to be penalized, based on flood rate, duration, and depth, in the objective function (Pinto 2013). The variable $f_{s,k}$ in this example denotes a function evaluated for each combination of scenario and assessment point.

A simplified gross pool market-clearing model is as follows.

Maximize$\sum_{i \text{ firms}}\sum_{j \text{ land types}}\sum_{b \text{ bid blocks}}\Big(Buybid_{i,j,b}\ buy_{i,j,b}$
$- Sellbid_{i,j,b}\ sell_{i,j,b}\Big) - \sum_{s \text{ scenarios}}\sum_{k \text{ assessment pts}}\phi_s \cdot C_k\left(f_{s,k}\right),$

Bids: $buy_{i,j,b} \le MaxBuy_{i,j,b}$, $sell_{i,j,b} \le MaxSell_{i,j,b}$, for each bid block b, land use j, and user i.

Rights: $cover_{i,j} = InitialPermit_{i,j} + \sum_{b \text{ bid blocks}}(buy_{i,j,b} - sell_{i,j,b})$, for each land type j and user i; dual price $p_{i,j}$.

Flood control: $\sum_{i \text{ firms}}\sum_{j \text{ land types}} H_{i,j,k} \cdot cover_{i,j} \le L_k + f_{s,k}$, for all scenarios s and assessment points k.

With the extensions described in Pinto (2013), the regulator can constrain (or at least penalize) flood rate, length, and depth.

How should the regulator set $C_k(f_{s,k})$? The penalty function $C_{s,k}(f_{s,k})$ is not generally convex, so implementers will have to approximate it. Apart from mathematical issues, this function should be set to the true cost of exceeding the capacity at the assessment point. Estimating these values will be important to getting the market right. But the presence of variable $f_{s,k}$ implies that the market manager is selling a right to exceed the flood capacity in scenario s at assessment point k. So the market manager is allowing excess impervious cover, and bidders for this impervious cover must value it higher than the expected flood cost. This method shows how a smart market could be adapted to situations where society is willing to take the risk of violating a constraint.

Unfortunately, the people who want the excess impervious cover are not necessarily the people exposed to the risk. Imagine wealthy homeowners in the hills wanting to build tennis courts, which increases flooding on poor people in the flood plain. The market correctly charges the wealthy people for their run-off, but does not necessarily compensate the people in the flood plain for their additional risk.

Government could compensate people in the flood plain in different ways. First, government could keep the extra market revenue, and agree to pay the costs when people in the flood plain are harmed. This approach implies that government is the insurer of last resort, which it usually is anyway. People in the flood plain could still be required to buy flood insurance, but they may feel justified in complaining, because the market manager has increased their risk and taken money to do so. Second, perhaps more interestingly, they could be given an initial allocation associated with $f_{s,k}$, so that they could sell the right. They should then be required to buy insurance associated with the flood risk. This would incentivize them to leave the flood plain, should the insurance cost be sufficiently high. We have not researched all of these issues, but it should be clear that the market makes these issues more salient and provides new ways to manage flood risk.

References

Collentine, D. (2005). Including non-point sources in a water quality trading permit program. *Water Science and Technology 51*(3–4), 47–53.

Hardin, G. (1968). The tragedy of the commons. *Science 162*, 1243.

Hung, M.-F., & Shaw, D. (2005). A trading-ratio system for trading water pollution discharge permits. *Journal of Environmental Economics and Management 49*(1), 83–102.

Kostel, J. A., Monchak, J., Tomer, M. D, Bingner, R. L, Bank Lentz, A. T, Ando, A. W, Brozovic, N., Raffensperger, J. F, Ranga Prabodanie, R. A., Wohlgezogen, F., Zajac, E., & Hey, D. (2014, September). *Feasibility assessment of a nutrient trading market in the big bureau creek watershed final report*, Prepared for U.S. Environmental Protection Agency, Targeted Watershed Grant WS-00E71101.

Lefkoff, L. J., & Gorelick, S. M. (1990). Benefits of an irrigation water rental market in a saline stream-aquifer system. *Water Resources Research 26*(7), 1371–1381.

Morgan, C. L., Coggins, J. S., & Eidman, V. R. (2000). Tradable permits for controlling nitrates in groundwater at the farm level: A conceptual model. *Journal of Agricultural and Applied Economics 32*(2), 249–258.

Pinto Rodriguez, A. A. (2013). *Market mechanisms to allow trading of impervious cover*. PhD thesis, University of Canterbury, Christchurch

Pinto Rodriguez, A. A., Raffensperger, J. F., Cochrane, T. A., & Dye, S. (2009, December 3–4). A proposed smart market for impervious cover runoff under rainfall uncertainty. In *Proceedings of the ORNSZ conference*, Christchurch, (pp. 190–199).

Pinto Rodriguez, A. A., Raffensperger, J. F., Cochrane, T. A., & Read, E. G. (2013). A proposed smart market design for sediment discharge. *Journal of Water Resources Planning and Management 139*(1), 96–108.

Prabodanie, R., Ranga, A., Raffensperger, J. F, & Milke, M. W. (2009, July 13–17). *Simulation-optimization approach for trading point and non-point source nutrient permits*, 18th World IMACS/MODSIM Congress, Cairns, Australia.

Prabodanie, R., Ranga, A., Raffensperger, J. F., & Milke, M. W. (2010). A pollution offset system for trading non-point source water pollution permits. *Journal of Environmental and Resource Economics 45*(4), 499–515.

Prabodanie, R. A., Ranga, John F., Raffensperger, E., Read, G., & Milke, Mark W. (2011). LP models for pricing diffuse nitrate discharge permits. *Annals of Operations Research*. doi: 10.1007/s10479-011-0941-0

Pritchard, G., Zakeri, G., & Philpott, A. (2010). A single-settlement, energy-only electric power market for unpredictable and intermittent participants. *Operations Research 58*(4), 1210–1219.

Raffensperger, J. F., & Cochrane, T. (2010). A smart market for impervious cover. *Water Resources Management 24*(12), 3065–3083.

Chapter 13
Critical Perspectives on Water Markets

Abstract This chapter describes a wide range of comments and objections about water markets generally, and our smart market design specifically. The first section describes our experience in collecting these comments. The remaining sections then group the comments, along with our responses to those comments. The comments included objections to water markets, such as "Pricing and trading water are ethically bad," and doubts about the feasibility of a smart market systems, such as "The problem cannot be fixed" and "It is more complicated than you think."

Keywords Fairness • Ethics • Fear • Communication

13.1 Listening to Stakeholders Before Market Development

We have given presentations about smart water markets all over the world. Our workshops often began by asking the audience to write objections to water trading on sticky notes and attach them to a wall near comments with a similar theme. We then invited people to share their concerns with the rest of the group. This took 30–45 min. Following this, we invited the audience to give us their hand-written lists of objections, which we collated and categorized.

The workshop audiences varied widely. The intended audience was mainly elected officials responsible for water management. These officials sometimes invited consultants, government staff, scientists, representatives of indigenous peoples, farmers, and members of the general public. Often a given person would be in more than one of these categories, such as the elected officials who were also farmers. Groups were of various sizes, from three or four to nearly a hundred. Participants tended to give similar responses between groups. The following categorization and analysis of responses should help readers better understand smart markets for water and to better manage the necessary stakeholder engagement.

Our experiences indicate that advancing plans for smart markets for water will be challenging because of pre-existing attitudes towards water markets. Reform of water use is a politically charged topic, even though the logic of using markets to protect the environment is strong. Any discussion with the public about water

J.F. Raffensperger, M.W. Milke, *Smart Markets for Water Resources*, Global Issues in Water Policy 12, DOI 10.1007/978-3-319-55008-4_13

markets will quickly find a strong resistance to the concept—even from those likely to benefit most from the introduction of a market.

Stakeholders and implementers should not use our replies to pressure people into accepting water markets. We intend our replies to help market proponents and skeptics understand each other's viewpoints. Economists may recognize many concerns about water trading as well-recognized problems of market design or market error.

Aside: Change Is Difficult and Necessary

Why allow markets for water? After all, some say, look at the mess markets have gotten us into—wars, famine, global recession, bankruptcy. Enron's exploitation of California's poorly designed electricity market is often waved as a red flag at the prospect of a market for water.

Poorly designed and operated markets and the complexities of modern society can induce fear of change in general, and of markets in particular. Most people do not wonder about or fear trading in baseball cards. The nature of the object immediately suggests an acceptable type of market. By contrast, the ubiquity of water and its many uses leave most people without a similar concept of how trading in water would work, and hence those people may object to "water trading" from fear of the unknown.

The lack of a market means letting business use up a precious resource without consideration of what is best for society. Whatever business you're in, imagine that a market for that business was not allowed to exist, and that governmental rules, political clout, and historical precedents determined who could access that good or service.

Some people don't want water markets because they believe that it will injure poor people. A common-pool smart market for water is not about charging poor people for drinking water. Nor is it about hurting disadvantaged people in developing countries. It is about managing the commercial use of water by business, especially the 75% of the world's fresh water that agriculture uses in the developed world. The smart market approaches developed in this book are about allocating a public resource that is precious. These approaches are also about finding new ways for society to resolve the tension between the environmental value of water and its value to a multitude of users who take that resource to make money. Preventing a water market does not remove the economic competition for water. Smart markets can protect the environment by shifting the competition over water's economic value to a fair and monitored market.

When water markets are mentioned, some people immediately think of expensive water projects. Maybe we can float a glacier and tow it to Shanghai? Maybe we can build a pipeline from Washington to Arizona? This misconception is a sign of just how much change is needed in how society

(continued)

perceives water problems: without markets today the big water projects can be seen as the only response to our ever-increasing water constraints.

Some people object to a market when they are really objecting to living under a constraint. They don't like markets because they don't like having to change their ways to get what they want. But no one can make rain. Everyone has to learn to live with the available water, and, to do that, society needs to manage it far better. Living sustainably *means* living under constraints on our behavior, so our children and the environment will have enough. With global warming and population growth, these constraints have suddenly gone critical. Society has to change and live under constraints in a smart way.

A few well-informed people object to water markets because water markets don't work well. In some places, water markets have allowed damage to the environment and injury to people who were not parties to the trade. To these well-informed people, we observe that damage has happened even without markets, so the bad experiences of some other types of markets should not be used to dismiss the potential of smart markets. We would also say that a common-pool smart market has not been tried. We have written this book as a plea for society to consider this alternative way of allocating water and to ask the skeptic to consider how this method will have fewer problems than others tried previously.

13.2 "Pricing and Trading Water Are Unethical"

Sample Comments

- "Many people feel that trading water is selling a necessity of life."
- "It puts a price on a sacred element."
- "It is commoditizing a precious part of nature for the sake of business."
- "It puts profit above cooperation."

Concerns about water trading partly reflect concerns about markets generally. The objection to a market economy might be on a spiritual basis or a philosophical one. These objections imply that a planned economy is better than a market economy. The current methods of water allocation are closer to those of a planned economy, and so individuals in favor of a planned economy will want to retain its planned nature. But history has painfully shown that planned economies do not work as well as markets.

Markets do not need to impinge on spiritual values or our desire to work cooperatively towards social goals. Smart markets for water can protect non-commercial water. Smart markets will not restrict communities' ability to work together on water issues—for example, on restoration of native vegetation near rivers. By separating the economic value of water from its social value, smart markets might make it easier for water users to participate cooperatively in social projects because it will be clearer that their participation is not because of some hidden agenda or a "greenwash" activity.

Sample Comments

- "It puts a price on water."
- "It views water as an 'economic commodity,' when it is something much more important."

Trading puts an explicit price on water, but this does not mean that it privatizes the resource.

Universally, it seems, the law recognizes water as shared public resource. Governments retain the power to control waters within their boundaries. Few countries allow absolute privatization of water, but, rather, they may allow privatization of the administrative right to use water. In this sense of the term "privatize," a smart market for water does not privatize the resource. Instead, it would make it much clearer to all water users that at its heart, water is a shared public resource, and that it needs a management method that more directly acknowledges its shared and public nature.

In some informal sense, however, users have already privatized the water. In fact, over-allocation has allowed people to claim more water than exists. When catchments have their water fully allocated, land prices reflect the value of water rights.

Even when water doesn't have a price, water rights change the value of the owner's assets. An industry with water rights might have the potential to expand production more than a competitor without water rights, increasing its company's share value. A city with water rights has the ability to develop new housing tracts, increasing taxable property in its boundaries, while a city without water rights does not. Water resources are in an awkward disequilibrium, where users see the resource as effectively privatized now, while other stakeholders see the resources as a non-privatized public good.

A good trading system will replace the implicit price with an explicit price that reflects society's marginal value for a given source of water (i.e., water from a given location), suitably weighted by environmental constraints to ensure sustainability. A trading system for water allocation would not restrict access when compared with non-trading systems. Oil, electricity, medical care, and food are all important, and the markets in developed countries enable people to have such things readily. The word "commodity" may come across as cold or impersonal. This word is the abstract language of economics, though everyone has feelings about how these materials and services impact our lives.

The 1992 International Conference on Water and the Environment in Dublin concluded that "Water has an economic value in all its competing uses and should be recognized as an economic good." The *lack* of pricing is causing many problems with water. Without pricing, a user who desperately needs water may be next door to a user who could give up water with little or no loss, but coming to terms is complicated. A proper market would allow these two users to communicate and come to an agreement to share the water, while fairly compensating the one who gives up water for the lost potential to use the water. The whole point of water trading is to give wiser incentives to society. For example, where and when water is scarce, its price should be high to incentivize conservation.

Sample Comments

- "People will profit on the trade of the water rather than the use of water."
- "The government allows personalized number plates (for example) to be traded with inflated prices—water trading would create more of the same."
- "Prices that result would not reflect any added value."

Markets do not convey why a price is what it is, but a properly regulated market does allocate goods better than rules and regulations. The high prices for personalized number plates simply reflect their high demand. As for profiting on the trade of the water, sellers would not voluntarily give up water unless someone paid them for that water above what they could earn with it from production. That is an ordinary part of doing business. The regulator should not allow traders to make money at the expense of those who are not in a smart market, and any market for water should be voluntary.

Sample Comment

- "Markets would place water ownership in corporations."

Corporations, especially hydroelectricity generators and large agricultural firms, already control water in industrialized economies. Large organizations may have more control because of their ability to lobby regulators (see Easter and Huang 2014, 3), and because of their stronger water rights and deeper pockets to pay the transaction cost of trading than those of small users. Smart market designs of the type described in this book should reduce the power of lobbying by allocating water through a market mechanism instead of through command and control. The smart market should reduce the transaction cost of trading, enabling small users to buy new allocations. The smart market should also create firmer water rights. In multiple ways, then, smart markets should increase the relative power of smaller water users.

13.3 "Trading Will Hurt the Environment"

Sample Comment

- "An attempt to provide a market for whale hunting just led to more-rapid destruction of whales."

The market for whale hunting did increase the destruction of whales. It did this by creating an incentive to hunt them when there was no strict regulation of the market. That is a great example of how not to protect the environment. The move toward strict regulation and allocations has done much to improve whale populations. Any market approach alone that does not carefully provide for and take into account critical environmental factors will also fail.

New Zealand's transferable quota system for fishery management has worked effectively to protect the resource. The rule of law, specific rights for users and the environment, and a well-designed and well-regulated market, as a package, effectively manage environmental resources. A good design should adjust allocations to maintain a sustainable resource.

The unchecked and unvalued use of water, as it is today, results in environmental degradation. Water markets have been bad for the environment in some places, such as Chile and Australia, but those markets are poorly regulated and do not have the needed central oversight. The smart market approach has explicit protections for the environment not seen in many other market approaches.

Sample Comments

- "Trading will lead to increasing pressure for use and the environment will lose out."
- "Current inefficiency in use results in more water in rivers, which, in over-allocated waters or those natural waters with insufficient flows, is beneficial to environmental interests."
- "A market would encourage the idea that any water that flows to sea is a loss."

Many users already believe that any water that flows to sea is a loss. We see greater environmental pressure without a market, because users cannot reallocate among themselves. In fully and over-allocated catchments, water users as a group often pit themselves against environmental groups. Users argue to take more water and environmental groups argue for more to be unused. Too often, this struggle has resulted in the environment losing out.

Users and environmental groups may have difficulty agreeing on the initial allocations for the environment, but once stakeholders agree (with a process for revising the agreement), the environmental requirements should be satisfied in a socially acceptable way. The market will be the pressure release valve that protects the environment.

A common misunderstanding of water markets is that the users gain by using water that otherwise goes into the environment, and so the gain comes at some cost to the environment. This is not the case. At no time would users take water at the expense of the environment, because the environmental bottom lines are hard constraints for the smart market model. Instead, the economic gains accrue during relative shortage, when the market reallocates water between users. During times of excess, the value of water would be low, and users would not gain much.

Reasoning that inefficiency protects the environment overlooks important situations. In certain catchments, the water authority allows users to use up to a certain limit of water, and so when the users use less than their limit, the remaining water benefits the environment. In some places and in some years, this inefficiency does help the environment. If this happens year after year, then the regulator could allocate more water, so the regulator has no water problem, and users need no market. More worrying and common is the situation where, in wet years, the users do not use all of their limit and the environment benefits to a small degree, but, in dry years, every user consumes their full limit, and the environment suffers dramatically. It seems a strange way to manage the health of the local environment by relying on wet years, and users not using all of their water; it would seem more prudent to decide on the appropriate amount of water in the environment in wet years and dry years, and then enforce that. A smart market moves societies away from the notion of letting the environment have what (if any) is left over, to one where the environment is protected by satisfying its needs as the first step in water allocation, not the last. A smart market matches water allocations to real water in every period, with explicit set-asides for the environment.

Sample Comments

- "Water trading would increase the amount of water used, because users could trade their unused allocation, as sleeper permits get traded to others."
- "Water markets would exacerbate environmental impacts of low flows in drought."

When Australia began trading water actively, users sold their unused "paper" allocations, and the environment did get hurt. The government then had to buy back rights in order to return water to the environment. This potential behavior is why allocations must match physical flows. The smart market can do this. But the regulator must not allow water users to trade their existing non-tradable rights one-to-one for new tradable rights. Current rights in fully or over-allocated catchments are not secure, and the regulator should reduce them, with the users in exchange getting the greater security of the new tradable rights.

A smart market would require more expense in monitoring water use, for example through meters. This additional regulation helps to manage the market and protect the environment. Currently, even without an active water market, government does not always monitor water use, which often results in damage to the environment. The problems of over-use have little to do with a market or a lack of a market. People are simply taking too much water.

Sample Comments

- "Environmental uses for water will be outbid."
- "Environmental interests don't consume water, yet can rarely afford to purchase, even if environmental interest is significant."

The smart market mechanism allows the regulator to require minimum environmental flow requirements by constraint. The environment should not need to buy water. Commercial users reliant on environmental flows, such as run-of-river hydroelectricity generators, would be assured of minimum environmental flows and the ability to buy allocations to increase the environmental flows as long as they do not violate environmental constraints.

Sample Comment

- "Trading will result in intensification of land use, causing adverse effects on water quality."

Yes, this is possible. The regulator should enforce rules to maintain water quality, whether or not a market is active. The regulator should include environmental constraints on water use that protect water quality. For example, lower limits on river flow might need to be set to ensure the concentration stayed below some critical value. This might result because lower flows from intensification of land use and other causes can concentrate contaminants in the water. We see the potential to develop a distinct, but related, market to control water quality (see Sect. 12.1).

Sample Comment

- "The market leaves no excess available 'just in case.'"

It is true that the smart market mechanism described here leaves no water anywhere "just in case." The mechanism leaves at least as much water for the environment as the regulator specifies. The regulator has various ways to be more conservative to address uncertainty, but, short of using a stochastic model (see Chap. 11), the modeling would be much more difficult with a range of scenarios. However, the market will specify water for shorter periods, closer to real-time, than the current command-and-control process, so the environment is more likely to get its specified requirement with the market than without. The smart market model will allocate all the water (to users and to the environment); but all water is already allocated during drought. Even under-allocated catchments should benefit from a smart market. If stakeholders want extra safety in the impact on the environment in a fully allocated catchment, stakeholders should be able to agree to increase the environmental flows requirements.

13.4 "A Water Market Will Lead to Unfairness, and Cultural and Social Loss"

Sample Comments

- "The cultural values held by indigenous peoples mean that water trading is unacceptable to them."
- "Seeing water as a 'commodity' may marginalize indigenous and non-Western perspectives."
- "Trading has a cultural impact—it takes water from the land."
- "There must be recognition of environmental, social, and cultural values in a trading system."

Society does not have to trade sacred water. Legally, users would trade permitted water allocation. Users trade the allocation rather than the water itself. Some indigenous cultures accept trading of water allocation. In New Zealand, Māori recognize a distinction between the sacred (*tapu*) right to water and the ordinary (*noa*) administrative right. Further, the sacred water can be set aside as an environmental requirement.

Trading might disproportionally benefit indigenous peoples with recognized water rights because they often hold very large investments in assets that depend on water. The environmental requirement constraints should incorporate indigenous and other cultural values via the democratic process.

Sample Comments

- "Your proposal has no provision for smaller water users."
- "People or animals that presently use water and are unable to pay for it may lose out."
- "Smallholdings that are uneconomic couldn't afford to pay much."

Nothing would stop small users from becoming involved. The smart markets for electricity have helped small electricity users enter the market. We expect that many small users will benefit from a smart market and will enter as bidders. All users, small or large, need to acknowledge that users will have different abilities to pay for water allocation. The water should go to the users with the highest ability to pay, as anything else is waste.

We recommend that a smart market start with the large water users and expand over time to include smaller users. As the Marlborough example showed (Fig. 5.1), many small users together can have small impact on the water resources. The regulator can omit them from market operations without large impacts on the value of the smart market, by giving small users fixed allocations. But this is to simplify implementation, not to discriminate against any user. Some small users would use water for domestic purposes, and others for watering small numbers of livestock. If users are taking large amounts of water for livestock, say in a feedlot in an arid environment, the operators of those facilities should face the true marginal cost to society of high water use.

Because the smart market has a lower transaction cost than existing allocation mechanisms, and the transaction cost will be about the same for users of any size, small users should be able to join and benefit.

Sample Comments

- "Business could monopolize the resource."
- "A wealthy operator could capture the market."
- "The water will be taken by the rich."

We do not expect that any users will monopolize the market. Schoengold et al. (2006) found that the price elasticity of agricultural water in California was about −0.79. That is, a 10% increase in price would result in a 7.9% decrease in demand. Such high elasticity is associated with highly competitive markets, which suggests that a water market is not likely to suffer from any user having market power.

We see other reasons that market power will not be an issue in a water market. To gain a monopoly on a spatially dispersed resource such as groundwater or combined ground and surface water, the monopolist would somehow have to access water bordering or under another user's land. An attempt to monopolize groundwater by pumping from one well will not work. Before a neighboring user would lose access to water, the hopeful monopolist would hit an environmental limit or have an empty well. The spatial nature of the resource allows a monopoly only over a local area at best. Even with a reservoir, one bidder would have difficulty in developing monopoly power. A shareowner of reservoir water could attempt to buy all the shares of the reservoir, but a monopolist has to own sufficient land to put the water to use; this is unlikely, as it raises a land ownership problem before it raises a water monopoly problem. In contrast to oil, unused water tends to evaporate or flow to the sea, thus giving the monopolist incentive to sell rather than lose the revenue. We do not say a monopoly would never happen. Rather we expect it to be rare. In these situations, the water regulator could impose an absolute upper limit on any user's ability to buy water.

Sample Comments

- "Trading will give some people a windfall, because they can sell a permit granted as a privilege by the community."
- "Implementing the market raises start-up issues of equity."

Windfall gains are possible but by no means certain. Users with currently underutilized water rights who can persuade water regulators to convert those into new tradable rights for the same quantity of water could receive large gains when the market starts. As part of implementation, the implementers could specify that initial tradable rights be based on actual past use. Even then, some users will find that they prefer to sell at the market price rather than continue with their historic rate of water use, thus making money on the sale. Any seller might be viewed as getting a windfall, but it is hard to see how the courts would see this as unfair. By preventing this sale, the regulator would also prevent reallocation, and would prevent people who need the water from getting it. Without a water market, the inefficiency of poorly utilized water would continue to the detriment of society.

The regulator, and the community, will have to address these start-up issues. Current users have received their existing permits by following proper procedures and so should expect a fair process of converting their existing rights into new tradable rights.

Sample Comments

- "People are not at the point of valuing water in ways that might be amenable to trading."
- "Trade advances a profit motivation versus other social concerns."
- "A trading regime must take into account all of the community values of water, i.e., social, environmental, recreational, tribal, public-ownership."
- "This should be a public resource for public good."

Current systems for water allocation may include some form of social optimization, but many people complain that the current system is bad, at least partly because reallocating water is difficult.

In response to the comment that people are not at the point of valuing water and that therefore there should not be trading: this is circular reasoning. People cannot value water for trading purposes now because trading is hard to do. People seem capable of valuing other goods where an effective market exists.

The social optimization argument ignores the fact that most water users are trying to make a profit. Hiding market forces within a framework of "social optimization" leads to less transparency and to greater potential for manipulation and for cynicism of citizens towards an allocation system.

Any water allocation mechanism can do only so much for society. The smart market system is intended to take into account all relevant information, including democratically chosen environmental flow requirements, the hydrology, users' existing permitted water allocations, and people's willingness to pay for water. A smart market for water would maximize the economic value for water. Social values can be effectively managed alongside a smart market (e.g., by financial support to certain parties) even if they are not considered within one.

13.5 "Trading Will Lead to Financial Loss"

Sample Comments

- "Some users will have limited ability to easily buy water."
- "Wouldn't water all be up for grabs, with arbitrary winners and losers?"
- "The market could lead to inefficient irrigation or use because of plentiful supply."
- "Sought-after areas are often controlled so land prices are kept high and new entrants have to pay up—the same could happen with controlled water."

A market should lead to greater efficiency, not less, and a lower transaction cost should increase the ability to buy water. Users who sell water allocation could re-enter the market as buyers later. So enabling more trading of water allocation will only enhance users' ability to get water. Trading would not result in arbitrary winners and losers, but, rather, in willing buyers and sellers, just as with markets for other goods.

Plentiful supply should result in lower prices, not necessarily inefficiency. More irrigation because of lower prices could mean less water uptake by crops and so less agronomic efficiency, but users would have to pay for the use of this extra water and would have a reason to be careful. The greatest potential benefit of smart markets is not where and when water is plentiful—many water allocation systems can work in those situations; the benefit is when water is in short supply. In a short-supply situation, the market would lead to an increase in prices, and more agronomic efficiency, unlike with current water allocation systems. The smart market, unlike other water allocation systems, is excellent at delivering economic efficiency. If the high price in a drought leads to some users having limited ability to buy water, because the price has risen, then that would be good. Users *should* have limited ability to buy water (i.e., users should face higher prices) when other users are willing to pay more for it because these other buyers can use it more productively.

Today's systems, where water allocations are fixed and tied to land, limit the ability of new entrants, assuming that a current system allows new entrants to take over water allocations, which many do not. Under a smart market, land prices will reflect the ability of the land to create economic value. Land values would be high when people want that land, because the land is productive, and not only because the land happens to have water rights. New entrants have to pay for land and water, just as they pay for farm equipment, seed, and fertilizer. Under a smart market, new entrants will look at the differences in land prices (reflecting differences in soil type, slope, road access, etc.) and differences in recent water prices. New entrants who demand certain land features and specific high or low water uses will be able to more precisely decide on where to purchase. Managing new entrants within a smart market will require care, but it will be easier than with existing systems.

Sample Comments

- "Trading wars may create inefficiencies."
- "Trading will lead to inefficient conflict between water users."
- "What if an upstream person buys, and thereby denies access downstream?"

Neighbors can be set against neighbors already, and this is worse when reallocating water is as complicated as it is now. By allowing co-management through a market, neighbors are more likely to get along. Trading should be cheaper than going to court.

Trading would increase efficiency, as water moves from low-value use to higher-value use.

In the smart market, an upstream person could not deny access to a downstream user. The upstream user could only bid higher, implying that the upstream user could use the water more productively.

Sample Comments

- "Some land may end up without water."
- "Water belongs to the land; trading takes it away from the land for anybody to use for any purpose."
- "If you own the land, what water goes with it?"

Requiring water to stay with land imposes rigidity, preventing reallocation of the water, and increasing people's risk. Yes, trading takes it away from the land for anybody to use for any purpose, as long as that purpose is more valuable than the seller's use. Water is likely to remain where it is drawn, because transporting water is expensive. Land can end up without water after trading, but probably only because that land is less productive than other areas.

The underlying, and perhaps unspoken, concern seems to be that property values will decrease for some once the regulator decouples water allocations from property rights. To an extent, this would happen. Certain parts of a catchment are likely to face higher market prices for extra water under a smart market, and this would likely lead to lower property values at that location. Some parts of a catchment will face lower water prices, resulting in higher land prices. The exact distribution of gains and losses would vary with varying environmental constraints and varying climate; still, implementers could estimate the effect in advance. Users might wish to devise a way to compensate each other for these effects as part of an implementation package. The likely gains for all users from introducing a smart market would allow various methods to financially compensate those who suffer an initial disadvantage from entering the market. Once a user is inside the market, however, other stakeholders should have no further opportunity for financial redress.

Sample Comments

- "Urban communities can afford to pay more than rural groups."
- "Water would shift to the non-productive sector, resulting in lower land productivity."
- "Trading may marginalize less commercial growers."
- "Different markets—e.g., industry versus community versus producer—have different abilities to pay."

Water needs to move to the highest-value uses. Anything else is waste. For example, if one user needs water for a valuable crop while another user is simply washing a car, they should trade, which would make both users better off.

Often, small users do not have the efficiency of scale that large users do. If these small users continue to get water, they are effectively wasting it in comparison with users who can use it more productively. The market should not be developed in a way that judges who is "small" and who is "large," or who is "inefficient" and who is "efficient," or who is "rich" and who is "poor." All of these factors, and many more, can influence the value of water, and a market works effectively when allowing people's own values to determine allocation. Alternative systems are inefficient because of the lobbying associated with the rules and definitions, and can often lead to hidden or explicit biases.

A water market is likely to accelerate changes in land uses. The government may need to help small water users manage change, but that should happen separately from the market. Under a market, land use changes will be based on clear economic signals. Those decisions will more likely be good decisions and longer-lasting ones. Without a market, the need for change will remain, but the ability to make good decisions related to change will be worse.

Users may want to impose restrictions on trade. Sometimes, they don't like how the other person might use water. Some New Zealand crop farmers, for example, told us that they would never sell water to a dairy farmer; those crop farmers think that dairy farming takes too much water and damages waterways. In California, some counties restrict sales of water to users outside the county. Often, these wishes for trade restrictions are based on genuine and valid concerns. The urgency for water allocation reform and the desire of water users for a smart market will need to be strong enough to overcome the inertia of the current system and its associated issues.

The introduction of a smart market will need a community's readiness to trade. People should have the opportunity to express their true concerns, and the policymakers must explain how the regulator and the market design will address those concerns. But the market should not suffer from needless restrictions.

13.6 "Uncertain Property Rights Prevent a Water Market"

Sample Comments

- "Who owns the water?"
- "Until property rights have been established in law, trading will not work effectively."
- "Tradability makes changes in rights tougher."

We do not address ownership of the water. A smart market can operate just the same whether society chooses to treat water rights as property rights or does not. Government could recognize water allocation as either a property right or an administrative right. As an administrative right, users would not trade water. Instead, they would trade the administrative right to use water. In either case, a governmental authority retains control of the resource to manage externalities from over-use of water.

The smart market requires conditions on the rights to make them tradable, and government should separate land ownership from water rights. But implementers should be able to base the tradable rights on existing rights, which the regulator has already granted.

Users tend to oppose any changes to their rights. People are concerned, often rightfully, that government's changes to their rights will cause them loss. Tradability itself actually enables changes to rights, by allowing people to more easily increase or decrease their rights.

Sample Comment

- "Indigenous peoples may claim water resources and may object to others profiting from trading water."

We do not think that we will be able to give a satisfying answer on this. Indigenous people have to take their claims to government and to society through the appropriate channels.

We note that people are already profiting from the use of water resources. We see no way to incentivize people to use water more efficiently or to give it up to others, other than by compensating them.

Sample Comments

- "Who is guaranteeing the supply? The farmer? The seller? The water regulator?"
- "Water of varying security of supply has varying values."
- "Trading will reduce water security."
- "If you are in an area with over-allocated water, then a trading regime could worsen security of supply."
- "Selling water locks you out of the market."
- "Users will not threaten their supply by trading."
- "People like to have their own security of supply, e.g., ponds, rather than rely on a market for security."

In a smart market system, the market manager confirms that the user can take the agreed allocation in the immediate period. Market rules have to address the rare case of water being much less available than expected, as well as the case of an environmental flow being below its minimum.

Beyond the immediate period, the market manager must plan to an uncertain forecast. Security of supply is whatever the clouds provide, combined with society's ability to be flexible in water use. Therefore, users' rights must be subject to adjustment, to match the actual inflows.

It is true that high-security water has higher value. Users may want a mechanism that allows them greater assurance of allocation. Implementers can design a security of supply mechanism through seniority of rights, or through the market. Users should be able to buy enough future rights to be sure that the market manager will not scale their rights below what they are willing to pay for.

In an over-allocated area, users already have poor water security, along with the risk that the regulator may cut their allocation. With the smart market, the users should have greater certainty of supply, because the market manager is using a system-wide calculator to be sure that the rights match the actual water.

Many concerns related to a loss of security from trading result from a preconception that trading water means trading water rights permanently. Many past efforts at water markets have focused on permanent trades. The focus of the smart market approach is on a spot market for water and not on permanent trades, though provision for permanent trades will be important for the ongoing fitness of a smart market approach. Because permanent water trades are what come to mind now for most water users as the only way to have additional water security, and because water users know how difficult those agreements are to transact, people rely on storage methods such as ponds for security. This viewpoint arises because trading is too hard, a problem that a smart market specifically looks to address.

We think that users do not want water security, per se. They want financial security. Trading would be voluntary. Users should start with rights similar to their existing rights. Having the option of trading should increase their security of both water and finances.

Sample Comment

- "A market will never get started because people will never agree on initial allocations."

The problem of initial allocations could hold up implementation of a smart market. The setting of initial allocations is inherently a political question. Somehow, society has to come to agreement on those allocations. Implementation should avoid disturbing current rights as far as is practical, in order to minimize opposition from those who fear that their allocations would diminish.

The regulator will need to make significant adjustments to initial allocations when catchments are over-allocated prior to implementation of a smart market, raising uncertainty about initial allocations. In the end, water users would need to choose between an existing right to a great deal of water that is rarely available, or a new variable right to water that is more secure in its supply.

13.7 "A Water Market Will Lead to Loss of Control"

Sample Comments

- "A market results in loss of power over who controls use."
- "Allocation of water would change from a process dominated by local government and landowner to a process dominated by economics and finance."
- "If water has a price, then society has pressure to maximize the monetary return."

Currently, people choose how to use water based on economics—money already dominates water allocation. Governments can crudely allocate water, but private individuals and corporations choose how to use it, and the vast majority of fresh water used worldwide is used to earn income. Users already face pressure to use water in a way that maximizes monetary return. Anything else could be considered waste. Government cannot and should not eliminate that pressure to use water efficiently, just as society wants people to use dental floss and plastic pipe efficiently. Because that dental floss and that plastic pipe are priced and sold within a well-regulated market, and because their externalities are managed, society can be confident that people use those commodities efficiently.

Because water is hard to manage, current markets do not manage it effectively. This diagnoses a problem. The "control" that water users and regulators have is a mask for inefficiency: there is too little monitoring of use and too little enforcement of conditions for water use. Control is so weak partly because users do not want to be controlled. Controls without markets lead to inefficient use of water as well as weak regulation of use. Water is too important for society to allow the current inadequate and inefficient forms of control to endure.

Sample Comments

- "The system gives no way to manage the resource."
- "The water regulator should not let individuals control surplus water."
- "The water regulator should be the water sharer, to maximize the best use of the water and not have individuals controlling surplus to meet their needs."

Under a smart market, the water regulator would retain the right to award permits to water users. The water regulator would retain control of water-related matters such as well permitting and construction, and the water regulator would oversee the market. But the market would choose water allocations.

Any water allocation system must give the regulator a way to manage the resource. The smart market must start with a strong hydrological model to ensure that a management capability is present. The regulator and market manager can use the model to examine situations where excess water use can damage the environment. Water regulators manage water this way now. The smart market models should have the best available hydrological science, and that science must be unbiased, able to withstand court challenge, and accepted by all stakeholders. Further, the water regulator must retain control over the model, and be able to improve it as the science improves.

What would change would be how the regulator uses the model to allocate water. The model would need to simulate every user's water take and future hydrological conditions.

The output of prices shown by location across the catchments will be of huge help in determining where water is most valuable, and how much should be spent on infrastructure such as dams and canals, or on new hydrogeological investigations.

To incentivize more-efficient use of water, users must have a greater sense of control, and the water regulator should not tell them what is efficient for them to do or what water is surplus to the users' needs. However, responsibility for the resource remains with the water regulator.

Sample Comment

- "What if we start a market and then decide it was a bad idea? We can't go back."

This would be a decision of government, just as the decision needed to start the market is a government decision. Government could dismantle and prohibit the smart market institution.

Sample Comment

- "Some water users will lie or cheat."

Markets work best when ruled by law. If people could steal without penalty, then no one would want to buy or sell. Potential buyers would find stealing cheaper than buying, while potential sellers would fear they would never be paid. A society with rampant theft, no law enforcement, and no trading, would not be society; it would be anarchy. It is true that people will always be tempted to steal, but that is a problem of law enforcement. A proper market for water would require careful laws and careful enforcement of those laws, just as our market for bread and other foodstuffs does.

Participants in the water market would have to meter their use, and submit to prudential requirements (such as creditworthiness) to join the market. As with any market, the regulator must enforce the rules. Users can lie about their water use without a water market, and the extra control that would come with a smart market will lessen the effect of lying and cheating.

Sample Comment

- "Users could trade outside of the market."

Yes, they could. But they do not trade much now, and trading would be much easier within the market because of the lower transaction cost. How might a black market work? Leaving aside the issue of illegal extraction of monitored water, a user might decide to capture rainfall on their property (say, through a large roof). The cost and visibility of water transport structures (pipes and canals) limits the potential to barter any unmonitored water take with other water users. Even when not limited by transport costs, trades outside the market would have limited effects because of the small water quantities involved.

13.8 "New Market Methods Are not Needed"

Sample Comments

- "No problem exists."
- "Trading would be useless, as we have an excess of water."
- "Existing users are happy with the system."

In a few places, people have no need or desire for trading. But in general, people do want to trade. Further, they are much more likely to want to trade if it were much easier. Many users wish they could change their water allocation. This is easy to test informally by simply picking a random price and asking existing users whether they would prefer to buy or sell at that price than to stay with their current water allocation. Both those who indicate that they would buy and those who indicate that they would sell are showing that they think that trading would have value.

Sample Comment

- "People can trade now."

Yes, people can trade now, but trading does not happen nearly often enough. Currently, trading has a transaction cost that is too high, so trading rarely happens. Reducing this transaction cost would reduce the cost to users, and they could decide themselves whether they wish to trade more or not.

Sample Comment

- "Before resorting to markets, we should review ways to use water more efficiently."

People have been improving water efficiency for many years. We think most users would say they use their water efficiently.

But water efficiency is relative to supply and price. Putting a lot of money into improving efficiency when water is plentiful could be a waste, because the water saved may not be worth the extra equipment. Users need to see the financial incentive to improve efficiency, in the form of a clear price signal from a market.

We agree that users should continue thinking about efficiency. But users also need a better mechanism for water reallocation and for water pricing, so that they know the right amount of efficiency to implement. All over the world, people have examined other ways to manage water and found them wanting. A new dam, for example, tends to reduce efficiency, because users have more water available. Society needs to try something new.

13.9 "The Problem Cannot Be Fixed"

Sample Comments

- "Users lack confidence in current management systems."
- "The water managers did a bad job because water is already over-allocated; what hope is there that the same people could run a market?"

The current system is hard to manage, even with competent managers and good data. Water managers face enormous pressure to both give more water to users for productive use, and to give less water to users so the environment benefits. The resource is complicated and variable. We understand why users would lack confidence in current management systems. A smart market would overhaul these systems, providing a simpler regulatory regime. It would also create a water market manager, distinct from the water regulator, to manage the new duties that would be required with a smart market.

Sample Comment

- "How can trades be fair when water is already over-allocated?"

Implementers should address over-allocation before the market begins. Within a smart market, users should be able to get more water only if it is available, and if someone else is willing to give it up. That seller should be compensated for it. Both would win, because the buyer would get additional water, and the seller would get a satisfactory price. Over-allocation has been reversed through government actions with (Australia) and without (Southern California) markets. Implementers should take special care when introducing markets into over-allocated catchments, but good market design can resolve over- allocation in a way that protects the environment while also allowing economic competition for water to happen.

13.10 "It Is more Complicated than You Think"

Sample Comment

- "The proposed model is too simplistic because it is a steady-state, linear model, when the problem is highly non-linear and dynamic."

Yes, the problem is non-linear and dynamic. The short time steps between auctions (roughly, weekly) are less than the timescales for non-linear hydrological processes (say, groundwater recharge). This allows a series of linear evaluations to approximate the dynamic, non-linear nature of the underlying physics.

The smart market models described in this book have linear components and objects that interact to model the hydrology. As better components become available, the regulator can insert them into the model. Some elements that we anticipate adding to the model would be weather forecasting and stochastic modeling, or trading in environmental impact quota. We have scientific concerns about these enhancements, but the concerns have partly to do with how much complexity users can tolerate. We see no strong scientific or technical impediment to implementation.

Sample Comments

- "We should not trade the resource until we understand it fully."
- "How can a market start to deal with such an uncertainty/information deficit? For example, we often don't understand groundwater resources until they are stressed, or environmental values may not be investigated until they are threatened."
- "The market implementers could get it wrong, and, after the market starts, we will find less water for everyone, especially the environment."
- "The smart market would need to be aquifer-specific, but many real aquifers are leaky and the connections between them can be significant. This aspect will make the market too complex to develop and operate."

Complex aquifers need management with or without a market. Water regulators currently allocate water under uncertainty and while lacking information. In that sense, a common-pool smart market system is no worse. It would require good hydrology models that incorporate the best available science, just as the current methods for setting maximum water use do. For water regulators to assess how much water use to permit in deeper aquifers they must assess the leakage between aquifers. A smart market would require a model no more sophisticated than the one used currently.

The smart market system would motivate continued improvement of our water resources. In general, society develops more-elaborate models when the costs of bad decisions become too high. The institution of markets for trading fish quota led to new research into reproduction patterns of marketable fish—sometimes leading to decreases in quota and sometimes to increases in quota. With a smart market, society will be in a better place to justify investments to improve its models.

With or without a smart market, the regulator needs some model to help allocate water in a way that is compatible with social and environmental targets. With a smart market, the market manager would use the model to simulate every trade, to ensure that flows satisfy environmental constraints.

Sample Comment

- "Your model is too simplistic because it does not consider recharge from irrigation."

This is not true. A smart market design can easily incorporate recharge. Water users may prefer to operate the smart market without considering recharge or with considering it in a simplistic manner at first. No matter how it is considered, the smart market will still function effectively. The smart market will respect environmental bottom lines, and society will receive greater value from water use. Including recharge in the model will simply change the spatial variability of prices.

Sample Comment

- "A smart market model for water allocation separates managing groundwater quantity from managing groundwater quality, yet the two are interrelated. For example, more irrigation in one part of a catchment, if not controlled, could lead to excessive nitrate flux that would not have occurred if the water had been used by the originally permitted user elsewhere in the catchment."

The physics of quantity flows and quality flows differ significantly. The effects of changing water quantity move much more quickly (hours, days) through an aquifer than the effects of changing water quality do (years, decades).

Further, the associated user behaviors are nearly independent. The commodity traded would connect to a single action. Users manage irrigation on a daily or even hourly schedule. Fertilizing occurs seasonally or annually. Thus, users will wish to trade a commodity associated with water quantity relatively frequently, perhaps daily. But they will wish to trade a commodity associated with water quality seasonally or annually, to match their actions.

Implementers could combine the quantity and quality markets by defining some kind of joint commodity, perhaps an allocation that has one value for water and another for nitrate. This is unnecessarily complicated, like requiring that a car owner can buy gasoline only when buying a car. Most likely, quantity and quality markets should operate separately, because the physics and demand timing of the two markets differ. They are related, just as car markets and gasoline markets are related, but they operate best separately.

Sample Comment

- "A large, new administrative system needs to be developed and implemented, which has great technology challenges—a much higher level of management would be needed."

The administrative system would be similar to any commercial website, with the addition of a significant calculation capacity. Robust administrative systems are already in place for commodities that require much higher technical sophistication, such as electricity.

Sample Comments

- "Trading needs to adapt in a way that ensures that new knowledge is not a barrier to change. Having a market will make incorporating new knowledge on the environment more difficult and so will be a barrier to change."
- "Water trading per se is not likely to be a good management tool unless it is linked to a useful set of knowledge (hydrology, climate, land use) that is dynamic and accounts for change."

Existing command-and-control approaches have these problems. Fixed water allocations become detached from new knowledge and changing water availability.

Implementers should write market rules that allow the regulator to update the hydrology model, and stakeholders should expect the regulator to do so. Implementers must create a legal framework to support scientific updates with equitable and efficient changes to allocations and environmental flow requirements.

Sample Comment

- "The users will want to vary the time between auctions."

In an active market, the length of any trade should reflect the hydrology and users' needs, so that users can obtain precisely what they need. The time between auctions can be flexible. There is no technical reason to fix the time between auctions.

Sample Comment

- "Markets will fail because they will create 'bottlenecks'—specific points with tight constraints that stop trades."

Currently "bottlenecks" exist in the sense that a catchment may have a critical point, say, near a stream, where further water use would damage the environment. The current command-and-control approaches do little to reduce such bottlenecks. Trading provides other options for users who wish to have more water—the market would compensate users for giving up water near the bottleneck. The optimization process will tend to prevent the formation of bottlenecks. Price signals will encourage users to come up with other solutions to the tight constraints, e.g., by changing land uses.

Sample Comment

- "The cost of operating the market will turn out to be too high."

The cost of operation should be not much more than that for an existing process, because in effect the same functions are done—granting of permits, managing hydrology models, and understanding the catchment. With a modest charge for each trade, the system could pay for itself. Other smart markets have been found to be cost effective to implement and operate.

Sample Comment

- "This has never been tried—will it work?"

The smart water market concepts developed here have not been implemented anywhere to our knowledge, at least not in a complete form. However, smart markets similar to it have been implemented for a wide range of other goods and services, all over the world.

Aside: Common-Pool Smart Markets for Water Are ....

- ... not ownership or sale of water, nor privatization. Users trade only the administrative permission to use water.
- ... not allowed to damage the environment. In fact, the market manager simulates every take to ensure that the environmental flow requirements are satisfied.
- ... not "free market" nor "just a free for all." They are smart markets, with water flows and bid accounting calculated by a computer, under complete control and supervision of a water regulator that has full authority to manage the resource.
- ... not about hitting poor and small water users with high water charges. The system is intended primarily for large commercial users, all of whom already have a quantity of allocated water and can sell from it or buy to add to it. Holders of small licenses can participate if they wish.
- ... not forced on anyone. Participation is voluntary, like with eBay. If you are satisfied with your license, you don't have to do anything. But realize that your water regulator can still reduce your license during drought.
- ... not going to promote monopolies. The proposed system is a rental/lease for a specific period of time, after which license returns to its original owner. A given user has access only to water available nearby. To have a true monopoly, someone would have to buy up all the land.

References

Easter, K. W., & Huang, Q. (Eds.). (2014). *Water markets for the 21st century: What have we learned? Global issues in water policy 11*. Dordrecht: Springer Science+Business Media.

Schoengold, K., Sunding, D. L., & Moreno, G. (2006). Price elasticity reconsidered: Panel estimation of an agricultural water demand function. *Water Resources Research 42*(9), W09411.

Chapter 14
Concluding Perspectives

Abstract Smart markets for water can benefit communities around the world. Some places will benefit more than others, and implementing the market will be easier in some places than in others. Overcoming the inertia of existing systems will be difficult. An initial focus on key users with small steps toward change will ease implementation. Every new idea leads to questions, and in this chapter we focus on some key questions, and give our thoughts on the way forward for this new water allocation method.

This concluding chapter reviews what's great about a smart market. Following sections discuss where to start, how to start, how much it would cost, and how to overcome political resistance. The final two sections discuss what can be done in the interim, and our thoughts about where to from here.

Keywords Implementation • Political resistance

14.1 What's So Great About a Smart Market?

In 2015, newspapers widely reported that some California households no longer had tap water, because local users had drawn down the local aquifer too far. A market would have prevented this: the homeowners could have bought water if they had no initial right, or they could have declined to sell if they had an initial right. Further, a smart market in particular could have guaranteed an initial right, preventing users from drawing down the aquifer level below the bottom of the wells that supplied tap water to those families.

Because water use has simultaneous effects everywhere, the solution is to coordinate water allocation to all users and the environment with a set of simultaneous equations. The smart market's rules represent the environmental flow requirements. Using a web page, users indicate how much water they want, and the price at which they would buy or sell, as in Fig. 14.1. Then a hydrology model calculates the effects of everyone's use on everyone else and the environment. With the list of required environmental flow requirements, the allocation model tracks all the simultaneous effects on the environment, now and into the future. The allocation model contains the rules corresponding to the environmental flow requirements, and thus it ensures sustainability into the future—indeed, forever. The

J.F. Raffensperger, M.W. Milke, *Smart Markets for Water Resources*, Global Issues in Water Policy 12, DOI 10.1007/978-3-319-55008-4_14

Fig. 14.1 With the smart market, Ann's transaction cost is mainly the time to type her bid into a web page

models can be independently audited. The modeling relies on 60 years of applied mathematics and improvements in communications technology.

The allocation model finds the lowest prices that balance the flow of water between those who give up water and those who receive more water, coordinating the different effects and the timing, while meeting the rules. Each user gets the mix of water and money that makes everyone best off. In this way, users with existing permits do not "pay for" water, because the regulator converted their existing permits to an appropriate corresponding allocation in the smart market. However, they do see a price for water. Rather than waste water, they will choose a price at which to sell it to someone else who can use it. Thus, the system is fair.

This is similar to a "cap and trade" system, but it has much less capping. Currently, if the regulator has capped users' abstractions, the users' own permits are the cap, yet users rarely trade with each other, even in the same catchment. Under a smart market, the cap would be catchment-wide, and thus far more flexible than the current "cap and use" system. Users could adjust their allocations every week, and have incentive to give up unused allocation. The water regulator retains its full authority over water allocation, and now becomes an enabler of a flexible water policy. A common-pool smart market serves to guarantee government's duty to protect the environment.

14.2 Where to Start?

In our conversations with users, we heard that users are waiting for government to act, believing that government must make some legal change to allow trading to occur. Because government is responsible for allocating water and approving trades, users have no way to create the market themselves. Any private agency would require government cooperation. So users are understandably waiting for government to make the first move.

But we have repeatedly heard government staff saying that they would be happy for users to start a market. The typical water agency may even deny that it *should* be in the trading business. The agency may point out that its commitments conflict with trading, such as when the agency has contracts with water users. The government agency may deny responsibility for the market, perhaps from an assumption that the government should not interfere with the free market. Government, following a laissez-faire approach, is waiting for users to act. Of course, this willingness lets the water agency off the hook for solving the problem. In fact, the water agency approves trades—it already is the principal player in trading.

Allocating water is the water agency's responsibility. The long list of prerequisites is almost entirely the responsibility of government. The private agent has no power to prescribe environmental flow requirements, or to adjust initial rights. Nothing will happen until government makes it happen. Implementation requires a committed and enthusiastic senior government person, supported by specialized scientific help.

Ultimately, the market will start where people agree to start it, and the market will have the constraints that stakeholders are willing to put in it. If the market starts with only some of the relevant constraints, then, when the market manager wants to add new constraints, participants would rightly raise new questions of initial rights. The solution is to have rules that establish processes for adding new complexities.

Assuming that someone in the local water agency is willing and enthused, the focus should turn to a particular location. Implementers should start by focusing on the catchments and the users in those catchments.

Catchments naturally partition land, and smart markets should respect their boundaries. If a water regulator does not exist to manage one catchment across multiple jurisdictions, government should appoint such a regulator. Government should strengthen and authorize them to manage a smart market. Each smart market should be associated with a specific hydrologically distinct catchment.

Implementers can develop smart markets where jurisdictional boundaries do not match catchment boundaries, or when users wish to trade between catchments and markets. Where a catchment overlaps two water regulators, implementers will have to choose who would supervise that catchment's market. With small catchments, a given water regulator may supervise more than one market. Implementers could consider trades between catchments, if users are currently trading large amounts of water between catchments with an inefficient system, or if users who currently cannot trade want to do so.

More benefits will result from a smart market that simultaneously considers the effects of a diverse set of users. The more uniform users are in how they use water and the value of water to them, the more difficult it is to find compromises that allow users to share water. Starting in a catchment with a mix of water users is preferable.

The users who will benefit most are those with the largest current takes. Thus, the major users of a smart market will be the major users of water—agriculture, industry, hydroelectricity generators, and possibly city water authorities acting on behalf of households. Implementers should first invite the largest users to enter the market. Other users will be welcome at any time, but they will have less incentive to join. Traders will need to apply formally to enter the market, and must register with the market manager. Traders must use water meters, guarantee payment for purchases, and be able to prove legal rights to water allocation that they sell. The regulator must approve their type of use in advance. Should a user wish to change type of use (e.g., from vineyard to dairy), the user would be expected to notify the market manager in advance, so the market manager can ensure that the physical model of water flow is adjusted for the expected change in use.

The available research suggests that individual households will have no direct interest in a water market. Some public education may be necessary to make clear that water trading is an activity for large commercial organizations, in particular, and is unlikely to affect households.

14.3 How to Start?

To apply the smart market concept to water, implementers could start with a trial catchment and then develop one catchment at a time. A water market could be implemented most easily for an existing users' group. The water regulator would need to approve the hydrological simulation as valid for permitted water allocation transfers.

The first implementation will be a bit like a first date, with uncertainty on everyone's part. Users are unlikely to fund the project because they may not believe that government will protect their interests. Users may be concerned that the local authority will reduce their permitted water allocations under the "use it or lose it" doctrine. Users may worry that a water market will cause them to lose money somehow. The local water regulator may not have the money to fund the project, and may be concerned that if something goes wrong, stakeholders will blame the regulator. The regulator needs financial support and political cover from higher levels of government.

Both users and the local authority are likely to want to see a complete functioning system before they agree to participate. A research project could play a pivotal role in developing a smart market system to a sufficient standard that stakeholders can run a simulated market and observe their roles and the effects of the market on their own operations. This is going to have to be funded by government or a charitable organization. Research-based development of a water market has many advantages. The research will be accountable to the funding body and the participants. The work is likely to be published in academic journals and thus refereed. Many research organizations have scientists with nearly the full range of disciplines required. The implementation team could include an educational institution to turn students into graduates who can fill the new positions that the market manager will have created.

We have found some value in workshop demonstrations of the software in action with a simplified catchment. The workshop allows individuals to see the impact of different bidding strategies on prices, and to see how prices can vary based on environmental constraints even with similar bids. However, these sessions can take time to run, and participants can still come away confused. Implementers should seek advice from experimental economists as to how to demonstrate the trading system.

We have found a challenge in having organizations take the step from expressing interest to becoming more involved in a trial implementation. Our perception is that the highly charged political atmosphere related to water allocation makes it difficult for individual organizations to push for change to a more market-based approach without appearing to be lobbying for changes that will benefit them to the detriment of others. We have found that currently competing interests will each recognize the need for markets privately, though not publicly, which implies that a coordinated trial could be easier to start than it would first appear. We find that interest is greater from larger organizations and from catchments where water quantity is more constrained, and a trial with those features would be more worthwhile to explore.

14.4 How Much Would It Cost?

A smart market for water would require modest, but significant, start-up costs and ongoing operation costs. Implementers will need to capitalize the market manager initially. Implementation requires a proper hydrology model, specification of the environmental requirements, and training of users. The regulator needs a proper hydrology model anyway. The additional costs would come with a need to rerun the model to generate a response matrix, and a need to forecast water availability and water use so all water users have similar information on which to base their bids. Implementers will incur additional start-up costs to develop a secure website, new water agreements with water users, and a new accounting system.

The market manager will need to cover ongoing costs such as accounting, web page design, programming, hydrological evaluation, and optimization software development. To cover these costs, users could pay a transaction fee on trades, either a fixed fee per trade or a volumetric rate that is added to the price. We do not know how large this transaction fee would be. It would restrict trade, depending on its size relative to trade volume, trade frequency, and allocation prices.

Because of economies of scale, the market will be most economic in a large catchment, with the largest water users, who already commonly face water constraints and take water for a variety of uses.

The first smart market system will be the most expensive, because of the need to establish new software and accounting methods, and the need to train people in the new methods. As the new technologies and skills diffuse out, later systems will be cheaper. The magnitude of cost would depend on the scale of the catchment, but would be similar to that of a feasibility study for an augmentation scheme (desalination, new canal plus purchase of rights, iceberg towing) for a similarly sized catchment.

Implementers should estimate the benefits of a smart market as well as the costs. Other smart markets have shown great economic benefits. The smart market would allow users to make trades that were previously impossible, as the market lowers the transaction cost for all trades. Trade increases economic value with no impact on the environment. In addition, users will receive other less obvious benefits. Users would have less need to fight legal battles for their water rights, and less need to invest in strategies to reduce their drought risk exposure, (e.g., using less-productive land for on-site reservoirs). These benefits may be difficult to estimate in advance, but we guess that the economic value of water to society will increase by 5–15%.

Partly because of the operator's role in protecting the environment, cash payments among traders in the market are unlikely to balance to zero (depending on the scaling regime). The market manager may need capital to pay sellers. In some cases, the seller may be selling for environmental protection, effectively selling water to the government. In such a case, the market manager is serving a custodial role on behalf of the government by buying water.

14.5 How to Overcome Political Resistance?

Smart markets for water resources would lead to a better environment and greater wealth. The issues likely to arise look superable. It is too bad that we cannot just walk into a politician's or a policymaker's office, in a rich or poor country, explain the many advantages, and get these new methods put into action.

We could say that the implementation of a smart market won't happen because people do not like change, or because the culture is not willing to accept smart markets for water, or because it is naïve to ignore political realities, or because policymakers need more education. All of these reasons have some merit, but none of them addresses the political and economic dynamics that limit change. The difficulties in implementing these new systems for water management arise from extractive economic and political water management institutions that inhibit change.

This argument and the terms themselves come from the book *Why Nations Fail: The Origins of Power, Prosperity and Poverty*, by Acemoglu and Robinson (2012).

They argue that economic development occurs when societies enter a virtuous cycle where inclusive economic and political institutions lead to increasingly inclusive institutions, and the nature of inclusive institutions leads to economic development. The emphasis in this theory is on the institutions, not on culture, education, power, or willingness to change, though the authors argue that these others are partial explanations because they partly result from inclusive institutions. The focus needs to be on whether institutions are "inclusive" so that they facilitate economic development.

Acemoglu and Robinson identify inclusive institutions as those that lead to

- broad participation in political processes by citizens and economic groupings;
- economic opportunity for a broad base of society;
- level playing field in terms of economic betterment by individuals;
- change to more-efficient methods of production; and
- ease of entrance for new economic actors.

They note that:

> To be inclusive, economic institutions must feature secure private property, an unbiased system of law, and a provision of public services that provides a level playing field in which people can exchange and contract; it also must permit the entry of new businesses and allow people to choose their careers.

They demonstrate with ample evidence that inclusive institutions lead to economic development. Economic development comes from new technology and good education, but these in turn result from inclusive institutions. They do not argue that free markets of themselves lead to economic development. In fact, they argue that a strong centralized state is an important condition—but not a sufficient condition—for economic development. New markets, like smart markets for water, will continue the virtuous cycle of development only when the growth of markets does not decrease the power of the state to enforce property rights, enforce other regulations, and ensure adequate infrastructure.

Political and economic institutions are called "extractive" when institutions give resources to a minority at the expense of the majority. Extractive institutions can lead to policies ranging from mildly extractive, such as subsidies, to highly extractive, such as slavery. A key advance of Acemoglu and Robinson's theory of economic development is to see the tight connections between economic and political institutions. With extractive institutions, the economic advantages that flow to elites happen because of obliging political institutions. At the same time, the political elite can extract benefits because of the way that economic institutions are developed. Extractive institutions are the opposite of inclusive ones in that they do not secure private property, can have biased law, and do not provide infrastructure services evenly (or of enough quality) among the population. Such institutions are characterized by narrow political participation and narrow economic opportunities.

Seen through this theoretical lens, the difficulties in implementing smart markets for water resources are clearer. Extractive institutions will inhibit change in water management because change would threaten those with privileges. The economic benefits of extraction support the political institutions that inhibit change, and in turn the political institutions limit the change for the benefit of the privileged. Even though people might say that wealthy societies today are generally inclusive, their parts related to water management tend to be more extractive than inclusive. The legal, regulatory, and political frameworks for water resources tend to support the status quo, to support subsidies, to discourage new entrants, and to discourage innovation. At the same time, societies in wealthy countries generally have the rule of law, strong private property protection, good infrastructure, and relative equality of opportunity. Acemoglu and Robinson show that "Fear of creative destruction is often at the root of the opposition to inclusive economic and political institutions." As difficult as it might be to change the extractive nature of the economic and political institutions specifically involved in water resource management, perhaps the extant inclusive institutions surrounding water management will make it easier.

This theoretical lens also helps us see why it would be foolish to try to implement smart water markets in poor countries. The difficulty is not that individual water users or their government are poor. The difficulty is that they have extractive institutions, and extractive institutions would throttle any attempts at smart water markets. The difficulties of implementing a smart market in a poor country are not an issue of a non-market culture or of weak education of water users, but instead are an issue of inappropriate institutions to support such innovation. Nor are they an issue of persuading policymakers in such countries to implement a market. No one would use a water market where laws, contracts, and respect of property are shaky. Until the institutional framework changes, efforts towards water markets in poor countries are set to fail.

Policymakers within extractive institutions are constrained in ways that limit their ability to change to something like smart water markets. They need to protect privileged economic interests so that extraction can continue. If they stray too far from this path, they are likely to be replaced by policymakers who will ensure that a system of extraction continues. Acemoglu and Robinson note that extractive institutions are stuck in a vicious cycle where more extraction leads to less efficiency in the system, which in turn requires ever greater levels of extraction for the privileged actors to maintain their position, and so on. The vicious cycle creates greater and greater inequality, which eventually threatens the stability of the state.

Fortunately, a virtuous cycle can lead away from the downward spiral. In the best case, economic development leads to a broader base of the population who acknowledge that change leads to prosperity, which leads to broader political support for further change, which leads to opening up of new economic opportunities, and so on. The virtuous cycle leads to more egalitarian societies without fear of change. They develop instead a constituency in favor of change because of the recognition that a rising tide of prosperity will lift all boats.

The existence of two potential cycles indicates that one or the other dominates. Unfortunately, this virtuous cycle does not seem to appear within the water sector of even wealthy economies, where otherwise generally inclusive institutions harbor mildly extractive water institutions. Either the broadening of support for change will eventually lead to change in water institutions, or the extractive demands of water institutions will become so great that they stop change in other sectors as part of a coalition against change.

If smart markets for water resources are to develop, society needs to break from the extractive to the inclusive cycle. How is that likely to happen? This analytical framework suggests that the change needs support from a broad range of social interests. This could be a mixture of economic, political and social groups: large water users hobbled by poor supply, potential new water users, advocates of lower government subsidies, environmental activists, and indigenous groups. Broad interest in the smart market, across the boundaries of multiple traditional political parties, would help the drive for change.

Advocates should also create or nurture key institutions, so society increases the trust needed for smart markets. This means improving regulatory and legal systems to make them faster and of higher quality. The media can play a role in breaking from one cycle to the other. Exposés of poor water management under existing institutions can help to build a constituency for change, which in turn builds a demand for media coverage, starting its own virtuous cycle that can lead to change.

Still, the analysis of Acemoglu and Robinson is sadly repetitive in showing the greater historic strength of extractive over inclusive institutions, and of vicious over virtuous cycles. The most effective way to bring about change might be to wait until what they call a "critical juncture" in history: a war, natural disaster, or long-term severe drought. Climate change will lead to significant changes in water availability and seasonality. When water users get 50% less water for 10 years, they might want a new system of water allocation. We hope this book will seed change.

In the meantime, how can policymakers fortify inclusive institutions that can institute smart water markets against slipping into becoming extractive ones that cannot?

- Build trust in the institutions needed to someday support a smart market.
- Ensure broad political participation so a coalition for change can develop.
- Incentivize cooperation between water users by forming water groups to decrease the tendency to use the courts to resolve disputes.
- Support the media that shine light on the inefficiencies and inequities in the water sector.
- Support the development of a society with a proper safety net, to ease fear of change.
- Centralize control of water resources to ensure integrated science-driven management.

The three conclusions seem to be that (1) smart markets will be difficult to implement, (2) our best chance for successful implementation is in wealthy countries, because of the inclusive institutions that made them wealthy in the first place, and (3) smart markets are more likely to be successful if implemented after a critical juncture that makes major change acceptable.

Aside: Advantages of the Smart Market for Water
For users:

- **Ease of trading**. Finding a trading partner is easy—it's the market manager. Negotiating is easy—just put in a bid on a website. The transaction cost for users is low.
- **Voluntary participation**. A user could make money in every trade, through either selling excess consented water allocation, or obtaining additional water for profitable production. A user need not participate, if doing so seems bothersome. Price information motivates participation.
- **Price information**. Users get new detailed public information on an important resource. Users can make better decisions about capital investment.
- **Fairness**. Stakeholders perceive the price associated with every trade as fair. Buyers understand why the price is what it is, and come away satisfied. Sellers feel adequately compensated.
- **Low costs**. The market can be operated without either taxing or subsidizing users. The market system pays its own costs of operating the market, and those costs are modest.
- **Market customization.** Users in each smart market implementation can have some options as to how the market system is run, such as how often: users can trade by day, week, or some other period, as they choose.

For the Environment:

- **More-Reliable Environmental Bottom Lines**. The market manager simulates future flows at every auction, checking that the allocated water will maintain adequate flows at every point in time. Allocations cannot violate environmental bottom lines, resulting in real-time, continuous enforcement. This approach is more precautionary than current methods.
- **More-Refined Environmental Requirements**. A smart market can readily include spatial and temporal variability in flows—to ensure good fishing for indigenous peoples, or extra spring surge for fish spawning, or greater flow for a kayaking event. The market manager treats the environment as a priority user, with firm rights, and with needs that can vary dynamically.
- **Faster Response**. A smart market relies on a large amount of real-time data, ensuring rapid response to changing environmental conditions without need to approach regulators or litigation in the courts.
- **State-of-the-Art Science**. The system relies on the best available science, and is easily updatable when better science becomes available.
- **Direct Intervention**. Any kind of interest group could easily participate in the market, whether non-profit environmentalist, sporting, or tourist,

(continued)

without the high transaction cost of the political process. Existing users would be compensated transparently. An environmental manager might ask, "Given a budget and the ability to purchase water in any week to increase river flow, how should that budget be spent, keeping in mind that the price of water changes every week?"

For Government:

- **High Accountability**. The system is auditable, with good records.
- **Less Bureaucracy**. Less bureaucracy is needed to manage water use applications. Water authorities should find their jobs easier, and that they can process consents more accurately, and at lower cost, than now.
- **Stewardship Mission**. The water regulator would have a narrower focus: on stewardship of the resource, rather than how it is used. The market has a clearer focus on managing non-commercial water. The water regulator would have a greater sense of control over the resource, and better information about the effects of society's use of those resources.
- **Better Stewardship Tools**. The system correctly accounts for physical externalities to the full extent of the available science, including effects on non-traders and the environment. It has an ability to take into account regional variation, and recent and forecasted weather.
- **Simpler Path to Sustainability**. There would be no need to "save water for the future": first, because the market system simulates and protects future water use; and, second, because new users will be able to enter as a normal market function at any time simply by leasing water. A catchment can never be over-allocated, because the legal rights match the physical water availability.
- **Support for Economic Growth**. The system provides an institution that supports economic growth by managing the total quantity of commercial water, rather than managing conflict over water allocations between users. Government can leave existing or potential new water users to explore economic development opportunities.
- **Clearer Impartiality**. The system has flexibility to allow different types of water users, such as agriculture and hydroelectricity generators, without choosing winners and losers.
- **Government as Enabler**. Government looks responsive, impartial, and fair. It gives opportunity, not rigid rules. It enables society to protect the environment, grow economically, and resolve conflict in more-civil ways.

14.6 What Can Be Done in the Interim?

Implementation will take an organization with a bold vision to drive it. In the meantime, how can stakeholders prepare the way for implementation for common-pool smart markets?

Some water regulators may already be well prepared for water markets. Others may be further away than they think. Luckily, good regulation of water without a market requires the same prerequisites as good regulation with a water market. To allow trading, a water regulator will have to

- get accounting systems organized;
- clean up its permit databases; and
- decide very clearly and very firmly the environmental flow requirements to be enforced.

All water users with flow above a certain amount should have meters. Better and more-accountable water management increasingly expects, if not demands, water metering. The regulator can reduce maximum unmetered flow over time. In this way, the regulator measures a greater quantity of water each year. All water users with flow above some amount should have real-time metering. The maximum unmonitored water use would decline as well. Smart markets will work better when more water use is metered, particularly when the smart market incorporates that water use in real-time.

Data and models must be public so trust can develop. Users or interest groups should see a clear process to describe the water resource and set the environmental bottom lines.

The regulator should standardize data to make the market more efficient, and to make it easier for other parties to test whether the market is fair or not. The data should be stored in standardized format, so that anyone may obtain that data, and at no charge, both by ordinary web page and by web services. (Texas and Utah have done this.) The regulator should use modern version control systems to track changes made to all data.

In addition, the public has a right to verify that the flows meet environmental standards. Users, researchers, and the public should have the ability to observe historical, up-to-date, and forecasted water flow information. The software should be made sufficiently user-friendly that laypeople can operate it. For example, anyone should be able to display data for a particular dam on their computer screen, and see that display updated as often as the data change. Similarly, anyone should be able to see real-time and forecasted data on rivers flows, to know when they would receive their water entitlement, or to know whether the environmental flow requirements are satisfied. The regulator should publish data on permitted water users to improve accountability. All data should be stored for historical analysis.

The information required is:

- remote sensing data for lake and reservoir levels,
- remote sensing data for river flow data, starting with the larger rivers, and progressively implemented to significant streams,

- frequent, preferably remotely sensed, data on water quality,
- climatology data, real-time and historical, including data that will work with the hydrology models,
- geological and seismic data stored in publicly available databases,
- environmental flow requirements and quality standards in every catchment,
- documentation associated with the hydrology models used to set water use limits, including model calibration.

Most importantly, implementers must communicate the problems with water allocations and the possible solutions to those problems, and must convince stakeholders that markets can do a better job of satisfying society's requirements for the environment while reducing users' risks. Implementers have to explain to the concerned public that when users face a price and an incentive to conserve, environmental flows will be more reliable.

14.7 Where to From Here?

A smart market is achievable as a result of a transforming and enabling confluence of technologies: hydrological simulation to inform a linear program, the Internet to transmit users' bids and the resulting prices and quantities, increased computation power to run the simulation and linear program, and real-time monitoring. This new mix of technologies will both enrich society and greatly improve the environment. One hard lesson to take away from the Australian water market experience is that implementers should expect to make mistakes. As long as implementers can go forward in good faith, anticipating mistakes and being willing to set them right, societies have a good chance at making progress. Worse than making mistakes in developing smart markets, is society dragging its feet when the technology exists to solve at least a major part of the water allocation crisis.

Even with an immediate full trial, much research remains to be undertaken. Twenty years from now, computing power will be thousands of times better than it is now. Sensors will be better. Communications will be better. Hydrologic understanding will be better. Smart markets will dramatically improve humanity's ability to get along with the environment. The millions of voices, with their many wants and needs, will find sustainable accord.

Perhaps this book has similarities to *Spot Pricing of Electricity*, written by Schweppe et al. in (1988); that book was written before electricity providers had considered smart market options for allocating electricity. Today, policymakers and researchers recognize smart markets as the preferred method for countries with advanced electricity control systems. We hope we provide enough here to make the first implementation of a smart market for water come sooner.

We see risks in operation of a smart water market. Government could neglect rule enforcement, allowing cheating, or neglect maintenance of the hydrology models. This system requires the rule of law, and it would work best in a democratic society. Further, as demand for water increases, and as water moves to its highest value, we expect the price of water to increase, though it will increase much more without trading. Once a smart market starts, society may blame trading itself for the high price of water, rather than society's ever-increasing demand for water.

With much of the GDP in many countries depending on water use, the adoption of a water market, through movement of water from lower- to higher-value use, holds the prospect of a measurable permanent gain to GDP.

In addition, a spot market would lead to more-reliable attainment of environmental bottom lines, clear and correct price signals for water where before there were no such signals, and fewer appeals and court cases associated with water. The hydrology simulation and optimization guarantee the sustainability of any water use, greatly reducing contention. The water regulator need no longer be the bearer of bad news by denying water permits. Instead, the water regulator can focus on the environment.

Currently, society makes decisions about how much water to set aside for non-commercial use, but government makes such decisions perhaps once every 10 years and the process is highly politicized. This resulting fixed number is always wrong, because a fixed number ignores natural variability and future science. Extractive institutions tend to produce endless conflict between interest groups. But a smart market for water can be part of the virtuous cycle of an inclusive institution, where participants recognize that markets can stimulate and reinforce trust, which in turn leads to more-efficient markets, and so on.

Our aim is not privatization of water, and not to take away users' rights, but to permanently solve key water allocation problems, while improving prosperity and the environment.

We hope we have inspired you to take on that challenge.

References

Acemoglu, D., & Robinson, J. A. (2012). *Why nations fail: The origins of power, prosperity and poverty*. New York: Crown Business.

Schweppe, F. C., Caramanis, M. C., Tabors, R. D., & Bohn, R. E. (1988). *Spot pricing of electricity, The Kluwer International Series in Engineering and Computer Science*. Boston: Kluwer Academic Publishers.

Appendix: Outline of Market-Clearing Code

Setting up software for the water market is a non-trivialexercise. Implementers will need the help of an experienced software development team. We have developed some programs and databases as a proof-of-concept. Software continues to improve, so the ideal specifications will change, too. For example, the code described below was written in a combination of PHP and Python, but could now be written almost entirely in Python, including the linear program (LP). In addition, the design of software items will need case-by-case consideration of the regulations, hydrology, and local users' needs. In particular, the data will depend on whether the catchment is predominantly a single reservoir, or surface water, or groundwater, or some combination. The example below is for groundwater.

We see the benefit in listing for others the set of software modules that we have used, and some of the reasoning behind them. If nothing else, this list could serve as a sort of checklist so that those who might attempt to replicate our work can be sure they have all the parts that they will need.

Key Modules

1. **GWM2K_DB**. An SQLite database that stores all the inputs to GWM2K for a given catchment. It also stores annual sets of rainfall data (mm/day).
2. **SetRainfall.py**. This allows the operator to choose a rainfall profile input to GWM2K. The script then starts up GWM2K.
3. **GWM2K**. Available from USGS. A hydrological simulation based on MODFLOW. Inputs are: rainfall (mm/day), and catchment data (well location, permeability, etc.). Implementers must run GWM2K once to set up the market. For a real water auction, it would need to be run occasionally, if nothing else to incorporate the effects of new rainfall inputs. It would be wise to avoid using the ArcGIS front end to GWM2K, and to instead run GWM2K via script.

J.F. Raffensperger, M.W. Milke, *Smart Markets for Water Resources*, Global Issues in Water Policy 12, DOI 10.1007/978-3-319-55008-4

4. **WaterAuctionDB**. An SQLite database containing all the relevant information to simulate the auction. Described in detail below.
5. **ReadMPS.py**. A Python script that extracts the response matrix and head limits from the "MPS" file produced by GWM2K, and writes to WaterAuctionDB.
6. **AuctionManager.php**. A set of PHP scripts that allow a researcher to schedule, run, and analyze auctions. AuctionManager.php can call SolveAuction.py, to trigger solution of the auction. AuctionManager.php is the operator's dashboard.

   AuctionManager.php should optionally be able to choose a rainfall forecast, via SetRainfall.py, and start GWM2K, and ReadMPS.py.

   AuctionManager.php should have an "auto-bid" feature, to allow the operator to re-run the auction with existing bids, with partially changed bids, or through Autobid.py.
7. **UpdateInitialAllocation.py**. A Python script that reads WaterAuctionDB, and calculates appropriate initial allocation for each user, based on each user's existing allocation, response coefficients, and the head limits. UpdateInitialAllocation.py writes the updated allocation values to WaterAuctionDB.
8. **SolveAuction.py**. A Python script that reads WaterAuctionDB, creates a gross pool linear program, solves it with open-source linear programming code (PuLP and CBC), and writes the solution to WaterAuctionDB.
9. **FrontEnd.php**. A set of PHP scripts that serve as user web pages to enable bidding. The scripts write the users' bids to WaterAuctionDB, read the auction-clearing prices and quantities from WaterAuctionDB, and write that information to the users' web pages.

   A key feature of these web pages could be a convenient means for users with different crop types to specify different default bids for water, for each period (day or week) remaining in the hydrological season.

   A login page is required. Users need to be able to view results of previous auctions.
10. **ViewPriceMap.php**. A PHP script that provides a visualization of prices, with the ability to output to an SVG file template, possibly with a background image. The SVG file (and possibly a background image) should be stored in the database. The operator should update prices (and quantities if the operator chooses) at the end of each auction.
11. **Autobid.py**. Autobid.py is a script that simulates user behavior and creates automatic bids. This code needs to allow almost arbitrary sophistication on the part of a simulated user.

Typical Order of Operation

1. The operator (the researcher or market manager) chooses a rainfall profile from GWM2K_DB, and starts SetRainfall.py, which loads the data into GWM2K and starts GWM2K. Using the modules described above, this operator does this at the command line, programmatically from some other script, and via AuctionManager.php.
2. After GWM2K finishes, the operator starts ReadMPS.py (optionally via AuctionManager.php), which reads the GWM2K output, and writes that to WaterAuctionDB.
3. The operator starts UpdateInitialAllocation.py, to adjust users' initial allocation (optionally via AuctionManager.php). This adjustment needs to be by assessment point, and available as the "initial allocation" on the user's web page.
4. Using AuctionManager.php, the operator then initializes an auction. An auction by definition is a period of time over which users can bid for water.
5. Each user opens their own FrontEnd.php web page, and places a bid for water, for each period remaining in the current hydrological season. At the end of the auction period, a script should update users' web pages to prevent any more bids.
6. SolveAuction.py runs to clear the market: it sets up the linear program, solves it, gets the solution (allocations and prices), and writes the solution to WaterAuctionDB.
7. FrontEnd.php then updates users' web pages with their allocated quantities and prices.

The Database: WaterAuctionDB

The database could be SQLite for portability and simplicity. The operator could store all data for a given catchment in a single directory, so the operator can change data simply by changing the directory. The database should have the following fields and features:

1. **Traders**: traderid, password, first name, last name, address, city, phone, email.
2. **TraderPermits (permanent):** traderid, wellid, permitid, permittotal, startdate, expirydate.
3. **TraderAllocations (permanent +/– leased):** traderid, auctionid, allocationadjusted, permitid, permittotal, dualprice, expirydate, lprowname, quantityallocated, quantitybought, startdate, wellid, AssessPointid.

   This row balances abstraction, allocation, and bids in the LP, for time period *effectdate*. The consent rows for this application are "zero-based" (i.e., a gross pool market) so all bids are non-negative, and settlement is done in a calculation after the solution of the LP. The right-hand side in the settlement equation is either *allocationadjusted* or *allocationtotal*, depending on the market manager's choice. To make bids zero-based, the operator can use a script, UpdateInitialAllocation.py, to adjust bid quantities.

This table also corresponds to a variable, which is the total amount of abstraction, *quantityallocated*, in the solution. Variable *quantityallocated* has an LP upper-bound of *wellcapacity*. Normally, this would be infinity.

4. **Wells**: wellid, traderid, permitid, layer, row, col, latitude, longitude, description (can be HTML media, including photos), wellcapacity. This is the ".decvar" file from GWM2K.
5. **Auctions**: auctionid, auctiontype, isauctionfinal, auctiondate, proportionalallowance, proportionalallowancefinal, netrevenue, description (can be HTML media, including photos), number-of-user-wells, Number-of-control-points, Number-of-auction-periods, Number-of-constraint-periods. Note: Number-of-auction-periods $\leq$ Number-of-constraint-periods.
6. **AssessPoints**: AssessPointid, layer1, row1, col1, layer2, row2, col2, latitude1, longitude1, latitude2, longitude2, AssessPointtype, name, description (can be HTML media, including photos).
7. **AssessPointEvent**: AssessPointeventid, AssessPointid, auctionid, dualprice, effectdate, head, lprowname, range_lower, range_ upper, slack.

   This is an assessment point row in the LP, associated with an assessment point and time period *effectdate*. The RHS is head. But GWM can give a separate variable for head, so we would get an equality plus an inequality. Therefore, we may need to have a separate bound on head. This would mean this table should be:

   AssessPointeventid, AssessPointid, auctionid, effectdate, equality-constraint-right-hand-side, head constraint-lower-bound, head constraint-upper-bound, lprowname, range_lower, range_ upper, slack, dualprice.
8. **TraderBids**: bidid, traderid, isbidautomatic, auctionid, biddate, effectdate, expirydate, pricebid, quantitybid.

   This table corresponds to bid variables in the LP. Each bid variable also has an upper bound. The consent rows for this application are "zero-based," so all bids are non-negative, and settlement is done in a calculation after the solution of the LP. For a given *traderid* and *auctionid*, each row corresponds to one block in the bid schedule.

   The operator could extend this table to have default bids, based on farm operation profile, by adding another table DefaultTraderBid like TraderBids, but without the date information.

   A smart market could have a table DefaultFarmTypeBid for default bids based purely on farm operation profile, independent of the *traderID*.
9. **CashTransactions**: cashtransactionid, auctionid, traderid, auctionid, transactiondate, cashadded, cashbalance, notes. The amountadded could be negative.
10. **AllocationTransactions**: watertransactionid, auctionid, traderid, transactiondate, futureauctiondate, wateradded, waterbalance, notes

At any auction, a *traderid* can buy or sell a quantity for any future period within the current auction, and a script should generate a *watertransaction* row for that future date, to indicate the *traderid*'s total right for the next auction. Also, the market manager may change users' quantities in response to water availability, which will generate additional rows in this table.

The balance is valid only for the period of the auction at *futureauctiondate*.

The size of this table will be something like 2*(#users)*(#auctions/year)^2, which could be a million rows for a few hundred users. An alternative structure, potentially less verbose, would be to add *expirydate* and use a single row for multiple future auctions: watertransactionid, auctionid, traderid, transactiondate, effectdate, expirydate, wateradded, waterbalance, notes.

The balance is valid for the period from the *effectdate* to the *expirydate*.

11. **ResponseMatrixInfo, about the response matrix**: ResponseMatrixCreationDate, cubic-meters-per-impulse-unit, notes
12. **ResponseMatrixCoefficients**

    This can be stored in two ways. First, as a time-independent response matrix. Zero *periods-lagged* means that the response is immediate: wellid, AssessPointid, periods-lagged, floating-point-impact-coefficient.

    Second, as a more detailed time-dependent matrix (which is what GWM does): wellid, AssessPointid, take-of-water-date, effect-on-control-point-date, floating-point-impact-coefficient

    A field for unit flow, e.g., *mlunitflow*, could indicate the rate at which the hydrologist tested the well to obtain the meters-per-unit-flow coefficient. That field may be useful to also store data on measurement uncertainty and degree of non-linearity, if those data are available. The *notes* field could store which scientist did the measuring, and by what method (e.g., MODFLOW simulation or eigenmodel).

    Most likely, the identifiers would be integers, and the time data would be *date-time* (not just *date*).
13. The operator may need to add other tables for riversegments and reservoirs when applicable.

Bibliography

ABC. (2015). *California's millennium drought turns attention to South Australia for expertise and experience*. http://www.abc.net.au/news/2015-05-03/california-millennium-drought-turns-attention-to-sa-expertise/6438004. 3 May 2015.

Acemoglu, D., & Robinson, J. A. (2012). *Why nations fail: The origins of power, prosperity and poverty*. New York: Crown Business.

Ahlfeld, D. P., & Mulligan, A. E. (2000). *Optimal management of flow in groundwater systems*. San Diego: Academic Press.

Ahlfeld, D. P., Barlow, P. M., & Mulligan, A. E. (2005). *GWM – A ground-water management process for the U.S. Geological Survey modular ground–water model (MODFLOW–2000)*. (U.S. Geological Survey Open–File Report 2005–1072). http://pubs.usgs.gov/of/2005/1072. Accessed 15 July 2006.

Alvey, T., Goodwin, D., Ma, X., Streiffert, D., & Sun, D. (1998). A security-constrained bid-clearing system for the NZ wholesale electricity market. *IEEE, 13*, 340–346.

Baliga, S., & Maskin, E. (2003). Mechanism design for the environment. In K. G. Mäler & J. R. Vincent (Eds.), *Handbook of environmental economics* (1st ed., pp. 305–324). Amsterdam/London: Elsevier Publishing.

Bastiaanssen, W. G. M., Molden, D. J., & Makin, I. W. (2000). Remote sensing for irrigated agriculture: Examples from research and possible applications. *Agricultural Water Management, 46*(2), 137–155.

Bate, R. (2002). Water – Can property rights and markets replace conflict? Chap. 15. In J. Morris (Ed.), *Sustainable development: Promoting progress or perpetuating poverty?* London: Profile Books.

Bauer, C. J. (1998). *Against the current: Privatization, water markets, and the state in Chile*. New York: Spring Science+Business Media.

Becker, N. (1995). Value of moving from central planning to a market system: Lessons from the Israeli water sector. *Agricultural Economics, 12*, 11–21.

Bidwell, V. J. (2005). Realistic forecasting of groundwater level, based on the eigenstructure of aquifer dynamics. *Maths and Computers in Simulation, 69*, 12–20.

Bjornlund, H. (2003). Efficient water market mechanisms to cope with water scarcity. *Water Resources Development, 19*, 553–567.

Bourbeau, B. Crainic, T. G., Gendreau, M., & Robert, J. (2005). Design for optimized multi-lateral multi-commodity markets. *European Journal of Operational Research, 163*(2), 503–529.

CA Water Commission. (2016). https://cwc.ca.gov/Documents/2014/02_February/February2014_Agenda_Item_10_Attach_1_TransferProcess.pdf. Accessed 13 Mar 2016.

J.F. Raffensperger, M.W. Milke, *Smart Markets for Water Resources*, Global Issues in Water Policy 12, DOI 10.1007/978-3-319-55008-4

Carlson, B., Chen, Y., Hong, M., Jones, R., Larson, K., Ma, X., Nieuwesteeg, P., Song, H., Sperry, K., Tackett, M., Taylor, D., Wan, J., & Zak, E. (2012). MISO unlocks billions in savings through the application of operations research for energy and ancillary services markets. *Interfaces, 42*(1), 58–73.

Clemen, R. T., & Reilly, T. (2001). *Making hard decisions with decision tools* (2nd ed.). Belmont: Duxbury/Thomson Learning.

Coase, R. H. (1960). The problem of social cost. *Journal of Law & Economics, 3*(1), 1–44.

Collentine, D. (2005). Including non-point sources in a water quality trading permit program. *Water Science and Technology, 51*(3–4), 47–53.

Cummings, R. G., Holt, C. A., & Laury, S. K. (2004). Using laboratory experiments for policymaking: An example from the Georgia irrigation reduction auction. *Journal of Policy Analysis and Management, 23*(2), 341–363.

Dalziel, P. C. (1987). Optimal water storage and pricing: The effect of monopoly. *New Zealand Economic Papers, 21*(1), 3–16.

Davids Engineering, Inc. (2015, October). *Presentation regarding measurement technology implemented in California's Reclamation District no. 108.* http://www.rd108.org/wp-content/uploads/2015/10/ RD108_2015_Measurement_Program_Board_Update_151015_v6bab.pdf. Accessed 24 July 2016.

Deshel, M. D. E. (2005). Optimal allocation of tradable pollution rights and market structures. *Journal of Regulatory Economics, 28*(2), 205–223.

Dinar, A. (Ed.). (2000). *The political economy of water pricing reforms*. New York: A World Bank Publication/Oxford University Press.

Dinar, A., & Subramanian, A. (1997). *Water pricing experiences: An international perspective*. World Bank technical paper, no. WTP 386. Washington, D.C.: The World Bank. http://documents.worldbank.org/curated/en/580151468768597034/Water-pricing-experiences-an-international-perspective

Dinar, A., Howitt, R. E., Murphy, J. J., Rassenti, S. J., & Smith, V. L. (1998). *How might future water markets look: The use of experimental economics to design markets for water*. Venice: World Congress of Environmental and Resource Economists.

Disegni-Eshel, D. M. (2005). Optimal allocation of tradable pollution rights and market structures. *Journal of Regulatory Economics, 28*, 205–223.

Draper, A. J., Jenkins, M. W., Kirby, K. W., Lund, J. R., & Howitt, R. E. (2003). Economic-engineering optimization for California water management. *Journal of Water Resources Planning and Management*, ASCE, *129*(3), 155–164.

Dye, S., Read, E. G., Read, R. A., & Starkey, S. R. (2012, June 14–15). Easy implementations of generalised stochastic CDDP models for market simulation studies. In *4th IEEE and Cigré international workshop on hydro scheduling in competitive markets, Bergen*.

Easter K. W., & Huang, Q. (Eds.) (2014). *Water markets for the 21st century: What have we learned?* (Global issues in water policy 11). Dordrecht: Springer Science+Business Media.

Easter, K. W., Rosegrant, M. W., & Dinar, A. (Eds.). (1998). *Markets for water: Potential and performance*. Norwell: Kluwer Academic Publishers.

EMRG. (2002). *1998–2002 final activity report*. Energy Modelling Research Group (EMRG). http://www.mang.canterbury.ac.nz/research/emrg/report9802.shtml. Accessed 5 Aug 2007.

Environment Agency. (2001). *Managing water abstraction map of CAMS areas*. Environment Agency, Bristol, UK.

Environment Agency. (2005). *The upper Ouse and Bedford Ouse catchment abstraction management strategy*. Environment Agency, Bristol, UK, https://www.gov.uk/government/uploads/system/uploads/attachment_data/file/289835/LIT7708_df73f8.pdf. Accessed 22 Mar 2017.

Environmental Protection Authority. (2003). *Hunter River Salinity Trading Scheme: Working together to protect river quality and sustain economic development*. Sydney: Environmental Protection Authority, Department of Environment and Conservation.

Erfani, T., Binions, O., & Harou, J. J. (2014). Simulating water markets with transaction costs. *Water Resources Research, 50*(6), 4726–4745.

FAA. (2015, January 6). *Fact sheet – Notice of proposed rulemaking: Slot management and transparency for LaGuardia Airport (LGA), John F. Kennedy International Airport (JFK), and Newark Liberty International Airport (EWR).* https://www.faa.gov/news/fact_sheets/news_story.cfm?newsId=18054. Accessed 14 May 2016. See also http://www.faa.gov/news/updates/?newsId=81144

Forni, L. (2010, September). *Economic and hydrologic models integration–new method: Sacramento Basin, California.* Master's thesis, Agricultural and Resource Economics, University of California, Davis.

Forsund, F. R. (2007). *Hydropower economics.* Boston: Springer Science+Business Media.

Freeze, R. A., & Cherry, J. A. (1979). *Groundwater.* Englewood Cliffs: Prentice-Hall.

Gallien, J., & Wein, L. M. (2005). A smart market for industrial procurement with capacity constraints. *Management Science, 51*(1), 76–91.

Garrick, D. E. (2015). *Water allocation in rivers under pressure: Water trading, transaction costs and transboundary governance in the western US and Australia.* Cheltenham: Edward Elgar Publishing, Cheltenham, UK.

Garrick, D., & Aylward, B. (2012). Transaction costs and institutional performance in market-based environmental water allocation. *Land Economics, 88*(3), 535–560.

Garrick, D., Whitten, S. M., & Coggan, A. (2013). Understanding the evolution and performance of water markets and allocation policy: A transaction costs analysis framework. *Ecological Economics, 88*, 195–205.

Gordon, H. S. (1954). The economic theory of a common-property resource: The fishery. *Journal of Political Economy, 62*(2), 124–142.

Gorelick, S. M., & Zheng, C. (2015). Global change and the groundwater management challenge. *Water Resources Research, 51*, 3031–3051. doi:10.1002/2014WR016825.

Greenwald, R. (1998). *MODMAN: An optimization module for MODFLOW version 4.0.* Freehold: GeoTrans.

Grether, D. M., Mark Isaac, R., & Plott, C. R. (1979). *Alternative methods of allocating airport slots: Performance and evaluation.* Prepared for the Civil Aeronautics Board, Polinomics Research Laboratories, Inc., Pasadena.

Griffin, R. C., & Characklis, G. W. (2002). Issues and trends in Texas water marketing. *Water Resources Update, 121*, 207–217.

Haimes, Y. Y. (1977). *Hierarchical analyses of water resources systems.* New York: McGraw-Hill.

Hanak, E., Lund, J., Dinar, A., Gray, B., Howitt, R., Mount, J., Moyle, P., & Thompson, B. (2011). *Managing California's water: From conflict to reconciliation.* San Francisco: Public Policy Institute of California.

Harbaugh, A. W. (2005). *MODFLOW-2005, the U.S. Geological survey modular ground-water model—the ground-water flow process.* U.S. Geological Survey Techniques and Methods 6-A16. http://pubs.usgs.gov/tm/2005/tm6A16/PDF.htm. Accessed 15 July 2006.

Harbaugh, A. W., & McDonald, M. G. (1996). *Programmer's documentation for MODFLOW-96, an update to the U.S. Geological Survey modular finite-difference groundwater flow model.* (Open-File report 96–486). U.S. Geological Survey; Branch of Information Services. https://pubs.er.usgs.gov/publication/ofr96486. Accessed 19 Mar 2017.

Hardin, G. (1968). The tragedy of the commons. *Science, 162*, 1243.

Hardin, G. (1972). *Exploring new ethics for survival.* New York: Viking Press.

Harou, J. J., Pulido-Velázquez, M., Rosenberg, D. E., Medellín-Azuara, J., Lund, J. R., & Howitt, R. E. (2009). Hydro-economic models: Concepts, design, applications, and future prospects. *Journal of Hydrology, 375*(3–4), 627–643.

Hey, D. L., & Philippi, N. S. (1999). *A case for wetland restoration.* New York: Wiley.

Hirshleifer, J., De Haven, J. C., & Milliman, J. W. (1960). *Water supply: Economics, technology, and policy.* Chicago: University of Chicago Press.

Hoffman, E., Rassenti, S., Smith, V, Dinar, A, & Howitt, R. (1993, November 21–23). Designing a smart, computer-assisted market for water allocation in Arid Lands. In *63rd annual conference of the southern economic association, New Orleans.*

Hogan, W. W., Read, E. G., & Ring, B. J. (1996). Using mathematical programming for electricity spot pricing. *International Transactions in Operations Research, 3*(3–4), 243–253.

Hollinshead, S. P., & Lund, J. R. (2006). Optimization of environmental water account purchases with uncertainty. *Water Resources Research, 42*(8), W08403.

Holt, C. A., Johnson, C. A., Mallow, C. A., & Sullivan, S. P. (2012). Water externalities: Tragedy of the common canal. *Southern Economic Journal, 78*(4), 1142–1162.

Howitt, R. E. (1995). A calibration method for agricultural economic production models. *Journal of Agricultural Economics*, Wiley Blackwell, *46*(2), 147–159.

Howitt, R. E. (2000). Is California's future hydraulically sustainable? *California Agriculture, 54* (2), 10–15. doi:10.3733/ca.v054n02p10.

Howitt, R. E. (2007, September 28–30). *Water market based conflict resolution*. Rosenberg International Forum on Water Policy, San Francisco. http://ciwr.ucanr.edu/files/186749.pdf

Howitt, R., Moore, N., & Smith, R. T. (1992, March). *A retrospective on California's 1991 emergency drought water bank*. California Department of Water Resources. Sacramento.

Hughes, S. http://geology.isu.edu/wapi/EnvGeo/EG7_water/EG_module_7pt2.htm. Accessed 22 Oct 2016.

Hung, M.-F., & Shaw, D. (2005). A trading-ratio system for trading water pollution discharge permits. *Journal of Environmental Economics and Management, 49*(1), 83–102.

Israel, M., & Lund, J. R. (1995). Recent California water transfers: Implications for water management. *Natural Resources Journal, 35*, 1–32.

Jacobs, J., Freeman, G., Grygier, J., Morton, D., Schultz, G., Staschus, K., & Stedinger, J. (1995). SOCRATES: A system for scheduling hydroelectric generation under uncertainty. *Annals of Operations Research, 59*(1), 99–133.

Jaikumar, R. (1980). Mathematical programming approaches to the design of auctions: A study of the auction of landing rights at congested airports. *Economic Forum, 11*, 29–47.

Jenkins, M. W., Lund, J. R., Howitt, R. E., Draper, A. J., Msangi, S. M., Tanaka, S. K., Ritzema, R. S., & Marques, G. F. (2004). Optimization of California's water supply system: Results and insights. *Journal of Water Resources Planning and Management, 130*(4), 271–280.

Kahneman, D., & Tversky, A. (1979). Prospect theory: An analysis of decision under risk. *Econometrica, 47*(2), 263–291.

Kirk, R. M. (1991). River-beach interaction on mixed sand and gravel coasts: A geomorphic model for water resource planning. *Applied Geography, 11*, 267–287.

Kostel, J. A., Monchak, J., Tomer, M. D., Bingner, R. L., Bank Lentz, A T., Ando, A. W., Brozovic, N., John F., Raffensperger, R. A., Prabodanie, R., Wohlgezogen, F., Zajac, E., & Hey, D. (2014, September). *Feasibility assessment of a nutrient trading market in the Big Bureau Creek Watershed final report*. Prepared for U.S. Environmental Protection Agency, Targeted Watershed Grant WS-00E71101.

Lachman, B., Resetar, S., Kolra, N., Schaefer, A., & Curtright, A. (2016). *Water Management, Partnerships, rights, and market trends* (RAND RR933). Santa Monica: RAND Corp.

Lefkoff L. J., & Gorelick, S. M. (1987). *AQMAN; linear and quadratic programming matrix generator using two-dimensional ground-water flow simulation for aquifer management modeling* (Water-resources investigations report 87-4061). Menlo Park: U.S. Geological Survey.

Lefkoff, L. J., & Gorelick, S. M. (1990). Benefits of an irrigation water rental market in a saline stream-aquifer system. *Water Resources Research, 26*(7), 1371–1381.

Lumbroso, D. M., Twigger-Ross, C., Raffensperger, J., Harou, J. J., Silcock, M., & Thompson, A. J. K. (2014). Stakeholders' responses to the use of innovative water trading systems in East Anglia, England. *Water Resources Management, 28*(9), 2677–2694.

Lund, J. R. (1993). Transaction risk versus transaction costs in water transfers. *Water Resources Research, 29*(9), 3103–3107.

Lund, J. R., & Ferreira, I. (1996). Operating rule optimization for the Missouri river reservoir system. *Journal of Water Resources Planning and Management*, ASCE, *122*(4), 287–295.

Lund, J. R. & Israel, M. (1995). Water transfers in water resource systems. *Journal of Water Resources Planning and Management*, ASCE, *121*(2), 193–205.

Mahakalanda, I., Dye, S., Read, E.G. and Raffensperger, J. F. (2012, December 10–11). *Intra-period market clearing for a multi-use catchment via CDDP*. Paper presented at the ORSNZ conference, Wellington.

Mahan, R. C., Horbulyk, T. M., & Rowse, J. G. (2002). Market mechanisms and the efficient allocation of surface water resources in southern Alberta. *Socio-Economic Planning Sciences, 36*, 25–49.

Markham, J. W. (1951). The nature and significance of price leadership. *The American Economic Review, 41*(5), 891–905.

McAfee, R. P. (2010). Four issues in auctions and market design. *Revista de Análisis Económico – Economic Analysis Review, 13*(1), 7–24. http://www.rae-ear.org/index.php/rae/article/view/117/226. Accessed 18 Mar 2017.

McCabe, K. A., Rassenti, S. J., & Smith, V. L. (1989). Designing 'smart' computer-assisted markets: An experimental auction for gas networks. *European Journal of Political Economy, 5*, 259–283.

McCabe, K., Rassenti, S., & Smith, V. (1990). Auction design for composite goods: The natural gas industry. *Journal of Economic Behavior and Organization, 14*, 127–149.

McCabe, K. A., Rassenti, S. J., & Smith, V. L. (1991). Smart computer-assisted markets. *Science, 254*(5031), 534–538.

McCabe, K., Rassenti, S., & Smith, V. (1994). Designing a real time computer assisted auction for natural gas networks. In W. Cooper & A. Whinston (Eds.), *New directions in computational economics* (pp. 41–54). Holland: Kluwer Academic Publishers.

McDonald, M. G., & Harbaugh, A. W. (1988). *A modular three-dimensional finite-difference ground-water flow model*. TWRI 6-A1, U.S. Geological Survey Reston. http://pubs.usgs.gov/twri/twri6a1

McGauran, P. (2006, April 28). *New Murray water plan a breakthrough for farmers*. http://www.maff.gov.au/releases/06/06047pm.html. Accessed 2 Oct 2006.

Montero, J. P. (1998). A simple auction mechanism for the optimal allocation of the commons. *The American Economic Review, 98*, 496.

Montgomery, W. D. (1972). Markets in licenses and efficient pollution control programs. *Journal of Economic Theory, 5*(3), 395–418.

Morgan, C. L., Coggins, J. S., & Eidman, V. R. (2000). Tradable permits for controlling nitrates in groundwater at the farm level: A conceptual model. *Journal of Agricultural and Applied Economics, 32*(2), 249–258.

Mulligan, K., Brown, C., Yang, Y.-C., & Ahlfeld, D. (2014). Assessing groundwater policy with coupled economic-groundwater hydrologic modeling. *Water Resources Research, 50*(3), 2257–2275.

Murphy, J. J., Dinar, A., Howitt, R., Rassenti, S. J., & Smith, V. L. (2000). The design of 'smart' water market institutions using laboratory experiments. *Environmental and Resource Economics, 17*(4), 375–394.

Murphy, J. J., Dinar, A., Howitt, R. E., Mastrangelo, E., Rassenti, S., & Smith, V. L. (2006). Mechanisms for addressing third-party impacts resulting from voluntary water transfers. In J. A. List (Ed.), *Using experimental methods in environmental and resource economics*. Cheltenham: Edward Elgar Publishing Ltd.

Murphy, J. J., Dinar, A., Howitt, R. E., Rassenti, S. J., Smith, V. L., & Weinberg, M. (2009). The design of water markets when instream flows have value. *Journal of Environmental Management, 90*(2), 1089–1096.

Murray–Darling Basin Authority. (n.d.-a 2). *Guidelines for water trading rules, managing surface water trade restrictions*. http://www.mdba.gov.au/sites/default/files/pubs/03_WTG-FREE-TRADE-OF-SURFACE-WATER_final.pdf. Accessed 30 Apr 2016.

Murray–Darling Basin Authority. (n.d.-b 1). *Guidelines for water trading rules, introduction.* http://www.mdba.gov.au/sites/default/files/pubs/01_WTG-REFERENCE_final.pdf. Accessed 30 Apr 2016.

National Water Commission. (2010). *The impacts of water trading in the southern Murray–Darling Basin.* National Water Commission, Commonwealth of Australia. http://www.nwc.gov.au/__data/assets/pdf_file/0019/10783/681-NWC_ImpactsofTrade_web.pdf. Accessed 30 Apr 2016.

National Watermarket. (2015). http://www.nationalwatermarket.gov.au/about-trade-delete/index.html. Accessed 30 Apr 2016.

Nemes, N., Plott, C. R., & Stoneham, G. (2008). *Electronic BushBroker exchange: Designing a combinatorial double auction for native vegetation offsets.* https://excen.gsu.edu/plott.pdf

Newlin Brad, D., Jenkins, M. W., Lund, J. R., & Howitt, R. E. (2002). Southern California water markets: Potential and limitations. *Journal of Water Resources Planning and Management, 128*(1), 21–32.

NSW DNR. Dept. of Natural Resources, New South Wales. http://www.dlwc.nsw.gov.au/water/recovery.shtml. Accessed 2 Oct 2006.

Ostrom, E. (2009). Social-ecological systems: A general framework for analyzing sustainability of social-ecological systems. *Science, 325*, 419–422.

Pidwirny, M. 2006 The hydrologic cycle. In *Fundamentals of physical geography* (2nd ed.). http://www.physicalgeography.net/fundamentals/8b.html. Accessed 18 Sep 2016.

Pinto Rodriguez, A. A. (2013). *Market mechanisms to allow trading of impervious cover.* PhD thesis, University of Canterbury, Christchurch.

Pinto Rodriguez, A. A., Raffensperger, J. F., Cochrane, T. A., & Dye, S. (2009, December 3–4). A proposed smart market for impervious cover runoff under rainfall uncertainty. In *Proceedings of the ORNSZ conference*, Christchurch, (pp. 190–199).

Pinto Rodriguez, A. A., Raffensperger, J. F., Cochrane, T. A., & Read, E. G. (2013). A proposed smart market design for sediment discharge. *Journal of Water Resources Planning and Management, 139*(1), 96–108.

Plagmann, M., & Raffensperger, J. F. (2007). A smart market for ground water using the eigenmodel approach. In *MODSIM Conference, Christchurch.*

Plott, C., & Zeiler, K. (2005). The willingness to pay–willingness to accept gap, the "endowment effect," subject misconceptions, and experimental procedures for eliciting valuations. *American Economic Review, 95*(3), 530–545.

Plott, C. R., & Zeiler, K. (2007). Exchange asymmetries incorrectly interpreted as evidence of endowment effect theory and prospect theory? *American Economic Review, 97*(4), 1449–1466.

Porse, E., et al. (2015). Pumping for the masses: Evolution of groundwater management in metropolitan Los Angeles. *GeoJournal.* doi:10.1007/s10708-015-9664-0, 17 pp, first online 7 August 2015.

Prabodanie, R. Ranga, A., Raffensperger, J. F., & Milke, M. W. (2009, July 13–17). Simulation-optimization approach for trading point and non-point source nutrient permits. In *18th World IMACS/MODSIM Congress, Cairns.*

Prabodanie, R., Ranga, A., Raffensperger, J. F., & Milke, M. W. (2010). A pollution offset system for trading non-point source water pollution permits. *Journal of Environmental and Resource Economics, 45*(4), 499–515.

Prabodanie, R. A., Ranga, John F., Raffensperger, E., Read, G., & Milke, Mark W. (2011). LP models for pricing diffuse nitrate discharge permits. *Annals of Operations Research.* doi: 10.1007/s10479-011-0941-0

Pritchard, G., Zakeri, G., & Philpott, A. (2010). A single-settlement, energy-only electric power market for unpredictable and intermittent participants. *Operations Research, 58*(4), 1210–1219.

Pulido-Velázquez, M. A. (1997). *Conjunctive use opportunities in Southern California.* Master's thesis, University of California, Davis.

Pulido-Velázquez, M., Andreu, J., & Sahuquillo, A. (2006). Economic optimization of conjunctive use of surface water and groundwater at the basic scale. *Journal of Water Resources Planning and Management, 132*(6), 454–467.
Raffensperger, J. F. (2011). Matching users' rights to available ground water. *Ecological Economics, 70*(6), 1041–1050.
Raffensperger, J. F., & Cochrane, T. (2010). A smart market for impervious cover. *Water Resources Management, 24*(12), 3065–3083.
Raffensperger, J. F., & Milke, M. W. (2005). A design for a fresh water spot market. *Water Science and Technology: Water Supply, 5*(6), 217–224.
Raffensperger, J. F., Milke, M. W., & Read, E. G. (2009). A deterministic smart market model for ground water. *Operations Research, 57*(6), 1333–1346.
Rassenti, S. J., Smith, V. L., & Bulfin, R. L. (1982). A combinatorial auction mechanism for airport time slot allocation. *Bell Journal of Economics, 13*, 402–417.
Read, E. G. (1979). Stochastic long-term scheduling models for a power system. In *Proceedings of 15th Annual Conference, Operational Research Society of New Zealand.* New Zealand Operational Research, special issue (pp. 41–52).
Read, E. G. (1989). A dual approach to stochastic dynamic programming for reservoir release scheduling. In A. O. Esogbue (Ed.), *Dynamic programming for optimal water resources system management* (pp. 361–372). New York: Prentice Hall.
Read, E. G, & Chattopadhyay, D. (1999). Electricity market models: Lessons from Australasian experience. In ORSNZ 99 (*Proceedings of the 34th Annual Conference, Operational Research Society of New Zealand*). http://orsnz.org.nz/conf34/PDFs/Chatterjee.pdf
Read, E. G., & Hindsberger, M. (2010). Constructive dual DP for reservoir optimization. In P. M. Pardalos, S. Rebennack, M. V. F. Pereira, & N. A. Iliadis (Eds.), *Handbook of power systems I* (pp. 3–32). Heidelberg: Springer.
Rosegrant, M. W., Ringler, C., McKinney, D. C., Cai, X., Keller, A., & Donoso, G. (2000). Integrated economic–hydrologic water modeling at the basin scale: The Maipo river basin. *Agricultural Economics, 24*(1), 33–46.
Sahuquillo, A. (1983). An eigenvalue numerical technique for solving unsteady groundwater continuously in time. *Water Resources Research, 19*, 87–93.
Salman, S. M. A. (ed.) (1999). *Groundwater: Legal and policy perspectives.* Proceedings of a World Bank seminar. World Bank technical paper, no. 456. World Bank, Washington, DC.
Samuelson, P., Dorfman, R., & Solow, R. M. (1958). *Linear programming and economic analysis.* New York: McGraw-Hill.
Schoengold, K., Sunding, D. L., & Moreno, G. (2006). Price elasticity reconsidered: Panel estimation of an agricultural water demand function. *Water Resources Research, 42*(9), W09411.
Schweppe, F. C., Caramanis, M. C., Tabors, R. D., & Bohn, R. E. (1988). *Spot pricing of electricity.* Boston: The Kluwer International Series in Engineering and Computer Science, Kluwer Academic Publishers.
Smith, V. L. (1982). On divestiture and the creation of property rights in public lands. *The Cato Journal, 2*(Winter), 563–629.
Smith, V., Hoffman, E., & Rassenti, S. (1995, October 1). *Environmental change and adaptive resource markets: Computer-assisted markets for water allocation.* Award Abstract #9513406. http://www.nsf.gov/awardsearch/showAward?AWD_ID=9513406. Accessed 14 May 2015.
Sprenger, C. (2015). An endowment effect for risk: Experimental tests of stochastic reference points. *Journal of Political Economy, 123*(6), 1456–1499.
Stanley, E. D., Honig, D. P., & Gainen, L. (1954). Linear programming in bid evaluation. *Naval Research Logistics, Quarterly, 1*(1), 48–54.
Starkey, S. R., Dye, S., Read, E. G. & Read, R. A. (2012, June 14–15). Stochastic versus deterministic market design: Some experimental results. In *4th IEEE and Cigré international workshop on hydro scheduling in competitive markets, Bergen.*

Stubbs, C. M. (2000, June). *Hydrologic-economic modeling of irrigated agriculture in the lower Murrumbidgee catchment: Investigations into sustainability*. PhD thesis, MIT.

Svorencík, A., & Maas, H. (Eds.). (2016). *The making of experimental economics: Witness seminar on the emergence of a field*. Cham: Springer International Publishing.

Teasley, R., Raffensperger, J., & Milke, M. (2010, May 16–20). *A smart market for ground and surface water with hydropower generation*. ASCE World Environmental and Water Resources Congress, Providence.

Tenkorang, F., & Lowenberg-DoBoer, J. (2008). On-farm profitability of remote sensing in agriculture. *The Journal of Terrestrial Observation, 1*(1), 6.

Thobani, M. (1997). Formal water markets: Why, when and how to introduce tradable water rights. *The World Bank Research Observer, 12*, 161–179.

Thompson Jr., B. H. (2005). Uncertainty and markets in water resources. *McGeorge Law Review, 36*, 117–138.

Tietenberg, T. H. (2006). The evolution of emissions trading: Theoretical foundations and design considerations. http://www.colby.edu/personal/thtieten/edintro.pdf. Accessed 19 March 2017.

Tilmant, A., & Kelman, R. (2007). A stochastic approach to analyze trade-offs and risks associated with largescale water resources systems. *Water Resources Research, 43*(6), W06425.

Tisdell, J., Ward, J., & Capon, T. (2004). Impact of communication and information on a complex heterogeneous closed water catchment environment. *Water Resources Research, 40* (9). doi:10.1029/2003WR002868/epdf

UNDP. (2011). *Fighting corruption in the water sector: Methods, tools and good practices*. New York: United Nations Development Programme, Bureau for Development Policy.

Vencorp. (2007). *Guide to the Victorian gas wholesale market*.www.vencorp.com.au/index.php?action=filemanager&doc_form_name=download&folder_id=466&doc_id=3468. Accessed 5 Aug 2007.

Vickrey, W. (1961). Counterspeculation, auctions and competitive sealed tenders. *Journal of Finance, 16*(1), 8–37.

Weeber, J. (2008). *Christchurch groundwater protection: A hydrological basis for the zone boundaries* (Technical report U08/21, 60 p). Christchurch: Environment Canterbury.

Young, R. A., & Loomis, J. B. (2014). *Determining the economic value of water: Concepts and methods* (2nd ed.). New York: Routledge Publishing.

Young, M., & McColl, J. (2005). Defining tradable water entitlements and allocations: A robust system. *Canadian Water Resources Journal, 30*(1), 65–72.

Yu, B., Tisdell, J., Podger, G., & Salbeb, I. (2003, July 14–17). *A hydrologic and economic model for water trading and reallocation using linear programming techniques*. MODSIM 2003: International Congress on Modelling and Simulation, Townsville.

Zetland, D. (2013). All-in-Auctions for water. *Journal. of Environmental Management, 115*, 78–86.

Zhu, T., Marques, G. F., & Lund, J. R. (2015). Hydroeconomic optimization of integrated water management and transfers under stochastic surface water supply. *Water Resources Research, 51*, 3568–3587. doi:10.1002/2014WR016519.

Index

J.F. Raffensperger, M.W. Milke, *Smart Markets for Water Resources*, Global Issues in Water Policy 12, DOI 10.1007/978-3-319-55008-4

H

I

U

The manufacturer's authorised representative in the EU is Springer Nature Customer Service Centre GmbH, Europaplatz 3, 69115 Heidelberg, Germany. If you have any concerns regarding our products, please contact ProductSafety@springernature.com

Printed and bound by CPI Group (UK) Ltd, Croydon, CR0 4YY
05/10/2026
02228127-0009